Jamal Berakdar
Concepts of Highly Excited Electronic Systems

Jamal Berakdar

Concepts of Highly Excited Electronic Systems

WILEY-VCH GmbH & Co. KGaA

Author

Jamal Berakdar
Max Planck-Institut für Mikrostrukturphysik
Halle, Germany
e-mail: jber@mpi-halle.de

This book was carefully produced. Nevertheless, authors, editors and publisher do not warrant the information contained therein to be free of errors. Readers are advised to keep in mind thar statements, data, illustrations, procedural details or other items may inadvertently be inaccurate.

1ˢᵗ edition

Library of Congress Card No.: applied for
British Library Cataloging-in-Publication Data:
A catalogue record for this book is available from the British Library

Cover Picture
Jamal Berakdar et al.: Calculated angular distribution of two equal-energy photoelectrons emitted from a copper surface.

Bibliographic information published by
Die Deutsche Bibliothek
Die Deutsche Bibliothek lists this publication in the Deutsche Nationalbibliografie; detailed bibliographic data is available in the Internet at <http://dnb.ddb.de>.

Printed in the Federal Republic of Germany
Printed on acid-free paper

Printing   Strauss Offsetdruck GmbH, Mörlenbach

Bookbinding   Großbuchbinderei J. Schäffer GmbH & Co. KG, Grünstadt

ISBN   3-527-40335-3

cm 7/17/03

for Gize & Sebastian

Preface

The understanding of elementary excitations in electronic systems is of a basic importance, both from a practical as well as from a fundamental point of view. For example, optical properties and electrical transport in materials are primarily governed by excitation processes. On the other hand, our main source of knowledge on the dynamics of elementary electronic systems, such as isolated atoms, ions or molecules, is their response to an external probing field, that excites the system. Generally, the corresponding quantum mechanical description in terms of excitation amplitudes, entails a thorough understanding of the relevant excitation spectrum of the system under study. For this purpose, efficient theoretical and calculational techniques have been developed, and their implementations have rendered possible a detailed insight into the nature and the dynamics of various electronic states in a variety of materials. This progress is driven, to a large extent, by recent breakthroughs in the experimental fabrication, characterization and spectroscopic techniques which, in addition to providing stringent tests of various aspects of current theoretical approaches, have also pointed out open questions to be addressed by theory. Particularly remarkable is the diversity of the electronic materials studied experimentally, ranging from isolated atoms to clusters and surfaces. It is this aspect which is emphasized in this presentation of some of the theoretical tools for the description of excited states of finite and extended electronic systems. The main goal is to highlight common features and differences in the theoretical concepts that have been employed for the understanding of electronic excitations and collisions in finite few-body (atomic) systems and large, extended systems, such as molecules, metal clusters and surfaces.

The complete work is divided in two parts. The first part, which is this present book, deals with the foundations and with the main features of the theoretical methods for the treatment of few-body correlated states and correlated excitations in electronic systems. The forthcoming second volume includes corresponding applications and the analysis of the outcome of theory as contrasted to experimental findings.

A seen from a quick glance at the table of contents, the book starts by reviewing the main aspects of the two-body Coulomb problem, which sets the frame and the notations for the treatment of few-body systems. Subsequently, we sketch a practical scheme for the solution of two-body problems involving an arbitrary non-local potential. Furthermore, an overview is given on the mainstream concepts for finding the ground state of many-body systems. Symmetry properties and universalities of direct and resonant excitation processes are then addressed. Starting from low-lying two-particle excitations, the complexity is increased to the level of dealing with the N-particle fragmentation in finite Coulomb systems. Having in mind the theoretical treatment of excitations in extended and in systems with a large number of active electrons, we introduce the Green's function theory in its first and second quantization versions and outline how this theory is utilized for the description of the ground-state and of many-particle excitations in electronic materials.

The topics in this book are treated differently in depth. Subjects of a supplementary or an introductory nature are outlined briefly, whereas the main focus is put on general schemes for the treatment of correlated, many-particle excitations. In particular, details of those theoretical tools are discussed that are employed in the second forthcoming part of this work.

Due to the broad range of systems, physical processes and theoretical approaches relevant to the present study, only a restricted number of topics is covered in this book, and many important related results and techniques are not included. In particular, in recent years, numerical and computational methods have undergone major advances in developments and implementations, which are not discussed here, even though the foundations of some of these techniques are mentioned. Despite these restrictions, it is nevertheless hoped that the present work will provide and initiate some interesting points of view on excitations and collisions in correlated electronic systems.

The work is purely theoretical. It should be of interest, primarily for researchers working in the field of theoretical few-body electronic systems. Nonetheless, the selected topics and their presentations are hoped to be interesting and comprehensible to curious experimentalists with some pre-knowledge of quantum mechanics.

The results of a number of collaborations with various friends and colleagues can be found in this book. I would like to take this opportunity to thank few of them. I am particularly indebted to M. Brauner, J. S. Briggs, J. Broad and H. Klar for numerous collaborations, discussions and advice on charged few-particle problems. I would also like to thank P. Bruno, A. Ernst, R. Feder, N. Fominykh, H. Gollisch, J. Henk, O. Kidun, and K. Kouzakov for fruitful collaborations, encouragements and valuable discussions on the theoretical aspects of electronic excitations and correlations in condensed matter. Furthermore, I am grateful to L. Avaldi, I. Bray, R. Dörner, A. Dorn, R. Dreizler, J. Feagin, A. Kheifets, J. Kirschner, A. Lahmam-Bennani, J. Lower, H. J. Lüdde, J. Macek, J. H. McGuire, D. Madison, S. Mazevet, R. Moshammer, Yu. V. Popov, A.R.P. Rau, J.-M.Rost, S. N. Samarin, H. Schmidt-Böcking, A. T. Stelbovics, E. Weigold and J. Ullrich for many insightful and stimulating discussions we had over the years on various topics of this book. Communications, discussions and consultations with E. O. Alt, V. Drchal, T. Gonis, J. Kudrnovský, M. Lieber and P. Ziesche are gratefully acknowledged.

Finally, it is a pleasure to thank Frau C. Wanka and Frau V. Palmer from Wiley-VCH for their competent help and support in the preparation of the book.

Jamal Berakdar

Halle, January 2003

Contents

1 The two-body Kepler problem: A classical treatment

This chapter provides a brief summary of the theoretical treatment of non-relativistic two-body Coulomb systems. An extensive account can be found in standard textbooks, e. g. [1]. The purpose here is to introduce the basic ideas and notations utilized in the quantum theory of interacting charged particles. Particular attention is given to the approach pioneered by W. Pauli [4] which utilizes the existence of an additional integral of motion due to the dynamical symmetry of Coulomb-type potentials, namely the Laplace-Runge-Lenz vector [2]. Let us consider a system consisting of two interacting particles with charges z_1 and z_2 and masses m_1 and m_2. In the center-of-mass system, the two-particle motion is described by a one-body Hamiltonian H_0 that depends on the relative coordinates of the two particles. Its explicit form is [1]

$$H_0 = \frac{1}{2\mu}\mathbf{p}_0^2 - \frac{z_{12}}{r_0},\tag{1.1}$$

where $z_{12} = -z_1 z_2$ and the reduced mass is denoted by $\mu = m_1 m_2/(m_1 + m_2)$. The vectors $\mathbf{r}_0$ and $\mathbf{p}_0$ are respectively the two-particle relative coordinate and its conjugate momentum. In addition to the total energy E, the angular momentum $\mathbf{L}_0 = \mathbf{r}_0 \times \mathbf{p}_0$ is a conserved quantity due to the invariance of Eq. (1.1) under spatial rotations. Furthermore, the so-called Laplace-Runge-Lenz vector [2]

$$\mathbf{A} = \hat{\mathbf{r}}_0 + \frac{1}{\mu z_{12}}\,\mathbf{L}_0 \times \mathbf{p}_0\tag{1.2}$$

is as well a constant of motion. This is readily deduced by noting that

$$\partial_t \mathbf{p}_0 = -z_{12}\hat{\mathbf{r}}_0/r_0^2$$

---
[1] Unless otherwise specified, atomic units are used throughout this book.

and therefore the time derivative of $\mathbf{A}$ vanishes, i.e.

$$
\begin{aligned}
\partial_t \mathbf{A} &= \partial_t \hat{\mathbf{r}}_0 + \frac{1}{\mu z_{12}} \mathbf{L}_0 \times (\partial_t \mathbf{p}_0) \\
&= \frac{\partial_t \mathbf{r}_0}{r_0} - \frac{[\mathbf{r}_0 \cdot (\partial_t \mathbf{r}_0)]\mathbf{r}_0}{r_0^3} - \frac{1}{r_0^3}[\mathbf{r}_0 \times (\partial_t \mathbf{r}_0)] \times \mathbf{r}_0 = 0.
\end{aligned}
\tag{1.3}
$$

For the derivation of the classical trajectories it is instructive to introduce the dimensionless

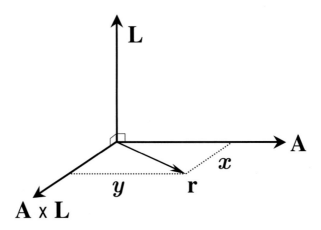

Figure 1.1: The motion of two particles interacting via a Coulomb-type potential takes place in the plane spanned by the vectors $\mathbf{A}$ and $\mathbf{A} \times \mathbf{L}$.

quantities

$$
\mathbf{r} = z_{12}\,\mathbf{r}_0/a, \ \mathbf{p} = \mathbf{p}_0\,a/(z_{12}\hbar), \ \text{and} \ \mathbf{L} = \mathbf{L}_0/\hbar,
$$

where the length scale is given by $a = \hbar^2/(\mu e^2)$ (for clarity, the electron charge e and $\hbar$ are displayed here). The Hamiltonian (1.1) transforms into $H = 2a/(z_{12}^2 e^2) H_0$, whereas in the scaled coordinates, the Laplace-Runge-Lenz vector ($\mathbf{A}$) has the form

$$
\mathbf{A} = \hat{\mathbf{r}} + \mathbf{L} \times \mathbf{p}.
\tag{1.4}
$$

Thus, the Hamiltonian H_0 (1.1) measured in the energy units $\epsilon = (z_{12}^2 e^2)/(2a)$ is

$$
H = H_0/\epsilon = p^2 - 2/r.
\tag{1.5}
$$

Since $\partial_t \mathbf{A} = 0 = \partial_t \mathbf{L}$ and $\mathbf{A} \cdot \mathbf{L} = 0$ we deduce that the relative motion of the two particles is restricted to a plane $\mathcal{P}$ defined by $\mathbf{A}$ and $\mathbf{L} \times \mathbf{A}$, as illustrated in Fig. 1.1. The relative position vector $\mathbf{r}$ in the plane $\mathcal{P}$ is uniquely specified by the components [see Fig. 1.1]

$$x = \mathbf{r} \cdot (\mathbf{L} \times \mathbf{A}), \quad \text{and} \quad y = \mathbf{r} \cdot \mathbf{A}.$$

The components x and y can be written in the form

$$y = r - L^2, \quad x = -L^2(\mathbf{r} \cdot \mathbf{p}). \tag{1.6}$$

Furthermore we conclude that since

$$L^2 = r^2 p^2 - (\mathbf{r} \cdot \mathbf{p})^2, \quad \text{the relation} \quad x^2/L^4 = r^2 p^2 - L^2 \tag{1.7}$$

applies.

Thus, for a given total (scaled) energy $E = p^2 - 2/r$ the components x and y of the position vector $\mathbf{r}$ are determined by the equation

$$\left[y E + (L^2 E + 1) \right]^2 - \frac{E}{L^4} x^2 = L^2 E + 1. \tag{1.8}$$

Further straightforward algebraic manipulations lead to

$$A^2 = L^2 E + 1 \geq 0.$$

Relation (1.8) is the defining equation for conic sections in the normal form:

- For $E < 0$ the motion proceeds along an *elliptical closed orbit* with $\mathbf{A}$ being along the main axis. The excentricity of the orbit is determined by $|\mathbf{A}|$.

- For $E > 0$ Eq. (1.8) defines a *hyperbola*.

- If $E = 0$ (and hence $p^2 = 2/r$) we conclude from Eq. (1.7) and from $r = y + L^2$ that

$$y = x^2/(2L^4) - L^2/2.$$

This equation describes a *parabola* with a curvature L^{-4}. If in addition $L \ll 1$ the parabola degenerates to an almost *straight line* along $\mathbf{A}$ starting from the origin [see Fig. 1.1].

2 Quantum mechanics of two-body Coulomb systems

2.1 Historical background

In a seminal work [4] W. Pauli applied the correspondence principle to introduce the hermitian Laplace-Runge-Lenz operator $\mathbf{A}$ and showed that $\mathbf{A}$ commutes with the total Hamiltonian H, i. e. $[H, \mathbf{A}] = 0$. Using group theory he utilized this fact for the derivation of the bound spectrum of the Kepler problem. Later on, V. A. Fock [5] argued that the degeneracy of the levels, having the same principle quantum numbers, is due to a "hidden" dynamical symmetry. I. e. in addition to the symmetry with respect to the (spatial) rotation group $O(3)$, the Kepler problem with bound spectrum possesses a symmetry with respect to a wider (compact) group $O(4)$ (rotation in a four-dimensional space). Shortly after that V. Bargmann [6] showed how the separability of the (bound) two-body Coulomb problem in parabolic coordinates is linked to the existence of the conserved quantity $\mathbf{A}$. J. Schwinger [7] utilized the rotational invariance with respect to $O(4)$ for the derivation of the Coulomb Green's function. It is this line of development which we will follow in the following compact presentation of this topic. A detailed discussion of the Coulomb Green's function in connection with the $O(4)$ symmetry is deferred to section 11.3.

The dynamical symmetry is related to the form of the Coulomb potential and persists in the n dimensional space. In fact, S. P. Alliluev [8] showed that the Hamiltonian of the n-dimensional (attractive) Kepler problem possesses a hidden symmetry with respect to the $O(n + 1)$ group. For $n = 3$ (e.g. the hydrogen atom) Fock showed [5] that the spectrum is described by the irreducible representations of $O(4)$. Those are finite dimensional since $O(4)$ is a compact group, i. e. a continuous group with finite volume. For the description of the discrete *and* the continuous spectrum one has to utilize the non-compact analog of $O(4)$, namely the Lorenz group (for more details and further references see Ref. [9]).

2.2 Group theoretical approach to the two-body problem

The classical vector (1.4) can not be translated directly into quantum mechanics as the Laplace-Runge-Lenz vector operator $\mathbf{A}$ because it would be non-hermitian. An acceptable definition for the (polar) vector operator $\mathbf{A}$ that satisfies $\mathbf{A} = \mathbf{A}^\dagger$, is [1]

$$\mathbf{A} = \hat{\mathbf{r}} + \frac{1}{2}\left(\mathbf{L} \times \mathbf{p} - \mathbf{p} \times \mathbf{L}\right). \tag{2.1}$$

Since for any vector operator, such as $\mathbf{p}$ the relation $\mathbf{L} \times \mathbf{p} - \mathbf{p} \times \mathbf{L} = i[\mathbf{L}^2, \mathbf{p}]$ applies, we can write $\mathbf{A}$ in the form

$$\mathbf{A} = \hat{\mathbf{r}} + \frac{i}{2}[\mathbf{L}^2, \mathbf{p}]. \tag{2.2}$$

With this definition of $\mathbf{A}$ one verifies the following commutation relations between the operators $\mathbf{L}$, $\mathbf{A}$ and H

$$\mathbf{L} \times \mathbf{L} = i\mathbf{L}, \tag{2.3}$$
$$\mathbf{A} \times \mathbf{A} = -iH\,\mathbf{L}, \tag{2.4}$$
$$\mathbf{L} \times \mathbf{A} + \mathbf{A} \times \mathbf{L} = i2\mathbf{A}, \tag{2.5}$$
$$[H, \mathbf{L}] = 0 = [H, \mathbf{A}], \tag{2.6}$$
$$\mathbf{A} \cdot \mathbf{L} = 0 = \mathbf{L} \cdot \mathbf{A}. \tag{2.7}$$

Furthermore, using Eq. (2.2) it is readily shown that

$$\mathbf{A}^2 = 1 + H(\mathbf{L}^2 + 1). \tag{2.8}$$

Eqs. (2.6, 2.7) state that $\mathbf{L}$ and $\mathbf{A}$ (and H) are conserved while Eq. (2.8) serves to derive Bohr's formula of the energy level scheme, as shown below. Introducing the normalized Laplace-Runge-Lenz operators as

$$\mathbf{N} = \mathbf{A}/h, \quad h = \begin{cases} 1/\sqrt{-H} & \forall E < 0, \\ 1/\sqrt{H} & \forall E > 0, \\ 1 & E = 0, \end{cases} \tag{2.9}$$

---

[1] Hereafter we use the shorthand notation $\sum_{ij}[\mathbf{A}_i, \mathbf{B}_j]\epsilon_{ijk} = (\mathbf{A} \times \mathbf{B})_k$ where $\mathbf{A}$ and $\mathbf{B}$ are vector operators. A vector operator satisfies the relation $\mathbf{L} \times \mathbf{A} + \mathbf{A} \times \mathbf{L} = 2i\mathbf{A}$ (same applies for $\mathbf{B}$ or any vector operator).

the Hamiltonain H is scaled out of Eqs. (2.4-2.7) which then simplify to

$$\mathbf{L} \times \mathbf{L} = i\mathbf{L} , \tag{2.10}$$

$$\mathbf{N} \times \mathbf{N} = \kappa\, i\mathbf{L} , \tag{2.11}$$

$$\mathbf{N} \cdot \mathbf{L} = 0 = \mathbf{L} \cdot \mathbf{N} , \tag{2.12}$$

$$\mathbf{L} \times \mathbf{N} + \mathbf{N} \times \mathbf{L} = 2\, i\mathbf{N}. \tag{2.13}$$

Here we adopt the definition $\kappa = \mathrm{sgn}(-E)$ and $\kappa = 0$ for $E = 0$. Eq. (2.8) becomes

$$-\kappa H \mathbf{N}^2 = 1 + H(\mathbf{L}^2 + 1), \quad \text{if} \quad E \neq 0, \tag{2.14}$$

$$\mathbf{N}^2 = 1 + H(\mathbf{L}^2 + 1), \quad \text{if} \quad E = 0. \tag{2.15}$$

The relations (2.10-2.13) coincide with the commutation relations between the generators of the homogeneous Lorenz group describing rotations and translations. For $E = 0$ (i. e. $\kappa = 0$) the Lorenz group (2.10-2.13) degenerates to the Galilean group. For a given positive energy $E > 0$ the continuum wave functions are the irreducible representations of the Lorenz group. This representation is infinite dimensional and unitary because the values of the orbital angular momentum l are not restricted for a fixed positive energy ($E > 0$). Furthermore, $\mathbf{L}$ and $\mathbf{N}$, the generators of the group, are hermitian for $H > 0$. In contrast, for the bound spectrum ($E < 0$) a finite number of (orbital angular momentum) states corresponding to a given principal quantum number n provides a finite dimensional, non-unitary representation. The non-unitarity is a consequence of the fact that the generators $\mathbf{N}$ are antihermitian for $H < 0$.

2.2.1   The bound spectrum

As well known, for $E < 0$ there is a one-to-one correspondence between the representations of the Lorenz group and the compact $O(4)$ group that have the six generators $(\mathbf{L}, \mathbf{N})$. On the other hand a reduction of the $O(4)$ algebra into two $O(3)$ algebras can be achieved upon introducing the operators

$$\mathbf{J}_{(1)} = \frac{1}{2}(\mathbf{L} + \mathbf{N}), \tag{2.16}$$

$$\mathbf{J}_{(2)} = \frac{1}{2}(\mathbf{L} - \mathbf{N}). \tag{2.17}$$

These satisfy the commutation relations of two independent angular momentum operators, namely

$$\mathbf{J}_{(1)} \times \mathbf{J}_{(1)} = i\mathbf{J}_{(1)}, \tag{2.18}$$

$$\mathbf{J}_{(2)} \times \mathbf{J}_{(2)} = i\mathbf{J}_{(2)}, \tag{2.19}$$

$$[\mathbf{J}_{(2)_\nu} \times \mathbf{J}_{(2)_\nu}] = 0, \quad \forall\, \nu = 1, 2, 3. \tag{2.20}$$

Thus, each of $\mathbf{J}_{(1)}$ and $\mathbf{J}_{(2)}$ can be regarded as the generators of rotations in three dimensions. From a group theory point of view the three components of $\mathbf{J}_{(1)}$ and of $\mathbf{J}_{(2)}$ are the members of two independent Lie algebras $SO_1(3)$ and $SO_2(3)$ [9]. As noticed by O. Klein [10], the Lie algebra implied by Eqs. (2.18-2.20) is then $SO_1(3) \times SO_2(3) = SO(4)$ which describes rotations in a four dimensional space. The representations of the Lie group $SO(4)$ are thus labelled by the two angular momenta $j_1 = 0, 1/2, 1, 3/2 \cdots$ and $j_2 = 0, 1/2, 1, 3/2 \cdots$. The eigenvalues of the Casimir operators $\mathbf{J}^2(1)$ and $\mathbf{J}^2(2)$ of the groups $SO_1(3)$ and $SO_2(3)$ are respectively $j_1(j_1 + 1)$ and $j_2(j_2 + 1)$. Due to the restriction (2.7) we deduce from (2.16, 2.17) that

$$\mathbf{J}^2_{(1)} = \mathbf{J}^2_{(2)} = \frac{1}{4}\left(\mathbf{L}^2 + \mathbf{N}^2\right), \tag{2.21}$$

i. e. $j_1 = j_2$. Furthermore, from Eq. (2.14) follows

$$H = -(\mathbf{N}^2 + \mathbf{L}^2 + 1)^{-1} = -(4\,\mathbf{J}^2_{(1)} + 1)^{-1}.$$

This means, employing the $|j_1 m_1\rangle \otimes |j_2 m_2\rangle$ representation, and taking into account the condition $j_1 = j_2 = j$, the energy eigenvalues E are $(2j + 1)^2$-fold degenerated and are given by

$$E = -\frac{1}{4j(j + 1) + 1} = -\frac{1}{(2j + 1)^2}, \quad j = 0, \frac{1}{2}, 1, \cdots,$$

$$E = -\frac{1}{n^2}, \quad n := (2j + 1) = 1, 2, 3 \cdots . \tag{2.22}$$

2.2.2   Eigenstates of two charged-particle systems

The states $|\Psi_{nlm}\rangle$ that form the representations of $SO(4)$ are obtained as follows. Since $|j_1 m_1\rangle$ and $|j_2 m_2\rangle$ are eigenvectors of angular momentum operators they can be regarded as spherical tensors (see appendix A.1 for definitions and notations of spherical tensors) of rank

j with components $m = -j \cdots j$ [note $j = j_1 = j_2 = (n-1)/2$, see Eq. (2.22)]. From $SO(4) = SO_1(3) \times SO_2(3)$ we deduce that $|\Psi_{nlm}\rangle$ is obtained via the tensor product[2]

$$
\begin{aligned}
&|\Psi_{nlm}\rangle \\
&= \sum_{m_1 m_2} \left\langle \frac{n-1}{2} m_1 \, \frac{n-1}{2} m_2 \middle| lm \right\rangle \left| j = \frac{n-1}{2} m_1 \right\rangle \otimes \left| j = \frac{n-1}{2} m_2 \right\rangle .
\end{aligned} \tag{2.23}
$$

Noting that $\mathbf{L} = \mathbf{J}_{(1)} + \mathbf{J}_{(2)}$ and $\mathbf{N} = \mathbf{J}_{(1)} - \mathbf{J}_{(2)}$ [cf. Eqs. (2.16, 2.17)] we can rewrite Eq. (2.23) in terms of the eigenvalues m and q of the components $\mathbf{L}_z$ and $\mathbf{N}_z$ with respect to an appropriately chosen axis z. Since $m = m_1 + m_2$ and $q = m_1 - m_2$ the state vector $|\Psi_{nlm}\rangle$ is readily expressed in terms of the common eigenstates $|\phi_{nqm}\rangle$ of the operators $\mathbf{L}_z$, $\mathbf{N}_z$ and H in which case Eq. (2.23) takes on the form

$$
|\Psi_{nlm}\rangle = \sum_q \left\langle \frac{n-1}{2} \frac{m+q}{2} \, \frac{n-1}{2} \frac{m-q}{2} \middle| lm \right\rangle |\phi_{nqm}\rangle . \tag{2.24}
$$

The explicit forms of the wave functions $\Psi_{nlm}(\mathbf{r})$ and $\phi_{nqm}(\mathbf{r})$ are given in the next section.

2.3   The two-body Coulomb wave functions

The quantum mechanical two-body Coulomb problem is exactly solvable in only four coordinate systems [3]:

1. In *spherical coordinates* the two-particle relative position $\mathbf{r}$ is specified by $\mathbf{r} = r(\sin\theta \cos\varphi, \sin\theta \sin\varphi, \cos\theta)$ where $\theta \in [0, \pi]$ and $\varphi \in [0, 2\pi]$ are the polar and the azimuthal angles. The chosen set of commuting observables is $\{H, \mathbf{L}^2, \mathbf{L}_z\}$.

2. In the *spheroconic coordinates* the coordinate $\mathbf{r}$ is given by $\mathbf{r} = r(\mathrm{sn}\alpha \, \mathrm{dn}\beta, \mathrm{cn}\alpha \, \mathrm{cn}\beta, \mathrm{dn}\alpha \, \mathrm{sn}\beta)$, where $r \in [0, \infty[$, $\alpha \in [-K, K]$, and $\beta \in [-2K', 2K']$, here $4K$ $(4iK')$ is the real (imaginary) period of the Jacobi-elliptic functions[3] [238]. The set of commuting observables that are diagonalized simultaneously is $\{H, \mathbf{L}^2, \mathbf{L}_x^2 + k'\mathbf{L}_y^2\}$.

---

[2] The tensor product of two spherical tensors $T_{q_1 m_1}$ and $T_{q_2 m_2}$ with ranks q_1 and q_2 and components $m_1 = -q_1, \cdots, q_1$ and $m_2 = -q_2, \cdots, q_2$ is the spherical tensor P_{km} where $P_{km} = \sum_{m_1 m_2} \langle q_1 m_1 \, q_2 m_2 | km \rangle T_{q_1 m_1} T_{q_2 m_2}$. For more details see appendix A.1.

[3] The Jacobi elliptic integrals are defined as $u = \int_0^\varphi \frac{d\theta}{\sqrt{1-k^2 \sin^2\theta}}$. $\mathrm{sn}(u) = \sin\varphi$, $\mathrm{cn}(u) = \cos\varphi$ and $\mathrm{dn}(u) = (1 - k^2 \sin^2\varphi)^{1/2}$. For the definition of the spheroconic coordinates k [k'] is deduced from the modulus of $\mathrm{sn}(\alpha)$ and $\mathrm{cn}(\alpha)$ [$\mathrm{sn}(\beta)$, $\mathrm{cn}(\beta)$ and $\mathrm{dn}(\beta)$].

3. In *spheroidal coordinates* the relative coordinate **r** is defined as
$\mathbf{r} = \frac{R}{2}\left(\sqrt{(\xi^2-1)(1-\eta^2)}\cos\varphi, \sqrt{(\xi^2-1)(1-\eta^2)}\sin\varphi, \xi\eta+1\right)$ where R is a real positive constant and $\xi \in [1,\infty[,\ \eta \in [-1,+1]$, and $\varphi \in [0,2\pi[$. In this case the chosen set of commuting operators is $\{H, \mathbf{L}^2 - 2R\mathbf{N}_z, \mathbf{L}_z\}$. For $R \to 0$ the spheroidal coordinates degenerates to the spherical coordinates whereas for $R \to \infty$ they coincide with the parabolic coordinates.

4. In the *parabolic coordinates* the preferred set of commuting operators is $\{H, \mathbf{L}_z, \mathbf{N}_z\}$. It is this coordinate system which will be discussed below and will be utilized in the next chapters of this book for the treatment of the few-body problem.

2.3.1 Spherical coordinates

The derivation of the normalized wave functions $\Psi_{nlm}(\mathbf{r})$ in spherical coordinates can be found in standard books on quantum mechanics [1]. Here we only give the final expression

$$\Psi_{nlm}(\mathbf{r}) = R_{nl}(r)Y_{lm}(\hat{\mathbf{r}}), \tag{2.25}$$

$$R_{nl}(r) = \frac{2}{n^2(n+1)!}\left[\frac{(n-l-1)!}{(n+l)!}\right]^{1/2}\left(\frac{2r}{n}\right)^l e^{-\frac{r}{n}} L_{n+l}^{2l+1}\left(\frac{2r}{n}\right). \tag{2.26}$$

The angular motion is described by the spherical harmonics $Y_{lm}(\hat{\mathbf{r}})$ whereas the radial part Eq. (2.26) is given in terms of the associated Laguerre polynomials[4] $L_b^a(x)$ [12, 99].

2.3.2 Parabolic coordinates

As mentioned above, as an alternative set of three commuting operators for the description of the two-body Coulomb problem one may choose $\{H, \mathbf{L}_z, \mathbf{A}_z\}$. The corresponding coordinate system in which the Schrödinger equation separates is the parabolic coordinates. Those are given in terms of the defining parameters of two systems of paraboloids with the focus at the origin and an azimuthal angle φ in the (x,y) plane. The relation between these coordinates

---

[4]The Laguerre polynomials are obtained according to the formula $L_n(z) = e^z \frac{d^n}{dz^n}\left[e^{-z}z^n\right]$. The associated Laguerre polynomials are given by $L_n^m(z) = \frac{d^m}{dz^m}L_n(z)$ and satisfy the differential equation

$$\left[z\partial_z^2 + (m+1-z)\partial_z + (n-m)\right]L_n^m(z) = 0. \tag{2.27}$$

and the cartesian coordinates (in which the position vector $\mathbf{r}$ is given by $\mathbf{r} = (x, y, z)$) is

$$
\begin{array}{ll}
x = \sqrt{\xi\eta}\cos\varphi, & \xi = r + \mathbf{r}\cdot\hat{\mathbf{z}}, \\
y = \sqrt{\xi\eta}\sin\varphi, & \eta = r - \mathbf{r}\cdot\hat{\mathbf{z}}, \\
z = \frac{1}{2}(\xi - \eta), & \tan\varphi = \frac{y}{x};
\end{array}
\tag{2.28}
$$

$$\xi \in [0,\infty[,\ \eta \in [0,\infty[,\ \varphi \in [0,2\pi].$$

The Laplacian Δ expressed in parabolic coordinates reads

$$\Delta = \frac{4}{\xi+\eta}\left(\partial_\xi\xi\partial_\xi + \partial_\eta\eta\partial_\eta\right) + \frac{1}{\eta\xi}\partial_\varphi^2. \tag{2.29}$$

Thus, the Schrödinger equation for an electron in the field of an ion with a charge $Z = 1$ a.u. is

$$
\begin{aligned}
(H - \bar{E})\phi &= 0, \\
\left(\mathbf{p}^2 - \frac{2}{r} - E\right)\phi &= 0,
\end{aligned}
\tag{2.30}
$$

where $E = 2\bar{E}$. In parabolic coordinates Eq. (2.30) has the following form

$$\left\{\frac{4}{\xi+\eta}\left[\partial_\xi\xi\partial_\xi + \partial_\eta\eta\partial_\eta + 1\right] + \frac{1}{\eta\xi}\partial_\varphi^2 + E\right\}\phi = 0. \tag{2.31}$$

Multiplying this equation by $(\xi + \eta)/4$ and upon inserting in (2.31) the ansatz

$$\phi = N_E\, e^{\pm i\, m\varphi}\, u_1(\xi)u_2(\eta), \tag{2.32}$$

where $m \geq 0$, and N_E is an energy dependent normalization constant, (2.33)

we obtain the two determining equations for functions $u_1(\xi)$ and $u_2(\eta)$ as

$$\partial_\xi(\xi\partial_\xi u_1) + \frac{E}{4}\xi u_1 - \frac{m^2}{4\xi}u_1 + c_1 u_1 = 0, \tag{2.34}$$

$$\partial_\eta(\eta\partial_\xi u_2) + \frac{E}{4}\eta u_2 - \frac{m^2}{4\eta}u_2 + c_2 u_2 = 0. \tag{2.35}$$

Here c_1 and c_2 are integration constants satisfying

$$c_1 + c_2 = 1. \tag{2.36}$$

Since the function u_1 have the limiting behaviour

$$\lim_{\xi\to 0} u_1 \to \xi^{m/2} \quad\text{and}\quad \lim_{\xi\to\infty} u_1 \to \exp(-\sqrt{-E}\,\xi/2) \tag{2.37}$$

it is advantageous to write down u_1 in the form

$$u_1 = \xi^{m/2}\exp(-\sqrt{-E}\,\xi/2)g_1(\xi). \tag{2.38}$$

From Eq. (2.34) one deduces for the unknown function g_1 the determining equation

$$\left[x\partial_x^2 + (m + 1 - x)\partial_x + \left(\frac{c_1}{\sqrt{-E}} - \frac{m+1}{2} \right) \right] g_1 = 0, \tag{2.39}$$

where $x = \xi\sqrt{-E}$. The solution of the differential equation Eq. (2.39) is the associated Laguerre polynomials $L_n^m(z)$, as readily verified upon a comparison with Eq. (2.27), i. e.

$$g_1(x) = L_{n_1+m}^m(x), \tag{2.40}$$

$$n_1 = \frac{c_1}{\sqrt{-E}} - \frac{(m+1)}{2} = 0, 1, 2, \cdots . \tag{2.41}$$

In an analogous way one expresses the function u_2 in terms of associated Laguerre polynomials as

$$u_2 = \eta^{m/2} \exp\left(-\sqrt{-E}\,\eta/2\right) L_{n_2+m}^m(\eta\sqrt{-E}), \tag{2.42}$$

$$n_2 = c_2/\sqrt{-E} - (m+1)/2 = 0, 1, 2, \cdots . \tag{2.43}$$

With this formula we conclude that the function, defined by Eq. (2.32), has the final form

$$\phi = N_E\, e^{\pm i\, m\varphi} u_1 u_2,$$

$$= N_E\, e^{\pm i\, m\varphi} \times$$

$$\times \xi^{m/2} \exp\left(-\sqrt{-E}\,\xi/2\right) L_{n_1+m}^m(\xi\sqrt{-E})$$

$$\times \eta^{m/2} \exp\left(-\sqrt{-E}\,\eta/2\right) L_{n_2+m}^m(\eta\sqrt{-E}). \tag{2.44}$$

Since $c_1 + c_2 = 1$ we conclude from Eqs. (2.41, 2.43) that

$$\frac{1}{\sqrt{-E}} = n = \frac{1}{\sqrt{-2\bar{E}}},$$

where the integer number n is identified as the principle quantum number and is related to n_1 and n_2 via

$$n := n_1 + n_2 + m + 1 = 0, 1, 2, \cdots , \quad \text{i.e.} \quad E = -\frac{1}{n^2}.$$

The connection between the separability sketched above and the Laplace-Runge-Lenz vector becomes apparent when the component $\mathbf{N}_z$ ($\mathbf{L}_z$) of $\mathbf{N}$ ($\mathbf{L}$), along a chosen quantization axis, is expressed in parabolic coordinates. This has been done by V. Bargmann [6] who showed that the states $|n_1 n_2 m\rangle$ are eigenvectors of $\mathbf{N}_z$, $\mathbf{L}_z$ and H.

The operator $\mathbf{N}$ is a polar vector (odd under parity operation). In contrast $\mathbf{L}$ is an axial vector and as such is even under parity. Therefore, as far as parity is concerned, the states

$|n_1 n_2 m\rangle$ are mixed, i.e. they have no well-defined parity. This is as well clear from the definition of the parabolic coordinates (2.28) that gives preference to the z direction (and therefore the parabolic eigenstates derived above are symmetrical with respect to the plane $z = 0$).

The presence of a preferential space direction in the definition of the parabolic variables makes this coordinate system predestinate for formulating problems that involve a direction determined by physical measurements, such as an external electric field or the asymptotic momentum vector of a continuum electron. A well-known demonstration of this statement is the separability in parabolic coordinates of the two-body Hamiltonian in the presence of an electric field $\mathcal{E}$ (the Stark effect). In this case one chooses the z axis to be along the field and writes down the Schrödinger equation as

$$(H - \mathcal{E}\,z - E)\,\phi = 0.$$

Expressing this relation in the coordinate (2.28) and making the ansatz (2.32) one obtains two separate, one-dimensional differential equations for the determination of u_1 and u_2, namely

$$\left[\partial_\xi \xi \partial_\xi + \frac{E}{4}\xi - \frac{m^2}{4\xi} + \frac{\mathcal{E}}{2}\xi^2 + c_1\right] u_1(\xi) = 0, \qquad (2.45)$$

$$\left[\partial_\eta \eta \partial_\eta + \frac{E}{4}\eta - \frac{m^2}{4\eta} - \frac{\mathcal{E}}{2}\eta^2 + c_2\right] u_2(\eta) = 0. \qquad (2.46)$$

These relations make evident the complete separability of the Stark effect in parabolic coordinates.

2.3.3 Analytical continuation of the two-body Coulomb wave functions

Another example involving a physically defined direction in space occurs in ionization problems. There, the wave vector $\mathbf{k}$ of the continuum electron is specified experimentally. Thus, a suitable choice for the space direction $\hat{\mathbf{z}}$, that enters the definition of the parabolic coordinates, is $\hat{\mathbf{z}} \equiv \hat{\mathbf{k}}$. In this case the parabolic coordinates are

$$\xi = r + \mathbf{r} \cdot \hat{\mathbf{k}}, \qquad (2.47)$$

$$\eta = r - \mathbf{r} \cdot \hat{\mathbf{k}}, \qquad (2.48)$$

$$\varphi = \arctan(y/x). \qquad (2.49)$$

Since we are dealing in case of $\bar{E} > 0$ with continuum problems one may wonder whether it is possible to utilize the wave function (2.32) (with the asymptotically decaying behaviour

(2.37)) to describe the two-particle continuum, i.e. whether Eq. (2.32) can be continued analytically across the fragmentation threshold. To answer this question we note that the associated Laguerre polynomials can be written in terms of the confluent hypergeometric functions[5] $_1F_1(a, b, z)$ [11, 12] as

$$L_n^m(z) = \frac{(m+n)!}{m!n!} \, _1F_1(-n, m+1, z). \tag{2.52}$$

Defining the (generally) complex wave vector $k = \sqrt{2\bar{E}}$ and choosing the phase convention for Eqs. (2.44) such that

$$\phi = N_E \, e^{\pm i m \varphi} u_1^* u_2 \tag{2.53}$$

we deduce the general solution of Schrödinger equation for one electron in the field of a residual ion with a unit positive charge (i.e. Eq. (2.32)) as (we recall the assumption that the electron-ion reduced mass is unity)

$$\begin{aligned}
\phi \quad = \quad & N_{k,m} \, e^{\pm i m \varphi} \, \xi^{m/2} \eta^{m/2} \, e^{i \mathbf{k} \cdot \mathbf{r}} \\
& _1F_1(-i\frac{c_1}{k} + \frac{1-m}{2}, 1+m, -ik\xi) \\
& _1F_1(i\frac{c_2}{k} + \frac{1-m}{2}, 1+m, ik\eta),
\end{aligned} \tag{2.54}$$

where $m \geq 0, \; N_{k,m} = N(k, \pm m), \; c_1 + c_2 = 1, \; k \in \mathbb{C}.$

Since $_1F_1(a, b, z)$ is analytic for all values of the complex arguments a, b, z, except for negative integer values of b one can use Eq. (2.54) for the description of the all bound and continuum states. Outgoing continuum waves characterized by the wave vector $\mathbf{k}$ are obtained from (2.54) upon the substitution $m \to 0, \; c_1 \to -ik/2$ in which case (2.54) reduces to

$$\phi_{out} \quad = \quad N_k^+ e^{i \mathbf{k} \cdot \mathbf{r}} \, _1F_1(-i\alpha_k, 1, ik(r - \hat{\mathbf{k}} \cdot \mathbf{r})). \tag{2.55}$$

---

[5]The confluent hypergeometric function $_1F_1(a, b, z)$ is the solution of the confluent Kummer-Laplace differential equation [11, 12]

$$z\, u'' + (b - z)u' - a\, u = 0. \tag{2.50}$$

$_1F_1(a, b, z)$ is also called Kummer's function of the first kind. The confluent hypergeometric function has the series representation [11, 12]

$$_1F_1(a, b, z) = 1 + \frac{a}{b}z + \frac{a(a+1)}{b(b+1)}\frac{z^2}{2!} + \cdots = \sum_{j=0}^{\infty} \frac{(a)_j}{(b)_j} \frac{z^j}{j!}, \tag{2.51}$$

where $(a)_j$ denotes the Pochhammer symbols; its form is inferred from (2.51). The power series representing the function $_1F_1(a, b, z)$ is convergent for all finite $z \in \mathbb{C}$. If a and b are integers, $a < 0$, and either $b > 0$ or $b < a$, then the series yields a polynomial with a finite number of terms. Except for the case $b \leq 0, \; b \in \mathbb{N}$, the function $_1F_1(a, b, z)$ is an entire function of z.

The incoming continuum wave is obtained by using $m \to 0$, $c_1 \to -ik/2$ in which case Eq. (2.54) yields

$$\phi_{in} = N_k^- e^{i\mathbf{k}\cdot\mathbf{r}} {}_1F_1(i\alpha_k, 1, -ik(r + \hat{\mathbf{k}}\cdot\mathbf{r})). \tag{2.56}$$

The parameter $\alpha_k = -\frac{1}{k}$ is generally referred to as the Sommerfeld parameter. For two particles with arbitrary masses m_1 and m_2 and charges Z_1 and Z_1, the Sommerfeld parameter α_{12} is defined as

$$\alpha_{12} = \frac{Z_1 Z_2 \mu_{12}}{k_{12}}, \tag{2.57}$$

where μ_{12} is the reduced mass and $\mathbf{k}_{12}$ is the momentum conjugate to the relative coordinate $\mathbf{r}_{12}$. The normalization constants $N_k^\pm$ can be obtained from the normalization of the asymptotic flux and read (a detailed derivation is given in chapter 8)

$$N_k^\pm = (2\pi)^{-3/2} e^{-\pi\alpha_k/2} \Gamma(1 \pm i\alpha). \tag{2.58}$$

It is readily verified that with this normalization the above wave functions satisfy

$$\langle \phi_{in,\mathbf{k}'} | \phi_{in,\mathbf{k}} \rangle = \langle \phi_{out,\mathbf{k}'} | \phi_{out,\mathbf{k}} \rangle = \delta(\mathbf{k} - \mathbf{k}'), \tag{2.59}$$

and

$$\phi_{in,\mathbf{k}}(\mathbf{r}) = \phi_{out,-\mathbf{k}}^*(\mathbf{r}). \tag{2.60}$$

In the above procedure we obtained the Coulomb continuum states from the bound states via analytical continuation in the complex k plane. One can now reverse the arguments and derives the bound state spectrum from the knowledge of the continuum states. This becomes of considerable importance when considering many-body systems for which approximate expressions for the continuum wave functions are known. Such a case we will encounter in chapter 8.

For bound states ($E < 0$) the wave vector k is purely complex and hence we write it in the form $k = -i/c$, $c \in \mathbb{R}$. Now we recall that the bound state spectrum is manifested as poles of the scattering amplitude in the complex k space. From the analyticity properties of the hypergeometric function one concludes that the poles in the complex k plane (which correspond to the bound states) originate from the analytically continued normalization factor

$$N^- = (2\pi)^{-3/2} e^{i\pi c/2} \Gamma(1 - c). \tag{2.61}$$

This function has isolated poles when $c = n \in \mathbb{N}^+$ is a positive integer, meaning that $k = -i/n = \sqrt{2E} \Rightarrow E = -1/2n^2$. The simultaneous description of the bound and the continuum spectrum is made possible by an analytical continuation into the complex k plane. This procedure can be utilized for cases involving a more general form of the potentials where an analytical treatment is not possible. The tools and methods for such a procedure are outlined in the next chapter.

3 One particle in an arbitrary potential

Due to the special functional form of the Coulomb potential we were able to derive the electronic quantum mechanical states analytically. In general, however, the electrons are subject to more complicated, non-local, energy-dependent forces. E. g., as shown in chapter 14, the propagation of a single particle or a single hole in a surrounding medium can be described by a Schrödinger-type differential equation involving a non-local, energy-dependent potential, called the self-energy. Therefore, we sketch in this chapter a general method for the treatment of the quantum mechanical properties of a non-relativistic particle in a non-local (energy-dependent) potential.

3.1   The variable-phase method

The quantum mechanics of a particle is fully described by its wave function derived from the Schrödinger equation, i.e. from the solution of a second-order differential equation. In numerous situations however, it suffices to know certain features related to the particle motion, e. g. for the derivation of the scattering cross section from an external potential the knowledge of only the scattering phase shifts is necessary. Therefore, it is useful to reformulate the problem of finding the wave function in a one which yields directly physical observables, such as the scattering phase shifts and the scattering amplitudes. This kind of an approach has been put forward by Calogero [13] and by Babikov [14] and is generally known as the variable-phase approach (VPA), or the phase-amplitude method. In the VPA the Schrödinger equation, as a second order differential equation (DE), is transformed into a pair of first-order, coupled DEs. The solutions of these DEs yield the so-called phase and amplitude functions. An important feature of the VPA is the decoupling of the two first order DEs [13, 14]. This is because the solution of some physical problems is obtained by treating only one of the two DEs. Solving both first-order DEs yields completely the same information carried by the wave function. Due to its versatileness and its close relation to the required physical quantities the VPA has found applications in various fields of physics [17, 18, 19, 20, 21]. In this chapter we sketch the basic elements of the VPA for local and non-local potentials.

3.2   Phase-amplitude equations for non-local potentials

Let us consider a non-relativistic, spinless particle with the energy k^2 subject to a hermitian, non-local potential $V(\mathbf{r}, \mathbf{r}') = V(\mathbf{r}', \mathbf{r})$. A quantum mechanical description of this particle requires the solution of the Schödinger eqaution[1]

$$\triangle \Psi(\mathbf{r}) + k^2 \Psi(\mathbf{r}) = \int d\mathbf{r}' V(\mathbf{r}, \mathbf{r}') \Psi(\mathbf{r}'). \tag{3.1}$$

Assuming isotropy of space, the potential $V(\mathbf{r}, \mathbf{r}')$ is a function of the scalar variables r^2, r'^2, $\mathbf{r} \cdot \mathbf{r}' = rr' \cos \theta$ only. In spherical coordinates, Eq. (3.1) separates into an angular and a radial part. The radial part $u_\ell(r)$ of the solution of the Schrödinger equation is derived from the relation

$$\frac{d^2}{dr^2} u_\ell(r) + \left(k^2 - \frac{\ell(\ell+1)}{r^2} \right) u_\ell(r) = \int_0^\infty dr' V_\ell(r, r') u_\ell(r'). \tag{3.2}$$

Here the non-local potential associated with the orbital angular momentum ℓ is given by

$$V_\ell(r, r') = V_\ell(r', r) = 2\pi rr' \int_{-1}^1 V_\ell(r, r') P_\ell(\cos \theta) d(\cos \theta),$$

where ℓ is an orbital quantum number. Instead of solving directly for u_ℓ one introduces in the VPA the functions $\alpha_\ell(r)$ and $\delta_\ell(r)$ such that

$$u_\ell(r) = \alpha_\ell(r) F_\ell(r), \tag{3.3}$$

where

$$F_\ell(r) = [\cos \delta_\ell(r) j_\ell(kr) - \sin \delta_\ell(r) n_\ell(kr)]. \tag{3.4}$$

The functions $j_\ell(kr)$ and $n_\ell(kr)$ are the Riccati-Bessel functions which are the regular and the irregular solutions in absence of the potential. Because two new functions have been introduced to mimic u_ℓ an additional condition on the derivative is required, namely

$$\frac{d}{dr} u_\ell(r) = \alpha_\ell(r) [\cos \delta_\ell(r) \frac{d}{dr} j_\ell(kr) - \sin \delta_\ell(r) \frac{d}{dr} n_\ell(kr)], \tag{3.5}$$

---

[1] For simplicity we use units in which $2m = 1 = \hbar$, $Z = 1$.

or one can write equivalently

$$\frac{d\alpha_\ell(r)}{dr}F_\ell(r) = \alpha_\ell(r)\frac{d\delta_\ell(r)}{dr}G_\ell(r),. \tag{3.6}$$

The function $G_\ell(r)$ is conveniently expressed in terms of the Riccati-Bessel functions as

$$G_\ell(r) = \sin\delta_\ell(r)j_\ell(kr) + \cos\delta_\ell(r)n_\ell(kr).$$

The functions $\delta_\ell(r)$ and $\alpha_\ell(r)$ are called respectively the *phase* and the *wave-amplitude functions*. The names as well as the physical significance of $\delta_\ell(r)$ and $\alpha_\ell(r)$ become obvious when we consider the cut-off potential $V_\ell^{(R)}$ obtained from the potential $V_\ell(r, r')$ terminated at the position R, i.e.

$$V_\ell^{(R)}(r, r') = V_\ell(r, r')\theta(R - r)\theta(R - r'), \tag{3.7}$$

where θ is the step function $[\theta(x > 0) = 1, \theta(x < 0) = 0]$. The functions $\delta_\ell(R)$ and $\alpha_\ell(R)$ are the (physically measurable) partial scattering phase shift $\widehat{\delta}_\ell^{(R)}$ and the asymptotic amplitude $\widehat{\alpha}_\ell^{(R)}$ of the wave function of the particle subjected to the potential $V_\ell^{(R)}$. This means that the function $\delta_\ell(R)$ and $\alpha_\ell(R)$ contain information pertinent not only to the potential in question but also to an infinite series of a certain class of potentials $[V_\ell^{(R)}(r, r'), \forall R\ ]$. Obviously the asymptotic value of $\delta_\ell(r)$ yields the scattering phase for the potential $V_\ell(r, r')$, i.e. $\delta_\ell(\infty) = \widehat{\delta}_\ell$.

To derive determining equations for the phase and for the amplitude functions we evaluate at first the derivative of the function F_ℓ as

$$
\begin{aligned}
\frac{dF_\ell(r)}{dr} &= \frac{d}{dr}\left[j_\ell(kr)\cos\delta_\ell(r) - n_\ell(kr)\sin\delta_\ell(r)\right], \\
&= \left[\frac{dj_\ell(kr)}{dr}\cos\delta_\ell(r) - j_\ell(kr)\sin\delta_\ell(r)\frac{d\delta_\ell(r)}{dr}\right. \\
&\quad \left. - \frac{dn_\ell(kr)}{dr}\sin\delta_\ell(r) - n_\ell(kr)\cos\delta_\ell(r)\frac{d\delta_\ell(r)}{dr}\right], \\
&= \left[\frac{dj_\ell(kr)}{dr}\cos\delta_\ell(r) - \frac{dn_\ell(kr)}{dr}\sin\delta_\ell(r)\right] - \frac{d\delta_\ell(r)}{dr}G_\ell(r).
\end{aligned}
\tag{3.8}
$$

Furthermore, for the first derivative of $u_\ell(r)$ we find

$$
\begin{aligned}
\frac{du_\ell(r)}{dr} &= \frac{d(\alpha_\ell(r)F_\ell(r))}{dr}, \\
&= \alpha_\ell(r)\frac{dF_\ell(r)}{dr} + \frac{d\alpha_\ell(r)}{dr}F_\ell(r) \\
&= \alpha_\ell(r)\frac{dF_\ell(r)}{dr} + \alpha_\ell(r)\frac{d\delta_\ell(r)}{dr}G_\ell(r), \\
&= \alpha_\ell(r)\left[\frac{dj_\ell(kr)}{dr}\cos\delta_\ell(r) - \frac{dn_\ell(kr)}{dr}\sin\delta_\ell(r)\right].
\end{aligned}
\tag{3.9}
$$

The second derivative of $u_\ell(r)$ is more complicated. Its explicit form reads

$$
\begin{aligned}
\frac{d^2 u_\ell(r)}{dr^2} &= \frac{d}{dr}\left(\alpha_\ell(r)\left[\frac{dj_\ell(kr)}{dr}\cos\delta_\ell(r) - \frac{dn_\ell(kr)}{dr}\sin\delta_\ell(r)\right]\right), \\
&= \frac{d\alpha_\ell(r)}{dr}\left[\frac{dj_\ell(kr)}{dr}\cos\delta_\ell(r) - \frac{dn_\ell(kr)}{dr}\sin\delta_\ell(r)\right] \\
&\quad + \alpha_\ell(r)\left[\frac{d^2 j_\ell(kr)}{dr^2}\cos\delta_\ell(r) - \frac{d^2 n_\ell(kr)}{dr^2}\sin\delta_\ell(r)\right] \\
&\quad - \alpha_\ell(r)\frac{d\delta_\ell(r)}{dr}\left[\frac{dj_\ell(kr)}{dr}\sin\delta_\ell(r) + \frac{dn_\ell(kr)}{dr}\cos\delta_\ell(r)\right], \\
&= \alpha_\ell(r)\frac{d\delta_\ell(r)}{dr}\left(\frac{G_\ell(r)}{F_\ell(r)}\left[\frac{dj_\ell(kr)}{dr}\cos\delta_\ell(r) - \frac{dn_\ell(kr)}{dr}\sin\delta_\ell(r)\right]\right. \\
&\quad \left. - \left[\frac{dj_\ell(kr)}{dr}\sin\delta_\ell(r) + \frac{dn_\ell(kr)}{dr}\cos\delta_\ell(r)\right]\right) \\
&\quad + \alpha_\ell(r)\left[\frac{d^2 j_\ell(kr)}{dr^2}\cos\delta_\ell(r) - \frac{d^2 n_\ell(kr)}{dr^2}\sin\delta_\ell(r)\right], \\
&= \alpha_\ell(r)\frac{d\delta_\ell(r)}{dr}\frac{W}{F_\ell(r)} \\
&\quad + \alpha_\ell(r)\left[\frac{d^2 j_\ell(kr)}{dr^2}\cos\delta_\ell(r) - \frac{d^2 n_\ell(kr)}{dr^2}\sin\delta_\ell(r)\right].
\end{aligned}
\tag{3.10}
$$

The function W has been introduced as an abbreviation for the expression

$$
\begin{aligned}
\frac{W}{F_\ell(r)} &= \frac{G_\ell(r)}{F_\ell(r)} \left[\frac{dj_\ell(kr)}{dr} \cos \delta_\ell(r) - \frac{dn_\ell(kr)}{dr} \sin \delta_\ell(r) \right], \\
&\quad - \left[\frac{dj_\ell(kr)}{dr} \sin \delta_\ell(r) + \frac{dn_\ell(kr)}{dr} \cos \delta_\ell(r) \right] \\
&= \frac{1}{F_\ell(r)} \left\{ G_\ell(r) \left[\frac{dj_\ell(kr)}{dr} \cos \delta_\ell(r) - \frac{dn_\ell(kr)}{dr} \sin \delta_\ell(r) \right] \right. \\
&\quad \left. - F_\ell(r) \left[\frac{dj_\ell(kr)}{dr} \sin \delta_\ell(r) + \frac{dn_\ell(kr)}{dr} \cos \delta_\ell(r) \right] \right\} \\
&= \frac{1}{F_\ell(r)} \left\{ n_\ell(kr) \frac{dj_\ell(kr)}{dr} - j_\ell(kr) \frac{dj_\ell(kr)}{dr} \right\} \left[\cos^2 \delta_\ell(r) + \sin^2 \delta_\ell(r) \right].
\end{aligned}
\tag{3.11}
$$

From this relation it is clear that W is the Wronskian of the functions $n_\ell(kr)$ and $j_\ell(kr)$ and hence W is equal to $-k$, i. e.

$$
W = n_\ell(kr) \frac{dj_\ell(kr)}{dr} - j_\ell(kr) \frac{dn_\ell(kr)}{dr} = -k.
\tag{3.12}
$$

Having derived the first and the second derivative for u_ℓ we can now write down the radial Schrödinger equation in the form

$$
\begin{aligned}
-\alpha_\ell(r) \frac{d\delta_\ell(r)}{dr} \frac{k}{F_\ell(r)(\delta_\ell(r))} &+ \left\{ \alpha_\ell(r) \left[\frac{d^2 j_\ell(kr)}{dr^2} \cos \delta_\ell(r) - \frac{d^2 n_\ell(kr)}{dr^2} \sin \delta_\ell(r) \right] \right. \\
+\alpha_\ell(r) \left(k^2 - \frac{\ell(\ell+1)}{r^2} \right) & \left. [j_\ell(kr) \cos \delta_\ell(r) - n_\ell(kr) \sin \delta_\ell(r)] \right\} \\
&= \int_0^\infty dr' V_\ell(r, r') \alpha_\ell(r') F_\ell(\delta_\ell(r')).
\end{aligned}
\tag{3.13}
$$

The curly bracket in this expression contains the solution of the free equation and hence vanishes. This leads to the differential equation for the phase function

$$
\frac{d\delta_\ell(r)}{dr} = -\frac{F_\ell(\delta_\ell(r))}{k} \int_0^\infty dr' V_\ell(r, r') \frac{\alpha_\ell(r')}{\alpha_\ell(r)} F_\ell(\delta_\ell(r')).
\tag{3.14}
$$

Once $\delta_\ell(r)$ has been derived one can obtain $\alpha_\ell(r)$ from the relation

$$
\frac{d\alpha_\ell(r)}{dr} F_\ell(\delta_\ell(r)) = \alpha_\ell(r) \frac{d\delta_\ell(r)}{dr} G_\ell(\delta_\ell(r)),
\tag{3.15}
$$

$$
\frac{\alpha_\ell(r')}{\alpha_\ell(r)} = \exp \left(-\int_{r'}^r ds \frac{d\delta_\ell(s)}{ds} \frac{G_\ell(\delta_\ell(s))}{F_\ell(\delta_\ell(s))} \right).
\tag{3.16}
$$

With this expression taken into account we arrive at the principal determining equation for $\delta_\ell(r)$ in the case of non-local potentials [23]:

$$\frac{d\delta_\ell(r)}{dr} = -\frac{1}{k} F_\ell(r) \int_0^\infty dr' V(r, r') F_\ell(r') \exp\left[-\int_{r'}^r \frac{d\delta_\ell(s)}{ds} \frac{G_\ell(s)}{F_\ell(s)} ds \right]. \qquad (3.17)$$

3.2.1   The local potential case

The case of local potentials is readily derived from Eq. (3.17) by imposing the restriction that

$$V(r, r') = V(r')\delta_\ell(r)(r - r').$$

Equation (3.17) reduces then to the well-established phase equation for problems involving local potentials, which has been derived in the original works of [13, 14]. The explicit expression for $\delta_\ell(r)$ in the local-potential case is

$$\frac{d\delta_\ell(r)}{dr} = -\frac{V(r)}{k} F_\ell^2(r) = -\frac{V(r)}{k} \left[j_\ell(kr) \cos \delta_\ell(r) - n_\ell(kr) \sin \delta_\ell(r) \right]^2. \qquad (3.18)$$

3.2.2   Numerical considerations

The equation (3.17) for the general case has to be solved with the initial condition

$$\delta_\ell(0) = 0$$

which corresponds to vanishing phase shift $\delta_\ell(0) = \widehat{\delta}_\ell^{(R=0)} = 0$, i. e. δ_ℓ vanishes in absence of the potential $V_\ell^{(R=0)}(r, r') = 0 \; \forall \; r, r'$. In this context we recall the statement made in the introduction with regard to the properties of the VPA, namely that the equation for the phase function does not contain the wave-amplitude function $\alpha_\ell(r)$, which is confirmed by Eq. (3.17).

Having solved for $\delta_\ell(r)$, one obtains the function $\alpha_\ell(r)$ upon an integration of Eq. (3.6) with the formal result being

$$\alpha(r) = \alpha(0) \exp\left(\int_0^r ds \frac{G_\ell(s)}{F_\ell(s)} \frac{d\delta_\ell(s)}{ds} \right). \qquad (3.19)$$

Furthermore, we note that the functions F, N and G are interrelated via the equations

$$dF = Nds - Gd\delta,$$

$$\Rightarrow \quad \int_{r'}^{r} \left(-\frac{G}{F}d\delta\right) = \int_{r'}^{r} \left(\frac{dF}{F} - \frac{Nds}{F}\right) = \ln\left(\frac{F(r)}{F(r')}\right) - \int_{r'}^{r} \frac{N(s)}{F(s)}ds.$$

$$(3.20)$$

Hence, the determining equation for the phase function can be written in the form

$$\frac{d\delta_\ell(r)}{dr} = \left(-\frac{1}{k}\right) F_\ell^2(r) \int_0^\infty dr' V_\ell(r,r') \exp\left[-\int_{r'}^{r} \frac{N_\ell(s)}{F_\ell(s)}ds\right],$$

$$(3.21)$$

where the function $N\alpha_\ell$ is given in terms of the derivatives of the Riccati-Bessel functions as

$$N\alpha_\ell(s) = \cos\delta_\ell(s)\frac{dj_\ell(ks)}{ds} - \sin\delta_\ell(r)\frac{dn_\ell(kr)}{ds}.$$

3.3   The scattering amplitude representation

The influence of an external potential on the motion of quantum particles is often described by means of the partial scattering amplitude (SA) $\widehat{\mathcal{F}}_\ell$. The SA is linked to the partial scattering phase via the relation [22]

$$\widehat{\mathcal{F}}_\ell = \frac{1}{k}\sin\widehat{\delta}_\ell e^{i\widehat{\delta}_\ell}.$$

In the same way one introduces in the VPA the scattering amplitude *function* as

$$\mathcal{F}_\ell(r) = \frac{1}{k}\sin\delta_\ell(r)\,e^{i\delta_\ell(r)}.$$

$$(3.22)$$

The significance of $\mathcal{F}_\ell(r)$ for physical problems follows from the properties of $\delta_\ell(r)$, namely the value of the scattering amplitude function $\mathcal{F}_\ell(r=R)$ yields the physical scattering amplitude $\widehat{\mathcal{F}}_\ell^{(R)}$ for a particle subjected to the cut-off potential $V_\ell^{(R)}$ [see Fig. 3.1 for an illustration]. The relations derived for $\delta_\ell(r)$ in the previous section lead to an integro-differential equation for the SA function that can be employed for the determination of all bound state energies of the particle in a local or in a non-local potential. For this purpose one introduces the auxiliary functions

$$\begin{aligned}
f_\ell(r) &\equiv k\mathcal{F}_\ell(r) = e^{i\delta_\ell(r)}\sin\delta_\ell(r), \\
\widetilde{F}_\ell(r) &\equiv F_\ell(r)e^{i\delta_\ell(r)} = j_\ell(r) + ih_\ell^{(1)}(kr)f_\ell(r), \\
\widetilde{G}_\ell(r) &\equiv G_\ell(r)e^{i\delta_\ell(r)} = n_\ell(r) + h_\ell^{(1)}(r)f_\ell(r).
\end{aligned}$$

$$(3.23)$$

Expressing $\delta_\ell(r)$ through $f_\ell(r)$ we find

$$\frac{d\delta_\ell(r)}{dr} = \frac{1}{2if_\ell(r)+1}\frac{df_\ell(r)}{dr}. \tag{3.24}$$

Taking Eqs. (3.4, 3.6) into account, the phase equation (3.17) yields a determining equation for the function $f_\ell(r)$ which has the explicit form

$$\frac{df_\ell(r)}{dr} = \left(-\frac{1}{k}\right)\sqrt{2if_\ell(r)+1}\,\widetilde{F}_\ell(r)$$

$$\times \int_0^\infty dr' V_\ell(r,r')\frac{\widetilde{F}_\ell(r')}{\sqrt{2if_\ell(r')+1}}\exp\left[-\int_{r'}^r \frac{\dot{f}_\ell(s)ds}{(2if_\ell(s)+1)}\frac{\widetilde{G}_\ell(s)}{\widetilde{F}_\ell(s)}\right]. \tag{3.25}$$

This equation is to be solved with the initial condition

$$f_\ell(0) = 0.$$

In case the potential is local, Eq. (3.25) reduces to (cf. Refs. [15, 16])

$$\text{if}\quad V(r,r') = V(r')\delta(r-r')\quad\Rightarrow\quad \frac{df_\ell(r)}{dr} = -\frac{1}{k}\cdot V_\ell(r)\widetilde{F}_\ell^2(r). \tag{3.26}$$

The partial SA associated with the orbital momentum ℓ describes stationary and quasi-stationary states characterized by the behaviour of SA for complex wave vectors k. The poles of the SA along the positive imaginary semi-axis ($k = i\kappa_n$, $\kappa_n \in \mathbb{R}^+$) correspond to the energies of stationary states of the discrete spectrum, i.e. $E_n = (i\kappa_n)^2 < 0$. Therefore, the energy spectrum is deduced from $f_\ell(r,k)$ via the condition

$$f_\ell(\infty;\kappa_n) = \infty. \tag{3.27}$$

For rewriting Eq. (3.25) in the case of $k = i\kappa$, $\kappa > 0$ we note that the Riccati-Bessel functions of imaginary arguments can be expressed through the modified Riccati-Bessel functions of real arguments $p_\ell(\kappa r)$ and $q_\ell(\kappa r)$, namely

$$j_\ell(i\kappa r) = \beta p_\ell(\kappa r), \tag{3.28}$$

$$n_\ell(i\kappa r) = \frac{i}{\beta}\left[\beta^2 p_\ell(\kappa r) - q_\ell(\kappa r)\right], \tag{3.29}$$

$$h_\ell^{(1)}(i\kappa r) = \frac{1}{\beta}q_\ell(\kappa r), \tag{3.30}$$

$$\beta := (i)^{\ell+1}. \tag{3.31}$$

In equation (3.25) the main integrand with respect to r' can be transformed into the form

$$
\begin{aligned}
&\exp\left[-\int_{r'}^{r}\frac{df_\ell(s)}{2if_\ell(s)+1}\,\frac{\hat{G}_\ell(s)}{\hat{F}_\ell(s)}\right]\\
&=\exp\left[-\int_{r'}^{r}\frac{df_\ell(s)}{2if_\ell(s)+1}\,\frac{i\left[\beta^2 p_\ell(\kappa s)-q_\ell(\kappa s)\left(if_\ell(s)+1\right)\right]}{\left[\beta^2 p_\ell(\kappa s)+q_\ell(\kappa s)\left(if_\ell(s)\right)\right]}\right]\\
&=\exp\left[-\int_{r'}^{r}\frac{1}{2}\frac{d(2if_\ell(s)+1)}{2if_\ell(s)+1}-\frac{q_\ell(\kappa s)\,d(if_\ell(s))}{q_\ell(\kappa s)\,if_\ell(s)+\beta^2 p_\ell(\kappa s)}\right].
\end{aligned}
\tag{3.32}
$$

At first we note that the factor involving the integral can be simplified as

$$
\begin{aligned}
\exp\left[-\int_{r'}^{r}\frac{1}{2}\frac{d(2if_\ell(s)+1)}{2if_\ell(s)+1}\right]&=\exp\left\{-\frac{1}{2}\ln\left[\frac{2if_\ell(r)+1}{2if_\ell(r')+1}\right]\right\}\\
&=\frac{\sqrt{2if_\ell(r')+1}}{\sqrt{2if_\ell(r)+1}}.
\end{aligned}
\tag{3.33}
$$

Comparing with Eq. (3.25) and using the identity (3.20) we obtain the Volterra integro-differential equation of the first-kind for the determination of the function [2] $if_\ell(r)$

$$
\begin{aligned}
\frac{d(if_\ell(r))}{dr}&=-\frac{2}{\beta^2\kappa}\left[if_\ell(r)q_\ell(\kappa r)+\beta^2 p_\ell(\kappa r)\right]^2\\
&\quad\times\int_0^r dr'_\ell V(r,r')\cosh\left\{-\int_{r'}^r ds\,\frac{if_\ell(s)\dot{q}_\ell(\kappa s)+\beta^2\dot{p}_\ell(\kappa s)}{if_\ell(s)q_\ell(\kappa s)+\beta^2 p_\ell(\kappa s)}\right\}.
\end{aligned}
\tag{3.34}
$$

Equivalently, one can substitute

$$
if_\ell(r)\equiv\beta^2 y_\ell(r)
$$

to obtain from Eq. (3.34) the relation for the *real* function $y_\ell(r)$

$$
\begin{aligned}
\frac{dy_\ell(r)}{dr}&=-\frac{2}{\kappa}\left[y_\ell(r)q_\ell(\kappa r)+p_\ell(\kappa r)\right]^2\\
&\quad\times\int_0^r dr' V_\ell(r,r')\cosh\left\{-\int_{r'}^r ds\,\frac{y_\ell(s)\dot{q}_\ell(\kappa s)+\dot{p}_\ell(\kappa s)}{y_\ell(s)q_\ell(\kappa s)+p_\ell(\kappa s)}\right\}.
\end{aligned}
\tag{3.35}
$$

Since the potential $V_\ell(r,r')$ is hermitian and because the functions $p_\ell(\kappa r)$ and $q_\ell(\kappa r)$ are real, the initial condition $y_\ell(0,\kappa)=0$ implies that $y_\ell(r,\kappa)$ is real everywhere. The bound energy spectrum $E=-\kappa^2$ is given by the poles of $y_\ell(\infty,\kappa)$.

---

[2]The upper dot on the function $f(x)$ denotes differentiation with respect to the argument, i.e. $\dot{f}(x)=\frac{df(x)}{dx}$.

For a numerical implementation it is advantageous to regularize Eq. (3.35) by introducing the inverse function

$$y_\ell = 1/\phi_\ell. \tag{3.36}$$

Equivalently, one may opt to use a tangent function regularization, i. e. one applies the substitution

$$y_\ell = \tan \gamma_\ell. \tag{3.37}$$

Upon using the procedure (3.36), and taking Eq. (3.35) into account, the problem reduces to the solution of the integro-differential equation

$$\frac{d\phi_\ell(r)}{dr} = \frac{2}{\kappa} [q_\ell(\kappa r) + \phi_\ell(r) p_\ell(\kappa r)]^2$$
$$\times \int_0^r dr' V_\ell(r, r') \cosh \left\{ - \int_{r'}^r ds \frac{\dot{q}_\ell(\kappa s) + \phi_\ell(s)\dot{p}_\ell(\kappa s)}{q_\ell(\kappa s) + \phi_\ell(s) p_\ell(\kappa s)} \right\}. \tag{3.38}$$

On the other hand, if the option (3.37) is more appropriate one has to deal with an equation for γ_ℓ that has the following form

$$\frac{d\gamma_\ell(r)}{dr} = -\frac{2}{\kappa} [q_\ell(\kappa r) \sin \gamma_\ell(r) + p(\kappa r) \cos \gamma_\ell(r)]^2$$
$$\times \int_0^r dr' V_\ell(r, r') \cosh \left\{ - \int_{r'}^r ds \frac{\sin \gamma_\ell(s) \, \dot{q}_\ell(\kappa s) + \cos \gamma_\ell(s) \, \dot{p}_\ell(\kappa s)}{\sin \gamma(s) \, q_\ell(\kappa s) + \cos \gamma_\ell(s) \, p_\ell(\kappa s)} \right\}. \tag{3.39}$$

For the solution of this equation the initial condition

$$\gamma_\ell(0, \kappa) = 0$$

has to be imposed. The eigenenergies are then provided by the zeros of $\phi_\ell(\infty, \kappa)$, or by the condition

$$\gamma_\ell(\infty, \kappa) = (2n - 1)\pi/2, \; n \in \mathbb{N}.$$

More details and an explicit scheme for the numerical implementation can be found in Ref. [23].

3.4 Illustrative examples

For an illustration let us consider a neutral atom. The quantity of interest is the scattering amplitude function $\mathcal{F}_0(r)$ for a zero orbital momentum $\ell = 0$.

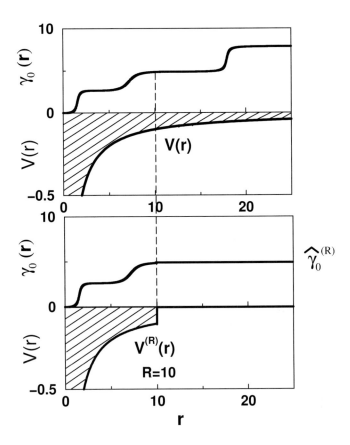

Figure 3.1: The arctangent of the scattering amplitude function $\gamma_0(r)$ for a particle subjected to the Coulomb potential $V(r) = -1/r$ (upper plot) or subjected to the Coulomb potential which is cut off at the radial distance $R = 10$ (lower plot). At the distance R the value of the function $\gamma_0(r = R)$ coincides with the arctangent of the scattering amplitude $\widehat{\gamma}_0^{(R)} = \arctan(\kappa\widehat{\mathcal{F}}_0^{(R)})$ of the particle that moves in the cut-off potential $V^{(R)}(r)$.

For a particle with the energy $E = -0.0556$ Hartree subject to an attractive Coulomb potential $V(r)$ the function $\gamma_0(r) = \arctan(\kappa\mathcal{F}_0(r))$ exhibits the behaviour shown in Fig. 3.1. The potential $V(r) = -1/r$ is finite everywhere and hence the derivative of $\gamma_0(r)$ is influenced by V at all radial distances. The physical, observable (arctangen) value of the scattering amplitude is given by the asymptotic (arctangent) value of the SA function, i.e. $\gamma_0(\infty) = \widehat{\gamma}_0 = \arctan\widehat{\mathcal{F}}_0$. For the potential $V^{(R)}$ obtained by cutting off $V(r)$ at the radial distance R, the value of the derivative of $\gamma_0(r)$ vanishes beyond the distance R. This means, the value of the

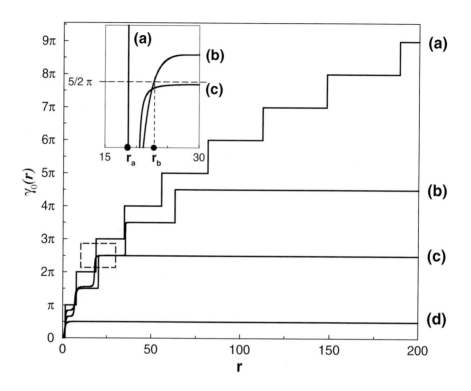

Figure 3.2: The function $\gamma_0(r)$ at different negative energies of a particle in an attractive Coulomb potential. The labels (a-d) refer to the different energies as follows: (a) $E = -0.5 \times 10^{-10}$ Hartree, (b) $E = -0.01$ Hartree, (c) $E = -0.0556$ Hartree, (d) $E = -0.5$ Hartree. The inset shows the behavior of $\gamma_0(r)$ near $\gamma_0(r) = \frac{5}{2}\pi$.

function $\gamma_0(r)$ coincides with the asymptotic value $\widehat{\gamma}_0^{(R)} \equiv \gamma_0(\infty) = \gamma_0(R)$, as illustrated in Fig. 3.1.

Figure 3.2 illustrates how $\gamma_0(r)$ behaves at different energies. We recall that we are dealing with s states only. The various energy levels occur when the value of the function $\gamma_0(r)$ is $(n - 1/2)\pi$. It might well be that this value is admitted at a finite radial distance R. In this case the eigenstate is associated with the cut-off potential $V^{(R)}$. From these considerations it is evident that the VP method provides, at the same time, the energy spectra of an infinite number of potentials $V^{(R)}$ where R is an arbitrary distance that falls within the range where the potential V is active.

The inset of Fig. 3.2 shows the behaviour of the function $\gamma_0(r)$ around its value $\gamma_0(r) = \frac{5}{2}\pi$ on an enlarged scale. The figure indicates the occurrence of eigenstates marked by (a) and (b) at finite radial distances and therefore, these eigenenergy states are associated with potentials which are cut off at corresponding radial distances. The inset reveals the behaviour of these states at the distances r_a and r_b. From the inset it is also clear that for the case labelled (c) (which corresponds to the 3s-state) the function γ_0 will reach $\frac{5}{2}\pi$ at an infinitely large distance and hence the corresponding energy eigenstate is an eigenstate of the Coulomb potential $V(r)$. The ground state energy E_d, associated with the potential $V(r)$, is indicated by the fact that $\lim_{r\to\infty} \gamma_0(r, E_d) \to \pi/2$.

4 Ground states of many-electron systems

In previous chapters we discussed the quantum mechanical methods and tools for the description of the bound and the excited states of a single particle in an external potential. With growing number of interacting particles an exact treatment becomes however intractable. Examples of this case are many-body systems with non-separable quantum Hamiltonians, such as large atoms, molecules, clusters and solids. Therefore, various concepts have been put forward to circumvent this situation. In particular, nowadays the ground state of a variety of many-body systems can be calculated efficiently by searching for the minimal energy of the system which singles out the ground state. In contrast, variational approaches to excitation processes [24] require more elaborate numerical procedures. On the other hand, the objects of physical interest are the transition amplitudes for certain reactions and response functions to external perturbations. These quantities are determined not only by the excited state achieved in the process, but also by the initial state, which in most cases is the ground state. In fact, weak perturbations (compared to the strength of interactions present in the probe) results is small fluctuations in the values of the relevant physical observables $<\mathcal{O}>$ around their ground state expectation values $<\mathcal{O}>_0$ [52, 84]. The calculations of $<\mathcal{O}>$ are efficiently performed by means of the so-called random-phase approximation (RPA) [52, 84], which will be discussed in chapter 14. In view of this situation its useful to give a brief overview on some of the ground-state methods, before addressing the treatment of excited and continuum states.

A widely followed rout for finding the ground-state of many interacting particles is to reduce the problem to the solution of a single particle moving in an effective (non-local) field. The Rayleigh-Ritz variational principle is then employed to find the ground state with the minimal energy. A prominent example is the Hartree-Fock (HF) method [25, 26]. In a HF approach the exchange interaction between parallel-spin electrons is correctly treated leading a non-local exchange potential for the single-particle motion. The correlation hole for antiparallel spin electrons is neglected however. More elaborate treatments are provided by the variational configuration interaction (CI) method (see e. g. Ref. [27, 28, 53] and references therein), and by the projective and size-extensive coupled cluster (CC) method [29, 28, 30, 53]. The Møller-Plesset many-body perturbation approach [31] and the wave-function based Monte Carlo method [33, 243] have as well proved to be powerful tools. On the other hand, density-based concepts [41, 43, 54, 55, 56], and in particular the local-density approximation within the density functional theory, proved to be exceptionally versatile and effective, and therefore they are now widely used in various branches of physics and chemistry. Furthermore, the Green's function method (see e. g. [52, 84, 85, 107] and references therein), which is based on the quasi-particle concept, offers a systematic and a diagrammatic tool for the treatment of the ground and the excited states of correlated systems. In chapter 14 the Green's function concept is covered, whereas is this chapter a brief overview is provided on the main ground state methods, mentioned above.

4.1   Time-scale separation

Generally, inhomogeneous electronic systems consist of heavy immobile ions surrounded by the much lighter electrons. While the electrostatic forces between the particles are generally comparable, the large difference between the electrons' and the ions' masses implies that the momentum transfer to the ionic motion is small. In addition, due to the large difference between the velocities of the ions and the electrons we can assume that, on the typical time-scale for the nuclear motion, the electrons may relax to the ground-state for a fixed configuration of the ions. This approximative separation of the electronic and the ionic motion has been pointed out the by Born and Oppenheimer (BO) in 1927 [34]. Our aim here is to illustrate the general idea of the BO approximation, which is utilized in later chapters of this book. Detailed discussions of the various aspects of the BO approximation can be found in numerous books on quantum chemistry (e. g. [35]).

Generally, the stationary state of a non-relativistic system consisting of N_e electrons and N_i less mobile ions[1] is described by the solution Ψ of the time-independent Schrödinger

---

[1]This approximation does not exclude the possibility of studying the dynamic of the ions, for once the electronic configuration is known, the ionic equation of motion can be solved, which yields the phonon spectrum (cf. Eq. (4.7)).

equation. At first, one writes the wave function Ψ is the form

$$\Psi\left(\{\mathbf{r}_i\},\{\mathbf{R}_\alpha\}\right) = \chi_{\{\mathbf{R}_\alpha\}}\left(\{\mathbf{r}_i\}\right)\Phi\left(\{\mathbf{R}_\alpha\}\right), \tag{4.1}$$

where $\{\mathbf{r}_i\}$ and $\{\mathbf{R}_\alpha\}$ are two sets of position vectors that specify, in an appropriately chosen reference frame, the coordinates of respectively the electrons and the ions. The charges and masses of the latter are denoted by Z_α and M_α respectively . The electronic wave function $\chi_{\{\mathbf{R}_\alpha\}}\left(\{\mathbf{r}_i\}\right)$ depends parametrically on the positions of the ions $\{\mathbf{R}_\alpha\}$ and is determined as a solution of the electronic part of the Schrödinger equation

$$\left[H_e - \mathcal{E}_e(\{\mathbf{R}_\alpha\})\right]\chi_{\{\mathbf{R}_\alpha\}}\left(\{\mathbf{r}_i\}\right) = 0, \tag{4.2}$$

where the electronic Hamiltonian H_e has the explicit form

$$\begin{aligned}
H_e &= \sum_{i=1}^{N_e}-\frac{1}{2}\nabla_{\mathbf{r}_i}^2 + \sum_{i,j>i=1}^{N_e}\frac{1}{|\mathbf{r}_i-\mathbf{r}_j|} - \sum_{i=1}^{N_e}\sum_{\alpha=1}^{N_i}\frac{Z_\alpha}{|\mathbf{r}_i-\mathbf{R}_\alpha|}, \\
&= \sum_{i=1}^{N_e}-\frac{1}{2}\nabla_{\mathbf{r}_i}^2 + \frac{1}{2}\sum_{i}^{N_e}\sum_{j\neq i}^{N_e}\frac{1}{|\mathbf{r}_i-\mathbf{r}_j|} + \sum_{i=1}^{N_e}V_{ext}^{(i)}(\mathbf{r}_i).
\end{aligned} \tag{4.3}$$

The energy eigenvalues $\mathcal{E}_e(\{\mathbf{R}_\alpha\})$ of the electronic Hamiltonian H_e are functions of the positions of the ions. The total wave function Ψ, given by Eq. (4.1), is the solution of the eigenvalue equation

$$\left[H_e - \sum_{i=1}^{N_i}\frac{1}{2M_\alpha}\nabla_{\mathbf{R}_\alpha}^2 + \sum_{\beta>\alpha=1}^{N_i}\frac{Z_\alpha Z_\beta}{|\mathbf{R}_\alpha-\mathbf{R}_\beta|}\right]\Psi\left(\{\mathbf{r}_i\},\{\mathbf{R}_\alpha\}\right)$$
$$= \mathcal{E}\Psi\left(\{\mathbf{r}_i\},\{\mathbf{R}_\alpha\}\right). \tag{4.4}$$

Taking Eq. (4.2) into account one concludes that

$$\chi\left[-\sum_{\alpha=1}^{N_i}\frac{1}{2M_\alpha}\nabla_{\mathbf{R}_\alpha}^2 + \sum_{\beta>\alpha=1}^{N_i}\frac{Z_\alpha Z_\beta}{|\mathbf{R}_\alpha-\mathbf{R}_\beta|} + \mathcal{E}_e(\{\mathbf{R}_\alpha\})\right]\Phi\left(\{\mathbf{R}_\alpha\}\right) - \mathcal{A}$$
$$= \mathcal{E}\Psi\left(\{\mathbf{r}_i\},\{\mathbf{R}_\alpha\}\right), \tag{4.5}$$

where the non-adiabatic coupling cross term $\mathcal{A}$ is given by

$$\mathcal{A} := \frac{1}{2M_\beta}\left\{\sum_\beta 2\left(\nabla_{\mathbf{R}_\beta}\Phi\right)\cdot\left(\nabla_{\mathbf{R}_\beta}\chi\right) + \Phi\nabla_{\mathbf{R}_\beta}^2\chi\right\}. \tag{4.6}$$

The adiabatic energy contribution $\mathcal{E}_e$ is generally dominant over the remaining non-adiabatic terms whose values can be calculated using perturbation theory [32].

Neglecting the non-adiabatic function Eq. (4.6) results in a separation of the electrons and the ions dynamics. The former is described by (4.2), whereas the ionic motion is treated by solving the Schrödinger equation

$$
\left[-\sum_{\alpha=1}^{N_i} \frac{1}{2M_\alpha} \nabla^2_{\mathbf{R}_\alpha} + \sum_{\beta>\alpha=1}^{N_i} \frac{Z_\alpha Z_\beta}{|\mathbf{R}_\alpha - \mathbf{R}_\beta|} + \mathcal{E}_e(\{\mathbf{R}_\alpha\}) \right] \Phi(\{\mathbf{R}_\alpha\})
$$

$$
= \mathcal{E}\Phi(\{\mathbf{R}_\alpha\}), \quad (4.7)
$$

i. e. the electronic energy $\mathcal{E}_e(\{\mathbf{R}_\alpha\})$ enters as a part of the adiabatic potential surface that dictates the motion of the ions[2]. The adiabatic technique of separating various degrees of freedom is of a general nature. In the context of the present work we will utilize the adiabatic idea, in particular in section 9.6.1, for separating out the laboratory-frame from the body-fixed motion in a three-body system.

4.2   Hartree-Fock approximation

Within the Hartree Fock theory [25, 26] the many-body eigenfunction Ψ is written as an antisymmetrized product (Slater determinant) of single-particle spin-orbitals $\psi_j(\mathbf{x}_j)$, where $\mathbf{x}_j$ stands for spin and spatial coordinates. The many-body problem is then reduced to the task of minimizing the energy functional

$$
E[\Psi] = \langle \Psi| \, \mathbf{H} \, |\Psi \rangle , \tag{4.8}
$$

which, according to the Rayleigh-Ritz principle, yields the ground-state wave function and the associated energy. In terms of the single-particle orbitals the minimization procedure is expressed as

$$
\frac{\delta}{\delta\psi} \left[<\mathbf{H}> - \sum_j \epsilon_j \int |\psi_j|^2 d\mathbf{r} \right] = 0, \tag{4.9}
$$

---

[2]In practice the use of Newtonian mechanics is generally sufficient for the description of the ionic motion (whilst the adiabatic potential surface is determined from quantal calculations of the electronic states). The relaxation of the nuclear positions to achieve the minimum-energy configuration is performed by means of molecular dynamic simulations [36, 37].

where ϵ_i are Lagrange multipliers introduced to ensure that the orbitals ψ_i are normalized. The key point is that Eq. (4.9) leads to a set of one-electron Schrödinger-type determining equations for the single particle orbitals, the well-known Hartree-Fock equations

$$
\left(-\frac{1}{2} \nabla_i^2 + V_{ion}(\mathbf{r}) \right) \psi_i(\mathbf{r}) + \sum_j \int d\mathbf{r'} \frac{|\psi_j(\mathbf{r'})|^2}{|\mathbf{r} - \mathbf{r'}|} \psi_i(\mathbf{r})
$$
$$
- \sum_j \delta_{\sigma_i \sigma_j} \int d\mathbf{r'} \frac{\psi_j^*(\mathbf{r'}) \psi_i(\mathbf{r'})}{|\mathbf{r} - \mathbf{r'}|} \psi_j(\mathbf{r}) = \epsilon_i \psi_i(\mathbf{r}),
$$

(4.10)

where σ_j are spin indices. The HF equation can be written in the form

$$
-\frac{1}{2} \nabla^2 \psi_i(\mathbf{r}) + V_{ion}(\mathbf{r}) \psi_i(\mathbf{r}) + \hat{U} \psi_i(\mathbf{r}) = \epsilon_i \psi_i(\mathbf{r}).
$$

(4.11)

Here $\hat{U}$ is a non-local potential operator whose action can be inferred from Eq. (4.10). V_{ion} is the local ionic (external) potential.

From a numerical point of view Eq. (4.11) is a one-electron equation involving a nonlocal potential. The eigensolutions can be found, for example by means of the methods outlined in chapter 3.

The HF-equation (4.10) contains four terms. The first and the second term of the right hand side of Eq. (4.10) describe the kinetic energy and the (external) electron-ion potential contributions to the total energy. The third term, also called the Hartree term, is due to the electrostatic potential arising from the charge distribution of the electrons. It should be mentioned that this term contains an unphysical self-interaction of electrons when $j = i$, which is removed by equivalent contributions from the fourth, or the exchange term. The exchange term is a result of the Pauli principle included in the assumption that the total wave function is an antisymmetrized product of single particle orbitals. The effect of accounting for exchange is that electrons with the same spin projections avoid each other leading to the so-called exchange hole. In contrast to exchange effects, the Hartree-Fock theory does not treat correctly the Coulomb correlations between the electrons since the electron is assumed to be subject to an average non-local potential arising from the other electrons. This shortcoming may lead to serious failures in predicting the electronic structure of materials [38]. Nevertheless, as will be discussed in chapter 14, the HF model provides a useful starting point for more elaborate many-body calculations.

4.2.1   Basis set expansion

For finding numerical solutions of the HF equations, it is customary to expand the orbitals
[28] in a basis set $\{\varphi_j\}$ of a finite size K, i. e. one writes

$$\psi_i = \sum_k^K c_{ij}\varphi_j. \tag{4.12}$$

With this ansatz for the HF orbitals the relation (4.10) transforms into a set of matrix equations for the expansion coefficients c_{ij} that can be solved, e. g. by iterative diagonalisation.
The basis must be sufficiently complete so that the Hartree-Fock orbitals are correctly reproduced. The suitable choice depends on the physical problem under study. E. g., for extended
(delocalized) electronic systems plane waves are useful, whereas for (molecular) systems with
localised electrons Gaussians basis sets proved to be most effective.

4.3   Configuration interaction

The configuration interaction (CI) method [27, 28, 53] is a variational-based approach that
goes beyond the Hartree-Fock ansatz of a single-determinant for the total wave function. In
contrast to Eq. (4.12), where the HF orbitals are written in terms of basis functions, the CI
method relies on a linear expansion of the exact many-body wave function Ψ in terms of
Slater determinants. This is done via the ansatz

$$\Psi = \sum_{k=0}^{\infty} c_k D_k. \tag{4.13}$$

In principle, the determinants D_k can be any complete set of N_e-electron antisymmetric functions. However, in practice D_k are often given by means of HF orbitals since the HF determinant D_0 is expected to be the best single-determinant approximation to the exact wave
function Ψ. This highlights the anticipation that the first term in the expansion is dominant,
i.e. $c_0 \approx 1$. This is because in many cases the HF energy is quite close to the total energy and
what is left is a small part that needs to be reproduced by a large number of configurations.
The number of determinants in the expansion (4.13) truncated at the term k_{max} is

$$k_{max} = \frac{K!}{N_e!(K-N_e)!}, \tag{4.14}$$

where K is the number of basis states ($K \gg N_e$, cf. Eq. (4.12)) (we recall N_e is the number of the electrons). Therefore, in practice the CI method can be applied to relatively small systems only. In addition, care should be taken as to where to truncate the expansion (4.13). Typically, the expansion is terminated after only double or quadruple excitations from the reference determinant D_0. For the ground state, this amounts to replacing occupied orbitals by an unoccupied one. The truncation scheme is called the CI singles-doubles (CISD) and CI singles-doubles-triples-quadruples (CISDTQ) method. Naturally, this procedure leaves asides all the other terms in the expansion (4.13). An additional known problem of the CI method is that the method is size-none-extensive. For finite reference spaces the CI scheme does not perform equally well when the size of the system varies. This shortcoming induces some difficulties when results for systems of different sizes are compared.

4.4   The coupled cluster method

The lack of size extensivity and the substantial computational cost of the CI method has led to the development of several related methods.

The coupled-cluster (CC) scheme [29, 28, 30, 53] is a non-variational method that resolves the problem of the size non-extensivity of the CI and yields in most cases very accurate results. Within the CC approach the wave function is obtained from the reference HF solution $\Psi_{HF} = D_0$ as

$$\Psi_{CC} = \exp(\mathbf{T})\Psi_{HF} \quad , \tag{4.15}$$

where $\mathbf{T}$ is an operator that generates the k fold excitations from a reference state

$$\mathbf{T} = \sum_k \mathbf{T}_k = \mathbf{T}_1 + \mathbf{T}_2 + \dots \quad . \tag{4.16}$$

For example, applying the operator $\mathbf{T}_2$ on the HF state generates excitations of pairs of occupied states ij to pairs of virtual states kl, i.e.

$$\mathbf{T}_2 \Psi_{HF} = \sum c_{ij}^{kl} D_{ij}^{kl} \ . \tag{4.17}$$

The expansion coefficients c_{ij}^{kl} have to be calculated self-consistently. The CC doubles wave function has thus the form

$$\Psi_{CCD} = (1 + \mathbf{T}_2 + \mathbf{T}_2^2/2 + \mathbf{T}_2^3/3! + \dots)\Psi_{HF} \ , \tag{4.18}$$

$$\Psi_{CCD} = D_0 + \sum t_{ij}^{kl} D_{ij}^{kl} + \frac{1}{2} \sum \sum t_{ij}^{kl} t_{mn}^{op} D_{ijmn}^{klop} + \dots . \tag{4.19}$$

It can be shown [29] that the CC singles-doubles calculations scale as the sixth power of the number of basis states. This sets a practical bound on the number of electrons that can be treated accurately with this method. For example, calculations including up to quadruple excitations scale as the tenth power of the number of states. Therefore, in practice, the CC expansion (4.17) is often truncated after including all double excitations.

4.5   Variational and diffusion Monte Carlo techniques

The variational Monte Carlo method (VMC) uses the Monte Carlo integration scheme [33, 243] and the variational principle to find out the best trial many-body function Ψ_t that minimizes the total energy (hereafter, for simplicity, we use N as the number of electrons)

$$E = \frac{\int \Psi_t^* H \Psi_t \, d\mathbf{r}_1 \dots d\mathbf{r}_N}{\int \Psi_t^* \Psi_t \, d\mathbf{r}_1 \cdots d\mathbf{r}_N} = \frac{\int |\Psi_t|^2 \left(\frac{H\Psi_t}{\Psi_t} \right) d\mathbf{r}_1 \cdots d\mathbf{r}_N}{\int |\Psi_t|^2 \, d\mathbf{r}_1 \cdots d\mathbf{r}_N} .$$

Expectation values of other observables can be found similarly. The position-dependent term $\frac{H\Psi_t}{\Psi_t}$ is called the local energy E_L and is central to both variational and diffusion Monte Carlo (DMC) methods [3]. From this quantity one calculates by means of, e.g. the Metropolis algorithm [338], the configuration sample from the probability density distribution and obtains the energy E by averaging E_L over the configuration set. In this way accurate energies of many-body wave functions are obtained [243]. The variance of E_L serves as an indicator for the accuracy of E (and the corresponding variational functions) but this does not guarantee the accuracy of other calculated observables.

4.6   Density functional theory

In contrast to the wave function-based HF and the post HF methods, the density functional theory (DFT) relies on the concept of electronic density. In its original version DFT deals with ground-state properties only. This section provides a brief overview on the DFT theory.

---

[3] In the DMC (and related) methods one converts the differential Schrödinger equation with certain boundary conditions into an integral equation and solves for the integral equation by stochastic methods. The motivation for this procedure is the formal similarity of the Schrödinger equation in imaginary time and the diffusion equation (see appendix A.4). This renders possible the use of a random process to solve the imaginary time Schrödinger equation [39, 40].

For more details the reader is referred to the extensive literature on this topic (see e.g. [54, 55, 56, 57]).

4.6.1   The Hohenberg-Kohn theorem

The basic idea of DFT relies on the following observation: the electronic Hamiltonian (4.3), within the BO approximation, consists of a static external potential $V_{ext} = \sum_{i=1}^{N} V_{ext}^{(i)}(\mathbf{r}_i)$ generated by the surrounding ions and of a remainder term

$$F = \sum_{i=1}^{N} \left[-\frac{1}{2}\nabla_{\mathbf{r}_i}^2 + \frac{1}{2}\sum_{j \neq i}^{N} \frac{1}{|\mathbf{r}_i - \mathbf{r}_j|} \right] ,$$

which is universal to all electronic systems with N number of particles. Thus, the ground state Ψ_0 of a specific system is determined once N and V_{ext} are given. This means that the ground state and the electronic density n_0 it generates, are functional of N and V_{ext}.

In 1964 Hohenberg and Kohn (HK) [41] proved the remarkable statement that, within an additive constant, an external potential V_{ext} is uniquely determined by the corresponding single-particle ground state electronic density. This important statement has the following consequence: For densities $n(\mathbf{r})$ which are ground-state densities for some external potential[4] V_{ext} the functional $\bar{\mathcal{F}}[n] = \langle \Psi|F|\Psi \rangle$ is determined uniquely. This is because $n(\mathbf{r})$ fixes, in addition to V_{ext}, the number of particles $N = \int d\mathbf{r}n(\mathbf{r})$ and hence the operator F and the associated wave function Ψ are determined. This leads to the second HK statement.

Let us define the functional

$$\mathcal{E}[n] = \mathcal{F}[n] + \int d\mathbf{r}\, n(\mathbf{r})V_{ext}(\mathbf{r}), \tag{4.20}$$

where $V(\mathbf{r})$ is an arbitrary external potential determined by $n(\mathbf{r})$ (and unrelated to V_{ext}) and $\mathcal{F}[n]$ is an unknown universal functional. The statement is then, for all v representable densities $n(\mathbf{r})$ the variational principle $\mathcal{E}[n] \geq E_0$ applies, where E_0 is the ground-state energy for N electrons in the external potential $V(\mathbf{r})$. This means, minimizing $\mathcal{E}[n]$ yields the ground state energy E_0.

Based on a constrained search formulation, Levy [58] gave the following arguments for the validity of the Hohenberg-Kohn theorem. Consider a conservative system consisting of

---

[4]Such densities are called v representable.

N particles interacting via the pair potentials v_{ij} and are subject to an external potential V_{ext}. The Hamiltonian is then

$$H = T + \frac{1}{2} \sum_{i \neq j} v_{ij} + V_{ext} = F + V_{ext},$$

where T stands for the kinetic energy operator. One considers the functional $\mathcal{F}$ defined as [58]

$$\mathcal{F}[n(\mathbf{r})] = \min_{n_\Psi(\mathbf{r}) \to n(\mathbf{r})} \langle\, \Psi|F|\Psi \,\rangle. \tag{4.21}$$

This means, given a set of functions $\{\Psi\}$ that yield the single-particle density[5] $n_\Psi(\mathbf{r}) = n(\mathbf{r})$, the functional $\mathcal{F}$ takes the minimal value in the set of the expectation values $\{\langle\, \Psi|F|\Psi \,\rangle\}$. Furthermore, let us assume Ψ_0 to be the ground state yielding the density $n_0(\mathbf{r})$ and having the energy E_0.

Considering an N electron state $\Psi_{[n]}$ that yields the density $[n]$ and minimizes $\mathcal{F}[n]$, then we obtain from the definitions of F and the energy functional $\mathcal{E}$

$$\mathcal{E}[n(\mathbf{r})] = \mathcal{F}[n(\mathbf{r})] + \int d\mathbf{r}\, n(\mathbf{r}) V_{ext}(\mathbf{r}) = \langle\Psi_{[n]}|F + V_{ext}|\Psi_{[n]}\rangle. \tag{4.22}$$

The variational principle sets a bound on the values of $\mathcal{E}[n(\mathbf{r})]$, namely

$$\mathcal{E}[n(\mathbf{r})] \geq E_0 = \langle\Psi_0|F + V_{ext}|\Psi_0\rangle. \tag{4.23}$$

This is valid for all densities obtainable from the N electron wave functions (i.e. N representable[6]). On the other hand, in the sense of Eq. (4.21), the ground-state wave function Ψ_0 is just a member of the set $\{\Psi_{n_0}\}$ and hence the relations apply

$$\mathcal{F}[n_0(\mathbf{r})] \leq \langle\Psi_0|F|\Psi_0\rangle,$$
$$\Rightarrow \quad \mathcal{F}[n_0(\mathbf{r})] + \int d\mathbf{r}\, n_0(\mathbf{r}) V_{ext}(\mathbf{r}) \leq \langle\Psi_0|F|\Psi_0\rangle + \int d\mathbf{r}\, n_0(\mathbf{r}) V_{ext}(\mathbf{r}),$$
$$\Rightarrow \quad \mathcal{E}[n_0(\mathbf{r})] \leq E_0. \tag{4.24}$$

Combining these findings with Eq. (4.23) we deduce

$$\mathcal{E}[n(\mathbf{r})] \geq \mathcal{E}[n_0(\mathbf{r})] = E_0. \tag{4.25}$$

---

[5]We recall that a given Ψ produces a unique $n(\mathbf{r})$. The reverse statement is however invalid because there is in general an infinite number of Ψ, that yield the same density $n(\mathbf{r})$.

[6]The requirement of the N representability of the density imposes the conditions [42] that $\int d\mathbf{r}\, |\nabla n^{1/2}(\mathbf{r})|^2$ is real and finite.

4.6.2   The Kohn-Sham equations

The HK theorem proofs the functional relation between physical observables and the ground-state density. Thus, instead of dealing with the many-electron wave function (which is generally dependent on $3N$ variables), it suffices to treat the ground-state properties using functions of only three variables, namely the single particle density. Thus, numerical efforts scale linearly with the system size. This great advantage goes on the expense that some expressions for the unknown universal functional $\mathcal{F}$ have to be determined. Once such approximations for $\mathcal{F}$ are available, one can perform the practical implementation according to the Kohn and Sham theory [43]. The correlated many-body system is formally mapped onto a fictitious system of non-interacting particles. Then a standard variational problem is formulated using Lagrange multipliers to account for certain constraints and the whole task reduces to the solution of a set of coupled differential equations. This is achieved by representing the single-particle density $n(\mathbf{r})$ by means of a set of auxiliary single-particle orbitals

$$n(\mathbf{r}) = \sum_{i=1}^{N} \psi_i^*(\mathbf{r})\psi_i(\mathbf{r}). \tag{4.26}$$

The next step consist of separating $\mathcal{E}[n(\mathbf{r})]$ into expressions related to the kinetic energy $\mathcal{T}[n(\mathbf{r})]$, the two-body interaction $\mathcal{W}[n(\mathbf{r})]$ and the external potential term, i.e.

$$
\begin{aligned}
\mathcal{E}\left[n(\mathbf{r})\right] &= \mathcal{T}\left[n(\mathbf{r})\right] + \mathcal{W}\left[n(\mathbf{r})\right] + \int d\mathbf{r}\, n(\mathbf{r}) V_{\text{ext}}(\mathbf{r}), \\
&= \mathcal{T}_s\left[n(\mathbf{r})\right] + \mathcal{W}_H\left[n(\mathbf{r})\right] \\
&\quad + \left\{ \mathcal{T}\left[n(\mathbf{r})\right] - \mathcal{T}_s\left[n(\mathbf{r})\right] + \mathcal{W}\left[n(\mathbf{r})\right] - \mathcal{W}_H\left[n(\mathbf{r})\right] \right\} \\
&\quad + \int d\mathbf{r}\, n(\mathbf{r}) V_{\text{ext}}(\mathbf{r}), \\
&= \mathcal{T}_s\left[n(\mathbf{r})\right] + \mathcal{W}_H\left[n(\mathbf{r})\right] + \mathcal{E}_{xc}[n(\mathbf{r})] + \int d\mathbf{r}\, n(\mathbf{r}) V_{\text{ext}}(\mathbf{r}).
\end{aligned} \tag{4.27}
$$

The functional $\mathcal{T}_s\left[n(\mathbf{r})\right]$ derives from the kinetic energy of the non-interacting electron gas that possesses the electronic density $n(\mathbf{r})$, i.e.

$$\mathcal{T}_s\left[n(\mathbf{r})\right] = -\frac{1}{2}\sum_{i=1}^{N}\int d\mathbf{r}\,\psi_i^*(\mathbf{r})\nabla^2\psi_i(\mathbf{r}), \tag{4.28}$$

whereas $\mathcal{W}_H\left[n(\mathbf{r})\right]$ is given by the Hartree energy

$$\mathcal{W}_H\left[n(\mathbf{r})\right] = \frac{1}{2}\int\int d\mathbf{r}d\mathbf{r}'\,\frac{n(\mathbf{r})n(\mathbf{r}')}{|\mathbf{r}-\mathbf{r}'|}.$$

Thus, many-body effects are contained in the exchange and correlation part

$$\mathcal{E}_{xc}[n(\mathbf{r})] = \mathcal{T}[n(\mathbf{r})] - \mathcal{T}_s[n(\mathbf{r})] + \mathcal{W}[n(\mathbf{r})] - \mathcal{W}_H[n(\mathbf{r})]. \tag{4.29}$$

Recalling that the reference system we chose, i.e. the non-interacting electron gas, has the same electronic density as the real system, we conclude that if the (exact) ground state can be written as a direct product of single-particle orbitals then the terms $\mathcal{T}_s$ and $\mathcal{W}_H$ describe the kinetic and the interaction energy. This is the reason for writing $\mathcal{E}$ in the form of Eq. (4.27). The variational principle for the Hohenberg-Kohn density-functional under the constraint $\int d\mathbf{r}\, n(\mathbf{r}) = N$ is formulated as

$$\delta \left\{ \mathcal{E}[n(\mathbf{r})] + \int d\mathbf{r} V_{ext} n(\mathbf{r}) - \mu \left[\left(\int d\mathbf{r}\, n(\mathbf{r}) \right) - N \right] \right\} = 0, \tag{4.30}$$

where μ is a Lagrange multiplier. Replacing the variation with respect to the density by a variation with respect to the orbitals ψ_i leads to the Kohn-Sham (KS) equations[7]

$$\left[-\frac{1}{2}\nabla_i^2 + V_{ext}(\mathbf{r}) + V_H(n,\mathbf{r}) + V_{xc}(\mathbf{r}) - \epsilon_i \right] \psi_i(\mathbf{r}) = 0,$$

$$\left[-\frac{1}{2}\nabla_i^2 + V_{eff}(\mathbf{r}) - \epsilon_i \right] \psi_i(\mathbf{r}) = 0, \tag{4.31}$$

where the exchange and correlation potential $V_{xc}(\mathbf{r})$ derives from

$$V_{xc}(\mathbf{r}) = \frac{\delta \mathcal{E}_{xc}[n(\mathbf{r})]}{\delta n(\mathbf{r})}. \tag{4.32}$$

The (density-dependent) effective one-particle Kohn-Sham!potential is given in terms of $V_{ext}(\mathbf{r})$, the Hartree potential term $V_H(n,\mathbf{r})$ and $V_{xc}(\mathbf{r})$ as $V_{eff}(\mathbf{r}) = V_{ext}(\mathbf{r}) + V_H(n,\mathbf{r}) + V_{xc}(\mathbf{r})$. Therefore, the KS equations depend on the density which is generated by the orbitals obtained from (4.31) (cf. 4.26). Thus, the Kohn-Sham equations have to be solved self-consistently[8]. The missing part in this solution procedure is $\mathcal{E}_{xc}$ (that yields V_{xc}). If one assumes $V_{xc} = 0$ the above procedure reduces to the Hartree approximation, whereas the assumption $V_{xc} = V_x$ leads to the HF approximation (V_x is the exchange potential) and the orbitals play the same role as in the HF theory. Generally, however, the orbitals ψ_i and the associated eigenvalues ϵ_i are just mathematical objects representing the physical ground-state. Unfortunately, the general form of the effective KS potential is unknown since $\mathcal{E}_{xc}$ has

---

[7]For finding the orbitals the Schrödinger-type equation has to be solved. The computational efforts (of diagonalizing the respective Hamiltonian) scale as the cube of the system-size, in contrast to the original linear scaling (with the system size) of finding the minimum of the HK functional only.

[8]The convergence to the ground-state minimum is guaranteed by the convex nature of the density-functional [44].

been evaluated only for few simple systems [54, 55, 56]. Formally however, one can show that in contrast to the non-local HF potential the KS potential is multiplicative and contains correlation effects (in as much as these effects are incorporated in $\mathcal{E}_{xc}$).

4.6.3   The local density approximation

The above approach to the many-body problem depends critically on the approximation to the generally unknown functional $\mathcal{E}_{xc}$. The most prominent of the approximations that have been put forward is the local density approximation (LDA) in which the value of an inhomogeneous electronic system with a density $n(\mathbf{r})$ is derived from the results for the exchange-correlation energy of the homogeneous electron gas that has the same density $n(\mathbf{r})$. This is done by assuming that, for inhomogeneous electronic systems, the contribution to the exchange-correlation energy originating from an infinitesimal spatial volume $\Delta\mathbf{r}$ is the same as in the case of a homogeneous electron gas with the same density $n(\mathbf{r})$ found in the local infinitesimal region $\Delta\mathbf{r}$, i.e.

$$\mathcal{E}_{xc}^{LDA}[n(\mathbf{r})] = \int d\mathbf{r}\, \epsilon_{xc}(n(\mathbf{r}))\, n(\mathbf{r})\,, \tag{4.33}$$

where $\epsilon_{xc}(n(\mathbf{r}))$ is the exchange-correlation energy per electron in a homogeneous electron gas with the electronic density $n(\mathbf{r})$. Accurate expressions for ϵ_{xc} are obtained from the Green's function Monte Carlo method [45]. Generally, one can say for systems with slow varying charge densities the LDA provides a good description of the ground-state properties. Unfortunately, in real systems the density generally varies rapidly. Nevertheless, the calculations using LDA are in surprisingly good accord with experimental findings for a wide range of materials [46]. This is traced back in part to the fact that the LDA functional satisfies certain sum rule for the exchange-correlation hole that must be fulfilled by the exact functional [46]. In fact, as shown by quantum Monte Carlo calculations, the LDA benefits from a cancellation of errors in the LDA exchange and correlation energies [47]. Therefore, care should be taken when improving on LDA. E.g. an improvement of only the exchange or the correlation term may invalidate the error cancellation which may adversely affect the performance of the theory. For strongly correlated systems the LDA is found be inaccurate (see e. g. [28] and references therein). E.g. the LDA calculations predict transition metal oxides, which are Mott insulators, to be either semiconductors or metals. In addition, the LDA does not describe correctly hydrogen bonding and van der Waals forces. Furthermore, within the LDA the asymptotic behaviour of the density is incorrect. The effective potential behaves as $V_{xc}^{LDA} \xrightarrow[r \to \infty]{} e^{-\gamma r}$ in contrast to the exact $V_{xc} \to -\frac{e^2}{r}$ which precludes a correct prediction of the properties of negative ions.

4.6.4   Gradient corrections

A way to improve on the LDA is to include gradient corrections by assuming $\mathcal{E}$ to be a functional of the density. Its gradient can be obtained by expanding the xc-energy in terms of gradients of the density

$$
\begin{aligned}
\mathcal{E}_{xc}[n] &= \mathcal{E}_{xc}^{[0]} + \mathcal{E}_{xc}^{[2]} + \mathcal{E}_{xc}^{[4]} + \dots, & (4.34) \\
&= \int d\mathbf{r} \left\{ \epsilon_{xc}^{LDA}(n(\mathbf{r})) + B_{xc}^{[2]}(n(\mathbf{r}))[\nabla n(\mathbf{r})]^2 + \dots \right\}. & (4.35)
\end{aligned}
$$

This procedure, which is termed generalized gradient approximation (GGA), improves the LDA description of the exchange and correlation hole only for short separations of the two interacting particles, for larger inter-particle distances it is oscillatory and of the wrong sign. In addition, the damping is poor and the exact sum rules for the exchange and correlation hole are not satisfied. To correct these deficiencies a real space cut-off may be carried out which removes the incorrect tails of the exchange and correlation hole, so that it matches as close as possible all the properties of the exact hole [59]. This results in a correction to the LDA that has the form

$$
\Delta\mathcal{E}_{xc}^{GGA} = \int d\mathbf{r} \, \epsilon_{xc}^{LDA}(n(\mathbf{r})) \, f_{xc}(n(\mathbf{r}), \frac{[\nabla n]^2}{n^{\frac{8}{3}}}) \, . \tag{4.36}
$$

The function f_{xc} ensures that the limits of low and high densities as well as of the low density gradients are correctly reproduced. In this way the exchange and correlation term V_{xc} has the correct asymptotic form which leads to a reasonable description of negative ions. While the GGA yields significantly improved results (over the LDA) for molecular systems, for solids the improvement is not systematic [48]. In addition, the van-der-Waals interaction can not be treated using the GGA.

4.6.5 Implicit orbital functionals

In this method $\mathcal{E}_{xc}$ is written in terms of KS orbitals (which are implicitly functionals of the electronic density). For the description of the exchange contribution one utilizes the Fock term

$$\mathcal{E}_x^{KS}[n] = \frac{1}{2} \sum_{i,j}^{occ} \iint d\mathbf{r}_1 d\mathbf{r}_2 \; \psi_i^*(\mathbf{r}_1)\, \psi_j^*(\mathbf{r}_2)\, v_{12}(\mathbf{r}_1, \mathbf{r}_2)\, \psi_i(\mathbf{r}_2)\psi_j(\mathbf{r}_1) \; , \qquad (4.37)$$

where v_{12} is the two-particle interaction. This procedure removes self-interactions from the term $\mathcal{W}_H + \mathcal{E}_x$ and leads to correct asymptotic behaviour. As far as exchange is concerned multiplicative effective potentials $V_x^{KS}(\mathbf{r})$ are obtained by means of the so-called optimized potential method (OPM) [60] by solving the integral equation

$$\int d\mathbf{r}' \; V_x^{KS}(\mathbf{r}')\, K(\mathbf{r}', \mathbf{r}) - Q(\mathbf{r}) = 0 \; . \qquad (4.38)$$

The kernel of this equation K and the inhomogeneous term Q are expressible in terms of the orbitals. Therefore, one has to iterate the KS equations and solve at the same time for the optimized-potential integral equation. For the treatment of the correlation part a perturbative approach similar to the Møller Plesset method has been suggested recently [61, 62]. Calculations showed that this scheme is capable of describing van-der-Waals bonding as well as atomic and ionic correlation energies [62]. An effective practical implementation of this approach for large systems is the subject of current research.

4.6.6 Self-interaction corrections

In DFT the question of the spurious self-interaction (SI) terms arises, i.e. the possibility of the electron interacting with itself. The self-interaction corrections (SIC) impose on the exact exchange-correlation functional the condition that the self-interaction energy of the electron cloud must be cancelled by an equivalent term in the the exchange-correlation energy. A procedure for removing SI from approximate expressions for the exchange-correlation functionals has been put forward by Perdew and Zunger (PZ) [117]. From Eq. (4.31) it follows that the KS total energy E_{tot}^{KS} is (for one spin-component)

$$E_{tot}^{KS} = \sum_{i=1}^{N} \langle \psi_i | -\frac{1}{2}\nabla_i^2 | \psi_i \rangle + \frac{1}{2} \int d\mathbf{r}_1 d\mathbf{r}_2 \frac{n(\mathbf{r}_1)\, n(\mathbf{r}_2)}{|\mathbf{r}_1 - \mathbf{r}_2|}$$

$$+ \int d\mathbf{r}\, V_{ext}(\mathbf{r})n(\mathbf{r}) + \mathcal{E}_{xc}[n(\mathbf{r})]. \qquad (4.39)$$

On the other hand the energy E_{tot}^{1e} of a one-electron system with a charge density distribution $n(\mathbf{r})$ is

$$E_{tot}^{1e} = \langle \psi| -\frac{1}{2}\nabla^2 |\psi\rangle + \int d\mathbf{r}\, V_{ext}(\mathbf{r})\, n(\mathbf{r}).$$

This imposes on the exact exchange-correlation functional the condition:

$$\frac{1}{2}\int \frac{n(\mathbf{r}_1)\, n(\mathbf{r}_2)}{|\mathbf{r}_1 - \mathbf{r}_2|}\, d\mathbf{r}_1 d\mathbf{r}_2 + \mathcal{E}_{xc}[n(\mathbf{r})] = 0.$$

Therefore, Perdew and Zunger (PZ) [117] suggested to remove the self-interaction from the Kohn-Sham total energy, by making the prescription

$$E_{tot}^{PZ} = E_{tot}^{KS} - \sum_{i=1}^{N}\left\{\frac{1}{2}\int d\mathbf{r}_1 d\mathbf{r}_2 \frac{n_i(\mathbf{r}_1)\, n_i(\mathbf{r}_2)}{|\mathbf{r}_1 - \mathbf{r}_2|} + \mathcal{E}_{xc}[n_i]\right\}.$$

The PZ correction scheme has the desirable properties that the correction term (enclosed in parentheses) vanishes if $\mathcal{E}_{ex}$ is exact. In addition, the exchange-correlation potentials possess the correct asymptotic (radial) behavior. On the other hand, the total energy and the exchange-correlation potentials become orbital-dependent. This is seen as follows. In the conventional DFT the KS-orbitals ψ_i are eigenfunctions of the same (Fock) operator

$$h^{(KS)} = -\frac{1}{2}\nabla^2 + V_H(\mathbf{r}) + V_{ext}(\mathbf{r}) + V_{xc}(\mathbf{r})$$

with an eigenenergy ϵ_i. In contrast, in the PZ scheme these orbitals are eigenfunctions of different Fock operators $h_i^{(PZ)}$ given by the equation

$$h_i^{(PZ)} = h^{(KS)} + V_i^{(PZ)}(\mathbf{r}),$$

and the potential $V_i^{(PZ)}(\mathbf{r})$ has the form

$$V_i^{(PZ)}(\mathbf{r}) = -\frac{\delta \mathcal{E}_{xc}[n_i]}{\delta n_i} - \int d\mathbf{r}' \frac{n_i(\mathbf{r}')}{|\mathbf{r} - \mathbf{r}'|}.$$

The orbital dependence of $h_i^{(PZ)}$ results in complications in the numerical self-consistent implementation. Therefore, one may choose to realize numerically the PZ-SIC within the optimized effective potential (OEP) scheme [118] resulting in eigen-equations that are formally identical to the KS equations (4.31). The resulting Fock operator $h_i^{(OEP)}$ has the form $h^{(OEP)} = h^{(KS)} + V^{(OEP)}(\mathbf{r})$, where $V^{(OEP)}(\mathbf{r})$ is such that the PZ functional is minimized.

4.6.7   Extensions of DFT

The original ground state DFT theory has been extended to deal with various systems and physical processes. Here, the prominent cases are mentioned.

To deal with situations where the spin polarization plays a role, such as in magnetic and relativistic systems a ground-state spin-polarized DFT [63, 64] has been developed. Furthermore, to address (equilibrium) statistical mechanical questions a finite temperature DFT has been put forward [65] as well as a density functional theory for superconductors [66]. Systems subject to an external time-dependent potential $V_{ext}(\mathbf{r}, t)$ can be in principle treated within the time-dependent DFT (TDFT) [67, 68]. In this case it has been shown [67, 68] that, for specified initial many-particle state, the time-dependent density $n(\mathbf{r}, t)$ determines the time-dependent external potential $V_{ext}(\mathbf{r}, t)$ up to a purely time-dependent function. This existence theorem has been generalized by Li and Tong [49] to multi-component systems with various kinds of particles. Furthermore, it has been shown that observables are expressible as functionals depending on the density and the chosen initial conditions. A time-dependent calculational scheme analogous to KS method can be developed such that

$$
\begin{aligned}
n(\mathbf{r}, t) &= \sum_{i=1}^{N} \psi^*(\mathbf{r}, t)\psi(\mathbf{r}, t), \\
i\partial_t \psi(\mathbf{r}, t) &= \left[-\frac{1}{2}\nabla_i^2 + V_{ext}(\mathbf{r}, t) + \int d\mathbf{r}' \, \frac{n(\mathbf{r}', t)}{|\mathbf{r} - \mathbf{r}'|} + V_{xc}(\mathbf{r}, t) \right] \psi(\mathbf{r}, t), \\
i\partial_t \psi(\mathbf{r}, t) &= \left[-\frac{1}{2}\nabla_i^2 + V_{eff}(\mathbf{r}, t) \right] \psi(\mathbf{r}, t), \qquad\qquad\qquad (4.40)
\end{aligned}
$$

where a splitting of the Hartree and the external potential term has been carried out in the same manner as in Eq. (4.31). In (4.40) the effective potential $V_{eff}(\mathbf{r}, t)$ is implicitly dependent on the initial conditions. From a practical point of view, it is not clear whether the approach leading to (4.40) will be useful for the time-dependent case as it has been for the time-independent DFT. For example, in the KS equations (4.40, 4.31) the relations have been rearranged such that the Hartree approach [which includes $V_{ext}(\mathbf{r}) + V_H(n, \mathbf{r})$, cf. (4.31)] is the leading order approximation (with respect to V_{xc}). For time-dependent problems it is not obvious that such a choice is reasonable, for the time-dependent HF theory was not particularly successful [69]. The question of which suitable (approximate) functionals to use is still open.

The analogous approximation to the LDA, the so-called adiabatic local density approximation (ALDA) has yielded useful results for some systems (e.g. laser ablation of atoms). In the ALDA one assumes the exchange and correlation term $V_{xc}(\mathbf{r}, t)$ to be approximated by $\frac{d}{dn}\epsilon_{xc}(n(\mathbf{r}))|_{n(\mathbf{r})=n(\mathbf{r},t)}$, where $\epsilon_{xc}(n(\mathbf{r}))$ is the exchange and correlation energy pro unit volume of the homogeneous electron gas. The ALDA is formally justified in case of slowly varying (in time) external potentials and slow varying densities. Extensions of the ALDA are provided by the linear response treatment [70] and the time dependent OPM. For small external perturbations the linear response approach (first order perturbation theory) is a reliable starting point and has led to satisfactory results for the excitation energies [70], even though a systematic application to extended systems is still outstanding.

5 Electronic excitations

In the previous chapter methods have been discussed that are capable of predicting ground state properties of electronic systems. In general, an adequate description of excitation processes entails the knowledge not only of the ground state but also of the excited states and of the way the external perturbation couples to the system under study. While this combined task is in general quite complicated to perform without severe approximations, valuable information can be gained from symmetry considerations of the transition amplitudes and the symmetry properties of the external perturbation field. In what follows we illustrate these statements in the case of excitation processes by ultraviolet radiations. Therefore, we specify first the properties of the radiation field and quantify its coupling to electronic systems. In a second step we inspect from a general point of view the symmetry properties of the electronic excitation probabilities. Subsequently, we address the question of obtaining appropriate expressions for the excited states.

5.1 Electric dipole transitions

We will be dealing with radiative processes where the photon density is large so that the electromagnetic field can be treated classically. We operate in the Coulomb gauge, i. e., $\nabla \cdot \mathbf{A} = 0$. In vacuum we can set $\Phi = 0$, where $\mathbf{A}$ and Φ are the field vector and the scalar potentials, respectively. It should be noted, however, that in a polarizable medium, such as near a metal surface $\mathbf{A}$ may change rapidly which invalidates the assumption $\nabla \cdot \mathbf{A} = 0$ unless the dielectric constant ϵ is unity [155, 156]. To avoid this complication the photon energies have to be well above the plasmon energies [157]. The monochromatic, plane-wave solution for $\mathbf{A}$ can be written in the form

$$\mathbf{A}(\mathbf{r}, t) = A\hat{\mathbf{e}}\Big\{ \exp\big[i(\mathbf{k} \cdot \mathbf{r} - \omega t)\big] + \exp\big[-i(\mathbf{k} \cdot \mathbf{r} - \omega t)\big] \Big\}$$
$$= \mathbf{A}_0 e^{-i\omega t} + \big(\mathbf{A}_0 e^{-i\omega t}\big)^* .$$

(5.1)

The wave vector $\mathbf{k}$ ist related to the photon frequency ω via $k = \alpha\omega$, where α is the fine-structure constant. The energy-density ρ of the (classical) radiation field averaged over the period $T = 2\pi/\omega$ is given by

$$\rho = \frac{\omega^2 A^2}{2\pi}. \tag{5.2}$$

Thus, the energy-flux density, i. e. the intensity I, is given by $I = \rho/\alpha$. The interaction of the classical radiation field with an N-electron system is described by the Hamiltonian

$$\mathcal{H} = \frac{1}{2}\sum_{j=i}^{N}\left[\mathbf{p}_j + \frac{1}{c}\mathbf{A}(\mathbf{r}_j, t)\right]^2 + V, \tag{5.3}$$

where V is the total potential of the undisturbed system and $\mathbf{p}_j$ are the one-particle momentum operators. For low-intensity field we can set $A^2 \approx 0$ (for $A \approx 0.01$ and photon energy of $50\ eV$ we arrive at a maximum intensity $I \approx 5 * 10^{17}\ W/m^2$). In addition, due to $\mathrm{div}\,\mathbf{A}{=}0$, $\mathbf{A}(\mathbf{r}_j, t)$ commutes with $\mathbf{p}_j$ and Eq. (5.3) reduces to

$$\mathcal{H} = H + W(\mathbf{r}_j, t), \tag{5.4}$$

where H is the Hamiltonian of the unperturbed system and the perturbation W is given by $W = \tilde{W} + \tilde{W}^\dagger$. With the solution (5.1) the perturbation $\tilde{W}$ has the explicit form[1]

$$\begin{aligned}
\tilde{W} &= \left\{\frac{A}{c}\sum_{j=1}^{N}\exp[i(\mathbf{k}\cdot\mathbf{r}_j)]\,\hat{\mathbf{e}}\cdot\mathbf{p}_j\right\}e^{-i\omega t}, \\
&= \tilde{W}_0 e^{-i\omega t}. \tag{5.5}
\end{aligned}$$

Let us assume the unperturbed system to be in the stationary state $|i\rangle$ with energy ϵ_i, i. e.,

$$(H - \epsilon_i)|i\rangle = 0. \tag{5.6}$$

Under the action of the perturbation $W(t)$, within the time lap τ, the system performs a transition into excited states $|f\rangle$ which lay within the interval β and $\beta + d\beta$, where β stands for collective quantum numbers that specify the final channel. In a time-dependent first-order perturbation treatment of the action of W, the transition probability dw_{if} amounts to [158]

$$dw_{if} = \sum_{\alpha_i}\left|\int_0^\tau dt\langle f|\tilde{W}_0 e^{i(E_f-\epsilon_i-\omega)t} + \tilde{W}_0^\dagger e^{i(E_f-\epsilon_i+\omega)t}|i\rangle\right|^2 d\beta. \tag{5.7}$$

---

[1] The diamagnetic term $\mathbf{A}^2/(2c^2)$ has been neglected and the relation $\nabla\cdot\mathbf{A}|i\rangle = (\mathrm{div}\,\mathbf{A})|i\rangle + (\mathbf{A}\cdot\nabla)|i\rangle = (\mathbf{A}\cdot\nabla)|i\rangle$ is utilized.

Here E_f is the energy of the final state and α_i denotes the unresolved quantum numbers which characterize the initial-state. From Eq. (5.7) it is clear that the system gains [loses] energy under the action of $\tilde{W}(t)$ $[\tilde{W}(t)^\dagger]$, i. e. $\tilde{W}(t)$ and $\tilde{W}(t)^\dagger$ corresponds, respectively, to photon absorption and induced emission. In what follows we consider photoabsorption, only, and define $E_i = \epsilon + \omega$. For a small characteristic interaction time τ the time integral in Eq. (5.7) is readily performed and we can define a transition rate $dP_{if} = dw_{if}/\tau$ which, after simple algebraic manipulations, can be deduced to be

$$dP_{if} = (2\pi)^2 \frac{I\alpha}{\omega^2} \sum_{\alpha_i} |\langle f|\tilde{W}_0|i\rangle|^2 \, \delta(E_f - E_i)\, d\beta. \tag{5.8}$$

We define the differential cross section $d\sigma/d\beta$ as the transition rate normalized to the incoming flux density I/ω, i. e.

$$d\sigma = \omega dP_{if}/I. \tag{5.9}$$

Considering moderate photon energies (say < 500 eV) one can operate within the dipole approximation in which case Eq. (5.9) reduces to

$$d\sigma = 4\pi^2 \frac{\alpha}{\omega} \sum_{\alpha_i} |M_{fi}|^2 \delta(E_i - E_f) d\beta, \tag{5.10}$$

where the dipole-matrix element is given by

$$M_{fi} = \sum_{j}^{N} \langle f|\hat{\mathbf{e}} \cdot \mathbf{p}_j|i\rangle. \tag{5.11}$$

Making use of the canonical commutation relations $-i[\mathbf{r}_j, H] = \mathbf{p}_j$ and assuming that $|i\rangle$ and $|f\rangle$ are eigenfunctions of the *same* Hamiltonian H, the *velocity form* Eq. (5.10) can be converted into the *length form*

$$d\sigma = 4\pi^2 \alpha\omega \sum_{\alpha_i} \left|\sum_{j}^{N} \langle f|\mathbf{r}_j|i\rangle\right|^2 \delta(E_f - E_i) d\beta. \tag{5.12}$$

In practice, $|i\rangle$ and $|f\rangle$ are derived using different approximate procedures for H and thus the velocity and length forms yield, in general, different predictions. Conversely, equivalent cross sections, calculated within the length and velocity forms, mean merely that same approximations have been made in the initial and final channel, say however nothing about the quality of these approximations. Nevertheless, it is desirable to choose $|i\rangle$ and $|f\rangle$ as eigenstates of the same (approximate) Hamiltonian to preclude spurious transitions in absence of the perturbation $\tilde{W}_0$. We note furthermore that regardless of the form in which the dipole operator is presented, it's mathematical structure is always a sum of single-particle operators.

5.2 Single-photoelectron emission

The calculations of the cross sections (5.12) entails the knowledge of the initial state $|i\rangle$, which is usually the ground state and can thus be evaluated by the methods discussed in chapter (4). For the determination of the excited state $|f\rangle$ different approaches are employed that will be discussed in the next chapters. In many important cases however, decisive information are extracted from symmetry considerations of the dipole matrix elements (5.12) and the symmetry of $|i\rangle$ and $|f\rangle$. Obviously this is a notable simplification and hence we will recall briefly some prominent cases.

5.2.1 One-electron photoemission from unpolarized targets

Let us consider the one-photon one-electron continuum transitions for unpolarized targets and linear polarized light. It has been shown by Yang [159] that if the spin of the photoelectron is not resolved and the residual ion is randomly oriented, the angular distribution of the photoelectrons, i.e. the cross section for the emission of a photoelectron under a solid angle Ω with a fixed energy E, is given by

$$\frac{d\sigma}{d\Omega} = \frac{\sigma_0}{4\pi}[1 + \beta_a P_2(\cos\theta)], \tag{5.13}$$

where σ_0 is the total cross section, $-1 \leq \beta_a \leq 2$ is the so-called asymmetry parameter, P_2 is the second Legendre polynomial and θ is the angle between the photoelectron momentum $\mathbf{p}$ and and the electric field direction (for linear polarization) or between $\mathbf{p}$ and the light propagation direction (for circular polarization). The argument for the validity of the relation (5.13) is straightforward: under the specified conditions the cross section is a scalar invariant under rotations. The directional dependence of σ is given by $\mathbf{p}$ and by the polarization direction $\hat{\mathbf{e}}$.

These two vectors can be considered as spherical tensors[2] of rank 1. On the other hand the scalar invariant formed by these two (spherical) vectors is the scalar product of two spherical tensors [143, 144] $\sum_m C_{lm}(\hat{\mathbf{e}})C_{lm}(\hat{\mathbf{p}})^* = P_l(\hat{\mathbf{e}} \cdot \hat{\mathbf{p}})$, where C_{lm} is a spherical harmonic normalized to $4\pi/(2l+1)$ [143] . Since $l \leq 2$ and $l = 1$ is excluded due to odd parity, one obtains the general expression (5.13) for the cross section.

We note that the helicity of the light does not enter into Eq. (5.13), i.e. the cross section (5.13) does not depend dynamically on the polarization of the photon. The determination of the initial and the final states wave function is necessary only when the actual value of β_a and/or σ_0 are needed. Otherwise, the only piece of information on $|i\rangle$ that is essential for the validity of (5.13) is that the initial state is an angular momentum eigenstate. On the other hand β_a and σ_0 and hence the entire cross section is completely specified by only two measurements, e.g. under the (magic) angle $P_2(\hat{\mathbf{e}} \cdot \hat{\mathbf{p}}) = 0$ one obtains σ_0.

5.2.2   Single photoemission from polarized targets

In a photoemission experiment that resolves the spin of the photoelectron or, if the target atoms are polarized, Yang's formula (5.13) breaks down, because more than two relevant directions are involved. As an example let us consider the case of polarized atomic targets with $\hat{\mathbf{a}}$ being a unit vector along the quantization axis. Furthermore, we assume the hyperfine structure of the target to be quantified by angular quantum numbers F_0. The population of the hyperfine states is conveniently described by the density matrix $\rho_{F_0 M_0 F_0 M_0'}$ (cf. Ref. [145]). To factor out the dependence of $\rho_{F_0 M_0 F_0 M_0'}$ on the magnetic sublevels M_0 the density matrix is expressed in terms of the state multipoles (also called the statistical tensors) ρ_{K0} (see Ref. [145] for details). This is done via the relation

$$\rho_{F_0 M_F F_0 M_F'} = \sum_K \left\{ (-1)^{K-F_0-M_0} \sqrt{\frac{4\pi}{2K+1}} \right.$$

$$\left. \langle F_0 - M_0 F_0 M_0' \mid K M_0' - M_0 \rangle \rho_{K0} Y_{K M_0' - M_0}(\hat{\mathbf{a}}) \right\}, \quad (5.15)$$

---

[2]The spherical components A_q of a cartesian vector operator $\mathbf{A} = (x, y, z)$ are

$$A_1 = -\frac{1}{\sqrt{2}}(x+iy), \quad A_0 = z, \quad A_{-1} = \frac{1}{\sqrt{2}}(x-iy).. \quad (5.14)$$

The scalar product of two spherical tensors T_l and S_l is given by $T_l \cdot S_l = \sum_m (-1)^m T_{lm} S_{lm}$, more details and definitions of spherical tensors are given in appendix A.1 .

where Y_{KQ} is a standard spherical harmonics and $\langle \cdots | \cdots \rangle$ stands for the Clebsch-Gordon coefficients. The structure of the angular distribution of the photoelectrons, emitted from the target described by the density matrix (5.15), is obtained by inspecting the scalar invariants that can be formed out of the three vectors $\hat{\mathbf{a}}$, $\hat{\mathbf{e}}$ and $\mathbf{p}$. The general form of the cross section is then deduced to be (see Ref. [146] for a detailed derivation)

$$\frac{d\sigma}{d\Omega} = 4\pi^2 \alpha \omega (-1)^{1+\lambda} \sum_{LKY} \langle 1\lambda 1 - \lambda | Y0 \rangle \, \rho_{K0} \, B(L, K, Y) \, \mathcal{B}_{Y0}^{LK}(\hat{\mathbf{p}}, \hat{\mathbf{a}}), \qquad (5.16)$$

where $\hat{\mathbf{p}}$ is the emission direction of the photoelectron. λ quantifies the polarization state of the light, i.e. $\lambda = \pm 1$ indicates right/left circular polarization whereas $\lambda = 0$ stands for a linear polarized photon. The angular function $\mathcal{B}_{Y0}^{LK}(\hat{\mathbf{p}}, \hat{\mathbf{a}})$ in Eq. (5.16) is a bipolar harmonic resulting of coupling (i.e., the tensor product) of two spherical harmonics associated with the directions $\hat{\mathbf{p}}$ and $\hat{\mathbf{a}}$ (cf. appendix A.1). The generalized asymmetry parameters B that appear in Eq. (5.16) are given by

$$B(L, K, Y) = (2F_0 + 1) \left\{ \begin{array}{ccc} K & J_0 & J_0 \\ I & F_0 & F_0 \end{array} \right\}$$

$$\sum_{lj\,Jl'j'J'} (-1)^{J_0 + I + F_0 + J_f + J - \frac{1}{2}} \sqrt{[(2l+1)(2l'+1)(2j+1)(2j'+1)}$$

$$(2J+1)(2J'+1) \langle l0l'0|L0 \rangle \left\{ \begin{array}{ccc} L & J' & J \\ J_f & j & j' \end{array} \right\} \left\{ \begin{array}{ccc} L & j' & j \\ \frac{1}{2} & l & l' \end{array} \right\}$$

$$\left\{ \begin{array}{ccc} L & J & J' \\ K & J_0 & J_0 \\ Y & 1 & 1 \end{array} \right\} \langle J_f(l\frac{1}{2}); J \| r \| J_0 \rangle \langle J_f(l'\frac{1}{2})j'; J' \| r \| J_0 \rangle^*.$$

$$(5.17)$$

In this equation L is the total orbital angular momentum, J is the total electronic angular momentum with a coupling scheme $(LS)J$, I is the nuclear spin, and F is the overall angular momentum with the coupling scheme $(JI)F$. $\langle \cdots \| \mathcal{O} \| \cdots \rangle$ is the reduced matrix element of the operator $\mathcal{O}$.

The symmetry properties of the Clebsch-Gordon coefficients [cf. Eq. (5.16)] indicate that the term $Y = 1$ is responsible for a circular dichroism in the cross section, i.e. a dependence of the cross section on the helicity of the photon, provided B and $\mathcal{B}$ are non-vanishing. In the simplest case $(K = L = 1)$ the angular function reduces to the expression $\mathcal{B}_{10}^{11}(\hat{\mathbf{p}}, \hat{\mathbf{a}}) \propto (\hat{\mathbf{p}} \times \hat{\mathbf{a}}) \cdot \mathbf{k}$, where $\mathbf{k}$ is the light propagation direction. This means that the circular dichroism vanishes if the vectors $\hat{\mathbf{p}}$ and $\hat{\mathbf{a}}$ are linearly dependent (i.e. parallel or antiparallel). A finite value of

B requires $J_0 \geq \frac{1}{2}$. One should note, however, that the total cross section does not reveal any dichroism since the angular integration over the emission direction of the photoelectron selects $L = 0$. Physically, the circular dichroism arises due to the existence of an initial target orientation described by $K = 1$ and realized, e.g. by optical pumping. In a photoemission process this orientation is transferred to the photoelectron continuum.

5.3   General properties of emitted dipole radiation

The dipole transition probabilities for the photoabsorption and for the induced photoemission are given by Eq. (5.7). In the previous section we analyzed the products (photoelectrons) of a photoabsorption process. This section provides a compact account of the theoretical description of the polarization and the angular distributions of emitted dipole radiations, as put forward originally by Fano and Macek [121]. Following the de-excitation of a certain sample photons are emitted. Starting from the expression (5.12) for the cross section one derives for the intensity of the emitted photons at a distance R from the source the expression

$$\begin{aligned}
\mathcal{I} &= \;<I>_i \\
I &= \; c\sum_{m_f}|\langle\, f|\hat{\mathbf{e}}^* \cdot \mathbf{D}|i\,\rangle|^2,
\end{aligned} \tag{5.18}$$

where $\mathbf{D} = \sum_i \mathbf{r}_i$ is the dipole operator, $c = \omega_{if}^4 \alpha^3/(2\pi R^2)$ and $\omega_{if} = (E_i - E_f)/\hbar$ is the frequency of the emitted light with E_i and E_f being the energies associated with $|i\,\rangle$ and $|f\,\rangle$. The symbol $<I>_i$ indicates the weighted average over initial-state degeneracies. The magnetic sublevels m_f of the final state are not resolved. The treatment is restricted by the requirement that $|i\,\rangle$ and $|f\,\rangle$ are angular momentum eigenstates.

Eq. (5.18) can be written as

$$I = c\,\langle i|(\hat{\mathbf{e}} \cdot \mathbf{D})P_f(\hat{\mathbf{e}} \cdot \mathbf{D})^\dagger|i\rangle. \tag{5.19}$$

Here we introduced the scalar projection operator $P_f = \sum_{m_f} |f\,\rangle\langle\, f|$ which is invariant under joint rotations of $|f\rangle$ and $\langle f|$. The operators $\mathbf{D}$ and $P_f\mathbf{D}$ are polar vector operators. We Recall that the relation

$$(\mathbf{a} \cdot \mathbf{b})(\mathbf{c} \cdot \mathbf{d}) = \frac{1}{3}(\mathbf{a} \cdot \mathbf{c})(\mathbf{b} \cdot \mathbf{d}) + \frac{1}{2}(\mathbf{a} \times \mathbf{c})(\mathbf{b} \times \mathbf{d}) + T_2(\mathbf{a}, \mathbf{c}) \cdot T_2(\mathbf{b}, \mathbf{d}) \tag{5.20}$$

applies where $\mathbf{a}, \mathbf{c}, \mathbf{b}, \mathbf{d}$ are vector operators and $T_2(\mathbf{a}, \mathbf{b})$ is a tensor of rang 2 constructed from the spherical vectors (see footnote 2 on p. 53) $\mathbf{a}$ and $\mathbf{b}$, i. e. $T_{2m}(\mathbf{a}, \mathbf{b}) =$

$\sum_{qq'} \langle\, 1q1q'|2m \,\rangle a_q\, b_{q'}$. Furthermore we note that the scalar product of two tensors is given by $T_2 \cdot S_2 = \sum_m (-1)^m\, T_{2m} S_{2m}$. Expressing the polarization vector $\hat{e}$ in terms of the ellipticity β of the light, i. e.

$$\hat{e} = (\cos\beta, i\sin\beta, 0) \tag{5.21}$$

one obtains for the intensity (5.19) using (5.20) (applied to $(\hat{e} \cdot \mathbf{D})P_f(\hat{e} \cdot \mathbf{D})^\dagger$) the following formula

$$
\begin{aligned}
I = c\Bigg[&\frac{1}{3}\langle\, i|\mathbf{D} \cdot (P_f\mathbf{D})|i \,\rangle - \frac{i}{2}\sin(2\beta)\langle\, i|[\mathbf{D} \times (P_f\mathbf{D})]_\mathbf{k}|i \,\rangle \\
&- \frac{1}{\sqrt{6}}\langle\, i|T_{20}(\mathbf{D},(P_f\mathbf{D}))|i \,\rangle + \cos(2\beta)\langle\, i|\bar{T}_{22}(\mathbf{D},(P_f\mathbf{D}))|i \,\rangle \Bigg], \\
= c\Bigg[&-\frac{1}{\sqrt{3}}\langle\, i|T_{00}|i \,\rangle - \frac{1}{\sqrt{2}}\sin(2\beta)\langle\, i|T_{10}|i \,\rangle \\
&-\frac{1}{\sqrt{6}}\langle\, i|T_{20}|i \,\rangle + \cos(2\beta)\langle\, i|\bar{T}_{22}|i \,\rangle \Bigg].
\end{aligned}
\tag{5.22}
$$

In this equation we utilized the relations

$$
\begin{aligned}
\hat{e} \cdot \hat{e}^* &= 1,\ \hat{e} \times \hat{e}^* = -i\sin(2\beta)\,(0,0,1) = -i\sin(2\beta)\,\hat{\mathbf{k}}, \\
T_{22}(\hat{e},\hat{e}^*) &= T_{2-2}(\hat{e},\hat{e}^*) = \frac{1}{2}\cos(2\beta),\ T_{20}(\hat{e},\,\hat{e}^*) = -\frac{1}{\sqrt{6}}, \\
T_{21}(\hat{e},\hat{e}^*) &= T_{2-1}(\hat{e},\hat{e}^*) = 0.
\end{aligned}
$$

The symbol $\mathbf{k}$ stands for the photon wave vector. In addition we defined

$$\bar{T}_{22}(\mathbf{D},(P_f\mathbf{D})) = [T_{22}(\mathbf{D},(P_f\mathbf{D})) + T_{2-2}(\mathbf{D},(P_f\mathbf{D}))]/2.$$

The tensor $T_{Kq}(\mathbf{D},(P_f\mathbf{D}))$ is built out of the spherical components of $\mathbf{D}$ and $(P_f\mathbf{D})$, i. e. (cf. appendix A.1)

$$T_{KQ}(\mathbf{D},(P_f\mathbf{D})) = \sum_{qq'} \langle\, 1q1q'|KQ \,\rangle\, \mathbf{D}_q(P_f\mathbf{D})_{q'}. \tag{5.23}$$

Therefore, the rang of tensors occurring in Eq. (5.22) indicates the transformation behaviour of the respective terms in Eq. (5.22). The first term behaves as scalar under rotations whereas the second term transforms as a vector. The third and fourth terms are alignment tensors. Considering the initial-state manifold $j_i m_i$, the Wigner-Eckart-theorem, stated by Eq. (A.8), can be utilized and yields in combination with (5.23) the relations

$$\langle i|T_{KQ}|i\,\rangle = \langle\, j_i m_i KQ|j_i m_i'\,\rangle\langle\, i\parallel T_K \parallel i\,\rangle, \tag{5.24}$$

$$= \sum_{qq'm_f} \langle\, 1q1q'|KQ\,\rangle\langle\, i|\mathbf{D}_q|f\,\rangle\langle\, f|(P_f\mathbf{D})_{q'}|i\,\rangle$$

$$= \langle\, i\parallel \mathbf{D}\parallel f\,\rangle\langle\, f\parallel\mathbf{D}\parallel i\,\rangle\, G, \quad \text{where}$$

$$G = \sum_{qq'm_f} \langle\, 1q1q'|KQ\,\rangle\langle\, j_f m_f 1q|j_i m_i'\,\rangle\langle\, j_i m_i 1q'|j_f m_f\,\rangle. \tag{5.25}$$

Thus, the K dependence of $\langle\, i\parallel T_K \parallel i\,\rangle$ is given by $\bar{G} = G/(\langle\, j_i m_i KQ|j_i m_i'\,\rangle)$ which is shown to be proportional to a re-coupling coefficient $\left\{\begin{matrix} j_i & 1 & j_f \\ 1 & j_i & K \end{matrix}\right\}$. Hence, from Eq. (5.24) we conclude that the K dependence of $\langle\, i|T_{KQ}|i\,\rangle/(\langle\, j_i m_i KQ|j_i m_i'\,\rangle)$ is given by this re-coupling coefficient. This in turn reveals the K dependence of the terms in the expansion (5.22). To express (5.22) in terms of the total angular molmentum J one relates the matrix elements (5.24) of the tensor T_{KQ} to those of the tensor $S_{KQ}(J,J)$. The object $S_{KQ}(J,J)$ is the result of a tensor product of the states $|Jm\,\rangle$ and $|Jm'\,\rangle$ (note that $|Jm\,\rangle$ can be considered as a spherical tensor of rang J with spherical components m) $S_{KQ}(J,J) = \sum_{q,q'}\langle\, 1q1q'|KQ\,\rangle|Jq\,\rangle|Jq'\,\rangle$. The same steps done in Eqs. (5.24-5.25) can be repeated for $\langle\, i|S_{KQ}|i\,\rangle$ and one deduces that the K dependence of $\langle\, i\parallel S_K \parallel i\,\rangle$ is given by a re-coupling coefficient so that

$$\frac{\langle\, i\parallel T_K \parallel i\,\rangle}{\langle\, i\parallel S_K \parallel i\,\rangle} = \lambda_K(j_i,j_f)\frac{\langle\, i\parallel T_0 \parallel i\,\rangle}{\langle\, i\parallel S_0 \parallel i\,\rangle}$$

and

$$\lambda_K(j_i,j_f) = (-1)^{j_f-j_i}\frac{\left\{\begin{matrix} j_i & 1 & j_f \\ 1 & j_i & K \end{matrix}\right\}}{\left\{\begin{matrix} j_i & 1 & j_i \\ 1 & j_i & K \end{matrix}\right\}}.$$

From these relations and from Eq. (5.24) we conclude that

$$\langle\, i|T_{KQ}|i\,\rangle = \langle\, i|S_{KQ}|i\,\rangle\frac{\langle\, i\parallel T_K \parallel i\,\rangle}{\langle\, i\parallel S_K \parallel i\,\rangle} = \lambda_K(j_i,j_f)\frac{\langle\, i\parallel T_0 \parallel i\,\rangle}{\langle\, i\parallel S_0 \parallel i\,\rangle}\langle\, i|S_{KQ}|i\,\rangle$$

which can be now inserted in Eq. (5.22) to obtain an expression for the light intensity in terms of the total angular momentum. In Ref. [121] a detailed discussion of the physical contents of the resulting terms in Eq. (5.22) is provided. E.g. the first term

$$I_{tot} := -c\frac{1}{\sqrt{3}}\langle\, i|T_{00}|i\,\rangle = -c\frac{1}{\sqrt{3}}\langle\, i\parallel T_0 \parallel i\,\rangle$$

is the unpolarized and angle-integrated total intensity. Since

$$\langle\, i \parallel S_0 \parallel i \,\rangle = -\frac{1}{\sqrt{3}} j_i(j_i + 1)$$

we can parameterize the intensity (5.22) in terms of the expectation values of the total angular momentum, namely

$$I = I_{tot}\Bigg[1 - \frac{3}{\sqrt{2}}\lambda_1(j_i, j_f)\sin(2\beta)\,\mathbf{O}_{\hat{\mathbf{k}}}$$
$$- \sqrt{\frac{3}{2}}\lambda_2(j_i, j_f)\,A_{20} + 3\lambda_2(j_i, j_f)\cos(2\beta)\,\bar{A}_2\Bigg], \quad (5.26)$$

where $\mathbf{O}_{\hat{\mathbf{k}}}$ is the component along $\hat{\mathbf{k}}$ of the orientation vector $\mathbf{O}$. $\mathbf{O}$ is proportional to the initial state expectation value of the components of the total angular momentum $\mathbf{J}$, i. e.

$$\mathbf{O} \propto \langle\, i|\mathbf{J}|i\,\rangle/[j_i(j_i + 1)].$$

The components of the alignment tensors A_{20} and $\bar{A}_2$ are respectively determined by

$$\langle\, i|3\mathbf{J}_z^2 - \mathbf{J}^2|i\,\rangle/[j_i(j_i + 1)], \quad \text{and} \quad \langle\, i|\mathbf{J}_x^2 - \mathbf{J}_y^2|i\,\rangle/[j_i(j_i + 1)].$$

The merit of Eqs. (5.26, 5.22) is that properties of the radiation field are completely decoupled from the dynamical target properties which are described by $\mathbf{O}_{\hat{\mathbf{k}}}$, A_{20} and $\bar{A}_2$. To illustrate the importance of this procedure let us ask which quantity describes the difference Δ_{LR} between the emission of left and right-hand (circular, or elliptically) polarized light?. From Eq. (5.21) it is clear that exchanging $\hat{e}$ by $\hat{e}^*$ amounts to the formal replacement $\beta \rightleftarrows -\beta$. The only quantity in Eqs. (5.26, 5.22) sensitive to this replacement is $\sin(2\beta)\,\langle\, i|T_{10}|i\,\rangle$ (or $\sin(2\beta)\,\lambda_1(j_i, j_f)\,\mathbf{O}_{\hat{\mathbf{k}}}$). Therefore, to evaluate Δ_{LR} it suffices to consider $\sin(2\beta)\,\langle\, i|T_{10}|i\,\rangle$.

5.4   Symmetry properties of many-body photoexcitations

In section 5.2.1 we derived Eq. (5.13) as a parameterization for the angular distribution of a single continuum photoelectron. In this section we seek a similar description for the cross section of multiple photo-excitation and multi-photoionization. As in the preceding section 5.3 we assume the initial target state to be randomly oriented. In addition the magnetic sublevels of the final-ion state and the spin-state of the photoelectrons are not resolved. The cross section for this process (5.12)

$$\sigma \propto \sum_{M_f} \frac{1}{2J_i+1} \sum_{M_i} |\langle f | \hat{\mathbf{e}} \cdot \mathbf{D} | i \rangle|^2,$$

$$= \sum_{M_f} \langle f | (\hat{\mathbf{e}} \cdot \mathbf{D})(\hat{\mathbf{e}} \cdot (P_i\mathbf{D}))^\dagger | f \rangle, \tag{5.27}$$

$$\text{where} \quad P_i = \frac{1}{2J_i+1} \sum_{M_i} |\Phi_i\rangle\langle\Phi_i| \delta(E_\gamma + \epsilon_i - E). \tag{5.28}$$

The dipole operator in length form is $\mathbf{D} = \sum_n \mathbf{r}_n$. The many-particle initial and final state vectors are $|i\rangle$ and $|f\rangle$, respectively. Since the initial state is randomly oriented the operator P_i is a scalar with respect to spatial rotations. Physically P_i describes the initial spectral density of the target.

Comparing Eq. (5.28) with Eq. (5.19) for emitted radiations from excited targets we note that the formal structure of these two equations is the same. Hence, we can write

$$\sigma \propto \left[\frac{1}{3}\langle f|\mathbf{D}\cdot\mathbf{D}'|f\rangle - \frac{i}{2}\sin(2\beta)\langle f|[\mathbf{D}\times\mathbf{D}']_\mathbf{k}|f\rangle \right.$$
$$\left. - \frac{1}{\sqrt{6}}\langle f|t_{20}(\mathbf{D},\mathbf{D}')|f\rangle + \cos(2\beta)\langle f|\bar{t}_{22}(\mathbf{D},\mathbf{D}')|f\rangle \right],$$
$$= \left[-\frac{1}{\sqrt{3}}\langle f|t_{00}|f\rangle - \frac{1}{\sqrt{2}}\sin(2\beta)\langle f|t_{10}|f\rangle \right.$$
$$\left. - \frac{1}{\sqrt{6}}\langle f|t_{20}|f\rangle + \cos(2\beta)\langle f|\bar{t}_{22}|f\rangle \right], \tag{5.29}$$

where $\mathbf{D}' := P_i\mathbf{D}$ and $\bar{t}_{22}(\mathbf{D},\mathbf{D}') = [t_{22}(\mathbf{D},\mathbf{D}') + t_{2-2}(\mathbf{D},\mathbf{D}')]/2$. The tensors t_{KQ} are given by

$$t_{KQ}(\mathbf{D},(P_i\mathbf{D})) = \sum_{qq'}\langle 1q1q'|KQ\rangle \mathbf{D}_q(P_i\mathbf{D})_{q'}. \tag{5.30}$$

One may suggest to repeat the steps (5.19-5.26) of the preceding section to arrive at a parameterization of the cross section analogous to the relation (5.28). This procedure is valid if the final excited (multi-particle) state is an eigenstate of angular momentum. The resulting parameterization is an obvious analogue of Eq. (5.26). Generally however, and in particular when dealing with continuum transitions (as in photoemission), the final state is not an angular-momentum eigenstate. Hence, the Wigner-Eckart-theorem, i.e. Eq. (5.25), does not apply (also compare Eq. (A.8)) and consequently a relation similar to (5.26) can not be obtained for photoemission. Two methods have been developed [147, 148, 149, 150, 151, 152, 153, 154] to remedy this situation:

1) one may expand the final state in angular momentum eigenstates (partial-wave expansion) in which case Eq. (5.25) applies to each term of the expansion.

2) One may express right from the outset the cross section (5.28) in terms of tensorial parameters and group these parameters according to their transformational properties.

Here we consider the first procedure, i. e. we expand the final state $|f\rangle$ in partial waves. To illustrate the significance of multiple excitations we consider both cases single and double photoionization. For single ionization the final state can be written as

$$|f_s\rangle = \sum_{lm} |\psi_{lm}\rangle C_{lm}(\hat{\mathbf{k}}_1) \tag{5.31}$$

where $C_{lm}(\hat{\mathbf{k}}_1)$ is a spherical harmonic in the notation of Ref. [143] and $\mathbf{k}_1$ is the wave vector of the photoelectron. For double ionization the analogous expansion is

$$|f_d\rangle = \sum_{l_a l_b l m} |\Psi^-_{l_a l_b l m}\rangle B^{l_a l_b}_{lm}(\hat{\mathbf{k}}_a, \hat{\mathbf{k}}_b). \tag{5.32}$$

The wave vectors of the two photoelectrons are $\mathbf{k}_a$ and $\mathbf{k}_b$ and $B^{l_a l_b}_{lm}(\hat{\mathbf{k}}_a, \hat{\mathbf{k}}_b)$ is a bipolar spherical harmonics [143] [$B^{l_1 l_2}_{lm}(\hat{\mathbf{x}}, \hat{\mathbf{y}}) = \sum_{m_1 m_2} \langle l_1 m_1 l_2 m_2 | lm \rangle C_{l_1 m_1}(\hat{\mathbf{x}}) C_{l_2 m_2}(\hat{\mathbf{y}})$].

The partial wave expansions can now be inserted in Eq. (5.29). For brevity and clarity let us consider only the term $-\frac{i}{2}\sin(2\beta)\langle f|[\mathbf{D} \times \mathbf{D}']_\mathbf{k}|f\rangle$ of this expansion. As explained in the preceding section this term describes the dependence Δ_{LR} of the cross section on the helicity of the light, i.e. $\Delta_{LR} = -i\sin(2\beta)\langle f|[\mathbf{D} \times \mathbf{D}']_\mathbf{k}|f\rangle$. According to Yang's formula [102] for single ionization Δ^s_{LR} should vanish, i.e. the relation must apply

$$\Delta^s_{LR} = -i\sin(2\beta)\langle f_s | [\mathbf{D} \times \mathbf{D}']_\mathbf{k} | f_s \rangle = 0. \tag{5.33}$$

Upon substituting (5.31) into (5.33), applying the Wigner-Eckart theorem, and performing the sum over m, we conclude

$$\Delta^s_{LR} = -i\sin(2\beta) \sum_{l'l} (-)^{l'} \begin{pmatrix} l' & l & 1 \\ 0 & 0 & 0 \end{pmatrix} C_{10}(\hat{\mathbf{k}}) \langle \psi_{l'} \| \mathbf{D} \times \mathbf{D}' \| \psi_l \rangle. \tag{5.34}$$

The key point is that only even or only odd values of l and l' contribute because of parity conservation. Assuming the initial state to be a parity eigenstate the $3-j$ symbol in (5.34) is then equal to zero because $l' + l + 1 = odd$. Moreover $l' = l$ because $(l', l, 1)$ satisfy a triangular relation.

In contrast to single photoemission, for one-photon double ionization (PDI) parity conservation does *not* imply the absence of dichroism. Parity conservation in DPI implies that $l_a + l_b$ is either even or odd. To explore the properties of quantity

$$\Delta_{LR}^d = -i\sin(2\beta)\langle f_d \mid [\mathbf{D} \times \mathbf{D}']_{\mathbf{k}} \mid f_d \rangle = 0$$

(which is generally referred to as the circular dichroism CD) we substitute the partial wave expansion (5.32) and apply the Clebsch-Gordon series for bipolar harmonics [144]

$$B_{lm}^{l_a l_b}(\hat{\mathbf{a}}, \hat{\mathbf{b}}) B_{l'm'}^{l'_a l'_b}(\hat{\mathbf{a}}, \hat{\mathbf{b}})^* = (-)^{l_a + l_b + l' + m'}\sqrt{(2l+1)(2l'+1)}$$

$$\sum_{L_a L_b K Q} (2L_a + 1)(2L_b + 1)\begin{pmatrix} l_a & l'_a & L_a \\ 0 & 0 & 0 \end{pmatrix}\begin{pmatrix} l_b & l'_b & L_b \\ 0 & 0 & 0 \end{pmatrix}\begin{Bmatrix} l & l' & K \\ l_a & l'_a & L_a \\ l_b & l'_b & L_b \end{Bmatrix}$$

$$\langle lml' - m' \mid KQ \rangle B_{KQ}^{L_a L_b}(\hat{\mathbf{a}}, \hat{\mathbf{b}}).$$

$$(5.35)$$

In the final channel one couples the angular momentum J_f of the residual ion with the angular momentum l of the electron pair to obtain the resultant angular momentum J. Now the Wigner-Eckart theorem can be applied and the summation over all magnetic quantum numbers can then be performed. The result for Δ_{LR}^d is

$$\Delta_{LR}^d = -i\sin(2\beta)\sum_{L_a L_b}\gamma_{L_a L_b} B_{10}^{L_a L_b}(\hat{\mathbf{k}}_a, \hat{\mathbf{k}}_b)$$

$$(5.36)$$

where

$$\gamma_{L_a L_b} = \sum_{l_a l_b l l'_a l'_b l' J J'} (-)^{l_a + l_b + l' + J' + J_f + 1}(2J' + 1)(2L_a + 1)(2L_b + 1)$$

$$\sqrt{\frac{(2l+1)(2l'+1)}{3}}\begin{pmatrix} l_a & l'_a & L_a \\ 0 & 0 & 0 \end{pmatrix}\begin{pmatrix} l_b & l'_b & L_b \\ 0 & 0 & 0 \end{pmatrix}$$

$$\begin{Bmatrix} l' & J_f & J' \\ J & 1 & l \end{Bmatrix}\begin{Bmatrix} l & l' & 1 \\ l_a & l'_a & L_a \\ l_b & l'_b & L_b \end{Bmatrix}$$

$$\langle J_f(l'_a l'_b)l'; J' \parallel \mathbf{D} \times \mathbf{D}' \parallel J_f(l_a l_b)l; J \rangle.$$

$$(5.37)$$

Eqs. (5.36) and (5.37) can be simplified if the assumption is made that the initial target state and the final ion state have a well-defined parity. In this case the state of the two continuum electrons is a parity eigenstate with a parity $\pi = (-)^{l_a + l_b}$, where l_a and l_b are the orbital angular momenta of the electrons. We note that many pairs of angular momenta (l_a, l_b)

will contribute to the two-electron continuum state such that $l_a + l_b$ is either *even* or *odd*. For example, double photoionization of He or H^- in the two-electron ground state has the symmetry (^1S) which results for the final state in a $^1P^{odd}$ symmetry with configurations $(l_a, l_b) = (s, p), (p, d), (d, f), \cdots$.

From (5.37) it follows that only pairs of (L_a, L_b) with $L_a = L_b$ contribute to the dichroism (5.36). Upon further inspection of the $3 - j$ symbols in (5.37) we conclude that the finiteness of the coefficients requires that the following relations have to be fulfilled

$$l_a + l_a' + L_a = even,$$
$$l_b + l_b' + L_b = even.$$

This means that

$$L_a + L_b = even$$

because $l_a + l_b$ *and* $l_a' + l_b'$ are both either *even* or *odd*. From the $9 - j$ symbol we deduce that the three numbers $(1, L_a, L_b)$ satisfy a triangular relation. Since the case $L_a = L_b \pm 1$ leads to odd values of $L_a + L_b$ we conclude that $L_a = L_b$. Eq. (5.36) can thus be written in the form

$$\Delta_{LR}^d = -i \sin(2\beta) \sum_L \gamma_{LL} B_{10}^{LL}(\hat{\mathbf{k}}_a, \hat{\mathbf{k}}_b). \tag{5.38}$$

Since only the diagonal elements γ_{LL} contribute to Eq. (5.38) one might expect that the dichroism is less sensitive a quantity to the description of the scattering dynamics than the cross sections.

5.4.1   Propensity rules for the dichroism in multiple photoionization

Parameterizing the cross sections in the form (5.22, 5.29) does not only disentangle the dependencies on the radiation field properties from the electronic dynamics but it also provides valuable information on the symmetry of the various terms in the expansion (5.29). To be specific let us analyze the main symmetry features of Eq. (5.38). Due to the Pauli principle the Δ_{LR}^d must be invariant under an exchange of the two electrons, i.e.

$$\Delta_{LR}^d(\beta, \mathbf{k}_a, \mathbf{k}_b) = \Delta_{LR}^d(\beta, \mathbf{k}_b, \mathbf{k}_a). \tag{5.39}$$

The dichroism Δ_{LR}^d is odd with respect to inversion of the helicity of the photon, i.e.

$$\Delta_{LR}^d(\beta, \mathbf{k}_a, \mathbf{k}_b) = -\Delta_{LR}^d(-\beta, \mathbf{k}_a, \mathbf{k}_b).$$ (5.40)

The angular functions $B_{10}^{LL}(\hat{\mathbf{k}}_a, \hat{\mathbf{k}}_b)$ determine the angular dependence of the dichroism, as seen from Eq. (5.38). These functions are explicitly given by

$$B_{10}^{LL}(\hat{\mathbf{k}}_a, \hat{\mathbf{k}}_b) = \sum_M \langle\, LML - M \mid 10 \,\rangle C_{LM}(\hat{\mathbf{k}}_a) C_{L-M}(\hat{\mathbf{k}}_b).$$ (5.41)

From this equation we deduce the following properties:

1. $B_{10}^{LL}(\hat{\mathbf{k}}_a, \hat{\mathbf{k}}_b)$ are purely imaginary. This follows from the relation

    $$C_{LM}(\hat{\mathbf{x}})^* = (-)^M C_{L-M}(\hat{\mathbf{x}})$$

    for spherical harmonics and the symmetry formula

    $$\langle\, L - MLM \mid 10 \,\rangle = -\langle\, LML - M \mid 10 \,\rangle$$

    for Clebsch-Gordon coefficients. The dichroism Δ_{LR}^d is a difference of cross sections and as such must be real. Therefore, we conclude that the coefficients γ_{LL} are real as well.

2. $B_{10}^{LL}(\hat{\mathbf{k}}_a, \hat{\mathbf{k}}_b)$ are parity-even in the solid angles associated with the momenta $\mathbf{k}_a, \mathbf{k}_b$ of the two photoelectrons, i.e. $B_{10}^{LL}(-\hat{\mathbf{k}}_a, -\hat{\mathbf{k}}_b) = B_{10}^{LL}(\hat{\mathbf{k}}_a, \hat{\mathbf{k}}_b)$ which follows from the parity of spherical harmonics given by $C_{LM}(-\hat{\mathbf{x}}) = (-)^L C_{LM}(\hat{\mathbf{x}})$.

3. From the symmetry of Clebsch-Gordon coefficients, we deduce furthermore that Eq. (5.41) is odd with respect to exchange of the electrons, i.e. $B_{10}^{LL}(\hat{\mathbf{k}}_b, \hat{\mathbf{k}}_a) = -B_{10}^{LL}(\hat{\mathbf{k}}_a, \hat{\mathbf{k}}_b)$. This relation implies that:

4. $B_{10}^{LL}(\hat{\mathbf{k}}_a, \hat{\mathbf{k}}_b)$ vanishes when the two electrons escape in the same direction and, due to relation (5.39), the functions γ_{LL} has to satisfy the condition

    $$\gamma_{LL}(k_a, k_b) = -\gamma_{LL}(k_b, k_a).$$ (5.42)

    This leads to a vanishing dichroism for emission of two electrons with equal energies.

5. B_{10}^{LL} vanishes when the electrons recede in a back-to-back configuration ($\hat{\mathbf{k}}_b \parallel -\hat{\mathbf{k}}_a$). This is concluded by considering the quantity

$$B_{10}^{LL}(\hat{\mathbf{x}}, \hat{\mathbf{x}}) = \sum_M \langle\, LML - M \mid 10\,\rangle C_{LM}(\hat{\mathbf{x}}) C_{L-M}(\hat{\mathbf{x}})$$

and substituting the expansion

$$C_{LM}(\hat{\mathbf{x}}) C_{L-M}(\hat{\mathbf{x}}) = \sum_K \langle\, LML - M \mid K0\,\rangle\langle\, L0L0 \mid K0\,\rangle C_{K0}(\hat{\mathbf{x}}).$$

The orthogonality of Clebsch-Gordon coefficients selects then the only value $K = 1$ for which, however, $\langle\, L0L0 \mid 10\,\rangle = 0$. For $\hat{\mathbf{k}}_b = -\hat{\mathbf{k}}_a$ we use $C_{LM}(-\hat{\mathbf{x}}) = (-)^L C_{LM}(\hat{\mathbf{x}})$ and repeat the arguments above.

6. The Δ_{LR}^d vanishes if the direction of the incident light $\hat{\mathbf{k}}$ and the electrons' vector momenta $\mathbf{k}_a, \mathbf{k}_b$ are linearly dependent. The above consideration assumes a co-ordinate frame with z axis being along $\hat{\mathbf{k}}$. Now let us select an arbitrary direction of $\hat{\mathbf{k}}_a$ described by the polar angles θ_a, φ_a. If the three vectors $\hat{\mathbf{k}}$ and $\mathbf{k}_a, \mathbf{k}_b$ are linearly dependent the spherical position of $\hat{\mathbf{k}}_b$ is determined by $\theta_b, \varphi_b = \varphi_a + N\pi$ with $N = integer$. The product of phase factors of the spherical harmonics is then real. For this reason also the bipolar harmonics (5.41) are real which contradicts the prediction that they are purely imaginary, except for the case when they are equal to zero.

7. The Δ_{LR}^d vanishes in a non coincidence experiment, i. e. if we integrate over one of the directions $\hat{\mathbf{k}}_a$ or $\hat{\mathbf{k}}_b$. This follows directly from the orthogonality of spherical harmonics and $L \geq 1$.

The above analysis does not provide a value for Δ_{LR}^d but it is nevertheless a valuable benchmark symmetry check for experimental and theoretical studies that calculate the magnitude of Δ_{LR}^d.

We recall that our derivation is based on the partial wave expansion (5.32) which has the advantage that the angular momenta in the expansion (5.32) can be related to those present in the initial states. As illustrated above, it allows a direct conclusion on the symmetry properties and propensity rules for the transition amplitude. The disadvantage of this procedure is as well clear from the above derivation: For many particle excitations the expansion (5.32) becomes rapidly more complicated and one has then to resort to a direct tensorial expansion of the cross sections in terms of rotational invariants.

5.5 Resonant photoexcitaions processes

In the previous sections we outlined briefly some of the main features of the electric dipole transitions that can be deduced from general symmetry considerations. In this section we illustrate the influence of the details of the excited states on the photoemission processes.

5.5.1 Single channel

We consider the case where, upon absorbing the photon, the ground state of the target A is elevated into a meta-stable bound state φ of A^* that couples to the continuum $\tilde{\psi}_E$ of A^+ and decays subsequently into A^+ and a photoelectron of energy E, i.e. we study the reaction

$$A + \hbar\omega \rightarrow A^* \rightarrow A^+ + e^-. \tag{5.43}$$

The questions of how to describe the photoexcited state Ψ_E and how the intermediate resonant state will show up in the spectrum of the photoelectron have been addressed by Fano [119, 120] using the following arguments. In the basis $(|\varphi\rangle, |\tilde{\psi}_E\rangle)$ the Hamiltonian H is not diagonal. Leaving aside the discrete state $|\varphi\rangle$ we determine the continuum states $\psi_E = \int_0^\infty c(E, E')|\tilde{\psi}_{E'}\rangle \, dE'$ such that $\langle\psi_{E'}|H|\psi_{E''}\rangle = E'\delta(E' - E'')$. Now we include in the basis set $|\psi_E\rangle$ the discrete state $|\varphi\rangle$ and determine the wave functions Ψ_E that satisfies $H\Psi_E = E\Psi_E$ by means of the basis-set expansion

$$|\Psi_E\rangle = a(E)|\varphi\rangle + \int_0^\infty b(E, E')|\psi_{E'}\rangle \, dE'. \tag{5.44}$$

Using this ansatz we conclude from the relations

$$\langle\varphi|H|\Psi_E\rangle = E\langle\varphi|\Psi_E\rangle \quad \text{and} \quad \langle\psi_{E''}|H|\Psi_E\rangle = E\langle\psi_{E''}|\Psi_E\rangle$$

that the expansion coefficients b are given by

$$b(E, E') = \left[P\frac{1}{E - E'} + Z(E)\delta(E - E')\right]V_{E'}a(E). \tag{5.45}$$

Furthermore, one deduces the relation

$$(E_\varphi - E)\,a(E) = -\int b(E, E')V_{E'}^* \, dE', \tag{5.46}$$

where $P(f)$ stands for the principle value of f. In Eqs. (5.45, 5.46) we introduced the definition

$$E_\varphi = \langle \varphi|H|\varphi \rangle, \tag{5.47}$$

$$V_E = \langle \psi_E|H|\varphi \rangle, \tag{5.48}$$

$$Z(E) = \left(E - E_\varphi - P \int \frac{|V_{E'}|^2}{E - E'} dE' \right) |V_E|^{-2}. \tag{5.49}$$

Asymptotically, i.e. for large distances r, the discrete and the continuum states behave as

$$\lim_{r\to\infty} \varphi(r) \to 0 \quad \text{and} \quad \lim_{r\to\infty} \psi_E \to \frac{1}{r}\sin(kr + \bar\delta(E)),$$

where k is the wave vector and $\bar\delta$ is a phase shift. From Eqs. (5.44, 5.46) we deduce thus the asymptotic behaviour for Ψ_E to be

$$\lim_{r\to\infty} \Psi_E(r) \to \frac{1}{r} \int \left[P\frac{1}{E - E'} + Z(E)\delta(E - E') \right] V_{E'} a(E) \sin(k'r + \bar\delta(E')) \, dE',$$

$$= \frac{1}{r}\sqrt{\pi^2 + Z^2} \left[\frac{Z}{\sqrt{\pi^2 + Z^2}} \sin(kr + \bar\delta) \right.$$

$$\left. - \frac{\pi}{\sqrt{\pi^2 + Z^2}} \cos(kr + \bar\delta) \right] V_E \, a(E)$$

$$= \frac{1}{r}\sqrt{\pi^2 + Z^2} \sin(kr + \bar\delta + \Delta) V_E \, a(E). \tag{5.50}$$

The phase Δ is determined by the relations

$$\cos \Delta = Z(\pi^2 + Z^2)^{-1/2}, \quad \text{and}$$

$$\sin \Delta = -\pi(\pi^2 + Z^2)^{-1/2}. \tag{5.51}$$

Thus, the phase shift Δ due to the resonant state is expressible as

$$\tan \Delta(E) = -\pi/Z(E). \tag{5.52}$$

From the normalization condition $\langle \Psi_E|\Psi_{E'} \rangle = \delta(E - E')$ and from Eqs. (5.44, 5.46) we conclude that

$$\langle \Psi_E|\Psi_{E'} \rangle = a^*(E)a(E') \left[\frac{f(E) - f(E')}{E - E'} \right.$$

$$\left. + P \int \frac{|V_{E''}|^2}{(E - E'')(E' - E'')} dE'' + \delta(E - E')Z^2(E')|V_{E'}|^2 \right], \tag{5.53}$$

where the function f is given by the equation

$$f(E) := P \int \frac{|V_{E'}|^2}{E - E'} dE'.$$

From the relation (5.53) we therefore obtain

$$\langle \Psi_E | \Psi_{E'} \rangle = a^*(E)a(E') \left[\pi^2 + Z^2(E')\right] |V_{E'}|^2 \, \delta(E - E'), \tag{5.54}$$

which upon integration over E' leads to the expression for $|a|^2$

$$|a(E)|^2 \quad = \quad \frac{1}{[\pi^2 + Z^2(E)] \, |V_E|^2}, \tag{5.55}$$

$$= \quad \frac{|V_E|^2}{[E - (E_\varphi + f(E))]^2 + \Gamma^2}. \tag{5.56}$$

In this equation we made use of (5.49) and introduced the quantity

$$\Gamma = \pi |V_E|^2,$$

which characterizes the width of a resonance located at

$$E_R := E_\varphi + f(E).$$

The resonance lifetime is thus

$$\tau = \hbar/\Gamma = \hbar/(\pi |V_E|^2).$$

The cross section (5.12) for the process (5.43) is $\sigma(E) = 4\pi^2 \alpha w |\langle \Psi_E | \hat{\mathbf{e}} \cdot \mathbf{r} | \phi_0 \rangle|^2$ where $|\phi_0 \rangle$ is the ground state and $|\Psi_E \rangle$ is represented by Eq. (5.44). From Eqs. (5.44, 5.45) it follows that Ψ_E can be written in the form

$$|\Psi_E \rangle \quad = \quad a(E) \left[|\phi \rangle + Z(E) V_E |\psi_E \rangle \right], \quad \text{where}$$

$$|\phi \rangle \quad = \quad |\varphi \rangle + P \int \frac{V_{E'}}{E - E'} |\psi_{E'} \rangle \, dE' \tag{5.57}$$

is the modified discrete state. From Eqs. (5.56, 5.52) we deduce that

$$a(E) = (\sin \Delta)/(\pi V_E) \quad \text{and} \quad Z(E) = -\pi \cot(\Delta).$$

Therefore, the dipole matrix element can be written as a sum of a resonant and an non-resonant term, namely

$$\langle \Psi_E | \hat{\mathbf{e}} \cdot \mathbf{r} | \phi_0 \rangle \quad = \quad \frac{\sin \Delta(E)}{\pi V_E} \langle \phi | \hat{\mathbf{e}} \cdot \mathbf{r} | \phi_0 \rangle - \cos \Delta(E) \langle \psi_E | \hat{\mathbf{e}} \cdot \mathbf{r} | \phi_0 \rangle$$

$$= \quad \langle \psi_E | \hat{\mathbf{e}} \cdot \mathbf{r} | \phi_0 \rangle \, (q \sin \Delta - \cos \Delta). \tag{5.58}$$

The dimensionless parameter

$$q = \frac{\langle \phi | \hat{\mathbf{e}} \cdot \mathbf{r} | \phi_0 \rangle}{\pi V_E \langle \psi_E | \hat{\mathbf{e}} \cdot \mathbf{r} | \phi_0 \rangle} \tag{5.59}$$

is called the shape parameter. From Eqs. (5.51) it follows that

$$(q \sin \Delta - \cos \Delta)^2 = \frac{(Z + q\pi)^2}{(\pi^2 + Z^2)}.$$

Introducing the dimensionless parameter ϵ, also called the reduced energy, as

$$\epsilon = \frac{(E - E_R)}{\Gamma} = Z/\pi,$$

we can thus write

$$(q \sin \Delta - \cos \Delta)^2 = \frac{(\epsilon + q)^2}{1 + \epsilon^2}.$$

With this expression we obtain the Fano-Beutler line profile [119] for the cross section σ as

$$\sigma(E) = \sigma_0 \frac{(q + \epsilon)^2}{1 + \epsilon^2}. \tag{5.60}$$

The cross section σ_0 represents a non-resonant background. Fig. 5.1 shows the quantity σ/σ_0 [cf. Eq. (5.60)] as a function of q and ϵ. As q increases the profile tends to a symmetrical Lorenzian-shape profile, whereas for $q = 0$ the profile exhibits a dip at $\epsilon = 0$. The latter case is called window resonance. In the general case however the profile is asymmetric.

5.5.2 Multi-channel resonant photoexcitations

The treatment of the photoexcitation [127] of a resonant state φ coupled to many continua ψ_E^i $(i = 1, \cdots, N)$ proceeds along similar lines. In brief, one performs in analogy to Eq. (5.44) for all the involved N continua the expansion

$$|\Psi_{jE}\rangle = a_j(E)|\varphi\rangle + \sum_{i=1}^{N} \int_0^\infty b_{ij}(E, E')|\psi_{iE'}\rangle \, dE'. \tag{5.61}$$

From $H\Psi_{jE} = E\Psi_{jE}$ one deduces that

$$\sum_{i=1}^{N} \int_0^\infty b_{ij}(E, E')\langle \varphi|H|\psi_{iE'}\rangle \, dE' = (E - E_\varphi)a_j(E), \quad \text{and}$$

$$(E - E')b_{ij}(E, E') = a_j(E)\langle \varphi|H|\psi_{iE}\rangle. \tag{5.62}$$

As a solution of these equations one writes

$$b_{ij}(E, E') = \frac{V_{iE'} \, a_j(E)}{E - E'} + \delta(E - E')\mathbf{C}_{ij}(E), \tag{5.63}$$

where V_{iE} is a matrix element given by

$$V_{iE} = \langle \varphi|H|\psi_{iE}\rangle.$$

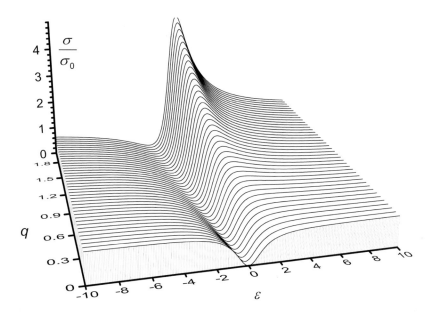

Figure 5.1: The Fano-Beutler Profile for different shape parameters q and different reduced energies ϵ.

On the other hand, the elements of the matrix $\mathbf{C}_{ij}(E)$ are unknown and need to be determined. To do so we conclude from Eqs. (5.62, 5.63) that

$$a_j(E) = \left\{ \sum_i V_{iE} \mathbf{C}_{ij}(E) \right\} \left\{ E - E_\varphi - \int \frac{\sum_k [V_{kE'}]^2}{E - E'} dE' \right\}^{-1}.$$

$$(5.64)$$

This relation combined with the normalization condition of the wave function (5.61) leads to the determining equation for the elements of $\mathbf{C}_{ij}(E)$

$$\sum_{jl} \mathbf{C}_{ij}^*(E) \mathbf{B}_{jl}(E) \mathbf{C}_{il} = 1,$$ (5.65)

where the matrix $\mathbf{B}_{jl}$ is given by

$$\mathbf{B}_{jl}(E) = \delta_{jl} + \frac{\pi^2 V_{jE} V_{lE} \sum_k [V_{kE}]^2}{\left(E - E_\varphi - \int \frac{\sum_k [V_{kE'}]^2}{E-E'} dE' \right)^2}.$$

Assuming that the channels are asymptotically decoupled we inspect the asymptotic behaviour of the wave function (5.61), as done in Eq. (5.50), and obtain a relation for $a_i(E)$ analogous to the single channel case [i.e. Eq. (5.54)], namely

$$\lim_{r \to \infty} \Psi_{iE} \quad \to \quad -\pi a_i(E) \sum_{j=1}^{N} V_{jE} \cos(k_j r + \bar{\delta}_j) + \sum_j \mathbf{C}_{ij}(E) \sin(k_j r + \bar{\delta}_j),$$

$$\Rightarrow \quad [-i\pi a_i(E) V_{jE} + \mathbf{C}_{ij}(E)] \, e^{i(k_j r + \bar{\delta}_j)} = 0, \quad i \neq j. \tag{5.66}$$

The energy-dependent function $a_i(E)$, $f_N(E)$ and $C_{ij}(E)$ are given by the relations

$$a_i(E) \quad = \quad \frac{\sum_j \mathbf{C}_{ij}(E) V_{jE}}{E - (E_\varphi + f_N(E))}, \tag{5.67}$$

$$f_N(E) \quad := \quad \int \frac{\sum_k [V_{kE'}]^2}{E - E'} dE',$$

$$C_{ij}(E) \quad = \quad i\pi V_{jE} a_i(E), \quad i \neq j. \tag{5.68}$$

The steps (5.53-5.56) can now be repeated to obtain the width of the resonance (which yields $\Gamma_N = \pi \sum_j^N (V_{jE})^2 = \sum_j^N \Gamma_j$).

An expression for the wave function (5.66) is obtained upon inserting Eqs. (5.67-5.68) in (5.66). This results in an expression that can be utilized for the calculation of the optical transition matrix elements (5.58). Formulas for the line profiles are obtained by defining, as done in Eq. (5.57), a modified discrete state ϕ_N and introducing the shape parameters q_N as

$$q_N = \frac{\langle \phi_N | \hat{\mathbf{e}} \cdot \mathbf{r} | \phi_0 \rangle}{\pi \sum_i^N V_{iE} \langle \psi_{iE} | \hat{\mathbf{e}} \cdot \mathbf{r} | \phi_0 \rangle}. \tag{5.69}$$

Analogously, one defines the reduced energy ϵ_N as the quantity

$$\frac{1}{\epsilon_N} = \frac{\pi \sum_j (V_{jE})^2}{E - [E_\varphi + f_N(E)]} = \frac{\pi \sum_j (V_{jE})^2}{E - E_{R_N}}. \tag{5.70}$$

The total cross section $\sigma_{tot}(E)$ in all channels normalized to the non-resonant background $\sigma_0(E)$ is then completely specified by the parameters q_N, ϵ_N and by the correlation index ρ_c

$$\sigma_{tot}(E)/\sigma_0(E) = 1 + \rho_c^2 \left[\frac{(q_N + \epsilon_N)^2}{1 + \epsilon_N^2} - 1 \right]. \tag{5.71}$$

The correlation index ρ_c is a relative measure for the coupling strength of the resonance to the individual continua, as readily deduced from its determining equation

$$\rho_c^2 = \frac{\left| \sum_j^N V_{jE} \langle \psi_{jE} | \hat{\mathbf{e}} \cdot \mathbf{r} | \phi_0 \rangle \right|^2}{\left[\sum_j^N (V_{jE})^2 \right] \left[\sum_j^N |\langle \psi_{jE} | \hat{\mathbf{e}} \cdot \mathbf{r} | \phi_0 \rangle|^2 \right]}. \tag{5.72}$$

Having sketched the essential steps for isolating and parameterizing the relevant quantities that describe the manifestation of resonant states, the question arises as how to calculate these quantities and what role is played by the inner-particle correlation in many-particle resonant states?. The next section gives an overview on some of the aspects of this intensively studied topic.

5.6   Few-body resonances

Resonances show up in various physical systems, such as in nuclear reactions, in electron-atom and electron-molecular scattering processes and in photoionization ([128, 162, 322] and references therein). The complexity and the various facets of resonance formation and decay in a many particle system are elucidated by one of the simplest few-body systems, namely helium. As illustrated in Fig. 5.2, doubly excited (two-electron) states lay above the first ionization threshold, i.e. it takes more energy to excite the two electrons to the next available level than to singly and directly ionize the target. While this observation is valid for two electron atoms, for more complex systems doubly excited states can as well be situated below the first ionization threshold.

As shown by Fig. 5.2, the doubly excited states of helium form an infinite number of perturbed Rydberg series converging to the single ionization thresholds, where one electron is left in an excited state of the residual ion. The doubly excited states are coupled to one or more continua and form thus resonances that decay by autoionization.

In 1963 Madden and Codling [163] made the first experimental observation of the full series of helium doubly excited levels with $^1P^o$ symmetry[3] in the $60 - 65$ eV energy range. Fano and co-workers [128] (see also [162]) pointed out that these early observations are manifestations of a breakdown of the independent-particle picture. Since $He^+(2s)$ and $He^+(2p)$ have the same energy one would expect three series with comparable intensity converging to the same threshold. In an independent particle model they are labelled by the configurations $2snp$, $2pns$, and $2pnd$. Experimentally, however, a single intense series was observed which is intercalated by a faint one. As an explanation, it has been suggested [120] that the electron-electron interaction will mix the series $2snp$, $2pns$, into a $2snp + 2pns$ (plus) and a $2snp - 2pns$ (minus) series. In this picture the excited electron pair is viewed as a joint (quasi) particle characterized by two quantum numbers. One quantum number is assigned to its inter-

---

[3]We recall the spectroscopic notation $^{2S+1}L^\pi$ in the LS coupling where L is the total orbital momentum, S is the total spin and π is the parity.

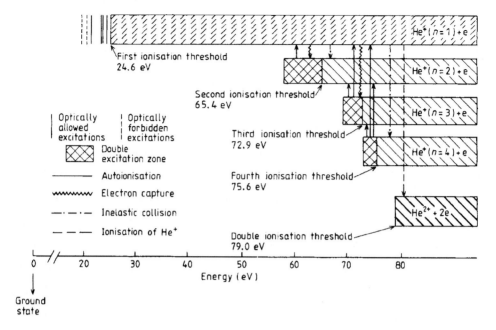

Figure 5.2: The term scheme of helium showing increasing number of channels and their interconnection by various processes (adopted from Ref. [128]).

nal motion whereas the other quantum number describes the motion of the center-of-mass of the pair. Since the optical transitions from the ground state to the plus levels are favored, only a single intense series is observed. The width of the measured lines is a measure for the rate of the autoionization of the excited levels.

5.6.1   Regularities and classifications of doubly excited states

These findings by Madden and Codling and by Fano and co-workers sparked a wealth of further experimental and theoretical investigations [162]. In particular, several classification schemes have been developed in order to understand the origin of the regularities in the occurrence of resonances and why resonances with the same total symmetry reveal drastically different characteristics. Here we briefly mention the adiabatic hyperspherical approach [128, 178, 164, 165, 166, 167], the widely used doubly excited symmetry basis scheme (DESB) [168, 169, 170] and the molecular orbital (MO) scheme [172] which contains and explains the DESB. A completely new approach for the treatment of the dynamics of two-electron atoms is offered by modern semi-classical concepts [173, 174]. In this method the

systematics of resonance series is ascribed to certain characteristic periodic orbits in a classical three-body system. Resonant and bound states are distinguished from the ground state via
the use of periodic orbit trace formulas.

The adiabatic hyperspherical approach [128, 178, 164, 165, 166, 167] was one of the first
models to be utilized. As will be outlined below, in the adiabatic hyperspherical approach the
overall size of the system is characterized by one coordinate, the hyperradius, which is treated
adiabatically, i.e. one assumes that the motion along the hyperradius is significantly slower
than along other coordinates. As in the case of the Born-Oppenheimer method (Sec.4.1) the
adiabaticity assumption leads to a quasi-separability of the problem. Thus, as long as the
adiabaticity hypotheses is viable the motion of the correlated electrons takes place in one-
dimensional (adiabatic) potentials. The lowest of these potential curves contains a Rydberg
series of bound states, whereas the potential curves of the excited channels carry the autoionizing, i. e. doubly excited Rydberg series. From a mathematical point of view, the adiabatic
assumption is no longer valid at very high excitations where a series of avoided crossing between the potential curves occur.

In the DESB approach features of the dynamical SO_4 symmetry for the two-body Coulomb
potential (cf. Sec. 2.2) are exploited by considering for the two electron case a coupled representation $SO_4 \otimes SO_4$. Doubly excited states are classified by the set of quantum numbers
$_N(K,T)_n^A$ that have the following meaning: N and n are respectively the principal quantum
numbers of the inner and outer electrons. Given a symmetry $^{2S+1}L^\pi$ there are several series
converging to the same threshold $He^+(N)$. Different series below a specified threshold N are
characterized by the approximate quantum numbers $(K,T)^A$, whereas a specific state within
the series is classified with the quantum numbers n, K, T. If the outer electron is excited further away from the inner core, the principle quantum number N of the inner electron is a good
quantum number. In contrast, n is, in general, not a good quantum number and therefore it is
conventionally replaced by an appropriate effective quantum number.

The quantum number T is obtained by quantizing the projection of the total angular mo-
mentum onto $\mathbf{B} := \mathbf{A}_1 - \mathbf{A}_2$, where $\mathbf{A}_{1/2}$ are the Laplace-Runge-Lenz vectors of the two
independent electrons (see Sec. 2.2). Specifically, the relation applies $(\mathbf{B} \cdot \mathbf{L})^2 |NnKTL^\pi\rangle = T^2(K+n)^2 |NnKTL^\pi\rangle$. The quantum number K describes the inter-electronic angular

correlation

$$\langle \cos \theta_{12} \rangle = -\frac{K}{N} + \frac{N^2 - 1 - K^2 - T^2 + 2l_1 \cdot l_2}{2Nn}, \tag{5.73}$$

where θ_{12} is the mutual angle between the position vectors of the two electrons. The ranges of the values of the quantum numbers K, T are given by

$$T = 0, 1, \cdots, \min(L, N-1),$$

$$K = -(N - T - 1), -(N - T - 1) + 2, \cdots, -(N - T - 1) + 2(N - T - 1).$$

For $n \gg N$ one deduces from Eq. (5.73) that $\langle \cos \theta_{12} \rangle \rightarrow -K/N$, meaning that for the highest value of K the electrons reside on average at opposite sides of the nucleus. On the other hand, for vanishing K, one obtains $\langle \theta_{12} \rangle \approx 90°$.

While K describes the angular correlation, the further quantum number A has been introduce [170, 178] to describe the radial correlation. The values of A are $-1, 0$ or 1. More precisely A behaves according to

$$A = \pi(-)^{T+S}, \quad \text{for} \quad K > L - N, \quad \text{eitherwise } A = 0.$$

This means A is not an independent quantum number. $A = \pm 1$ quantifies the nodal structure of the wave function for equidistant electrons from the origin ($r_1 = r_2$). States with $A = -1$ have a node, whereas states with $A = 1$ have an antinode. For $A = 0$ the state is highly asymmetric with one electron being close to the residual ion and the other electron being far away from the inner core, i.e. radial correlations are small in this case. Furthermore, the investigation of the vibrational structure [171] has led to the quantum number n_ν [179] which is given in terms of N, K, T as

$$n_\nu = (N - K - T - 1)/2.$$

This quantum number n_ν counts the nodes in the θ_{12} coordinate.

Using the set of the DESB quantum numbers, the structure of the doubly excited states can be made comprehensible. In addition, this scheme enabled a clear understanding of the autoionization rates to different channels and the photoexcitation strengths to different states as well as the existence of propensity rules for autoionization and photoexcitations [175, 176, 177], e.g. for photoexcitation experiment (investigating high-lying $^1P^o(N = 3 - 8)$ of H^-) the propensity rules $\Delta A = 0, \Delta n_\nu = 0$ apply [175, 176].

The DESB quantum numbers are combinations of the parabolic quantum numbers $[N_1 N_2 m]^A$ (cf. Sec. 2.2) that describe a polarized state in the single-electron N–manifold resulting from the removal of one electron whilst the other remains bound [176]. This interrelation is given below. The link between the adiabatic hyperspherical method and the algebraic approach has been established by Lin [178]. He was able to relate in a unique way the approximate quantum numbers from the algebraic approach to the hyperspherical potential curves.

In the MO approach [172] the three-body problem is formulated in a Jacobi coordinate system, i.e. one introduces the two position vectors $\mathbf{r}$ and $\mathbf{R}$, defined as

$$\mathbf{r} = (\mathbf{r}_1 + \mathbf{r}_2)/2, \quad \mathbf{R} = \mathbf{r}_1 - \mathbf{r}_2 \equiv \mathbf{r}_{12}.$$

The inter-electronic distance r_{12} is then treated adiabatically[4]. The advantage of this treatment is that for fixed (adiabatic coordinate) r_{12}, the wave function is separable in prolate spheroidal coordinates[5] [180]

$$\lambda = (r_1 + r_2)/r_{12}, \quad \mu = (r_1 - r_2)/r_{12}.$$

The (molecular) wave function is then written as a product of orbitals characterized by the molecular quantum numbers n_λ, n_μ that count the number of nodes along the coordinates λ and μ. In addition to the coordinates λ and μ a further coordinate φ is needed that describes the projection of the angular momentum of the electrons' center of mass onto the inter-electronic axis. The corresponding quantum number is usually referred to as m. Thus, an adiabatic MO state, which builds a Rydberg series, is specified by the quantum numbers (n_λ, n_μ, m) or equivalently by the parabolic quantum numbers $[N_1 N_2 m]^A$ (cf. Sec.2.2). The different members of each Rydberg series can be constructed as vibrational states in the corresponding molecular potential. A single member of a Rydberg series, is specified by a vibration-like quantum number $\bar{n} = n - N$ for even n_μ and $\bar{n} = n - N - 1$ for odd n_μ. The connection between the (independent) MO quantum numbers n_λ, n_μ, m, the DESB quantum numbers K, T, A, n_ν, and the parabolic quantum mumbers is [182]

---

[4]We note here that treating r_{12} adiabatically does not mean that the electron-electron interaction is weak. In fact it is the strong repulsion between the two particles which stabilizes this coordinate [181].

[5]In the limits $r_{12} \to 0$ or $r_{12} \to \infty$ the spheroidal surfaces coincide with spherical or paraboloidal surfaces.

DESB	parabolic	MO
$A =$	A	$= (-)^{n_\mu}$,
$T =$	m	$= m$,
$K =$	$N_2 - N_1$	$= [n_\mu/2] - n_\lambda$,
$N =$	$N_1 + N_2 + |m| + 1$	$= n_\lambda + [n_\mu/2] + |m| + 1$,
$n_\nu =$	N_1	$= n_\lambda$.

The symbol $[x]$ means the integer part of x. Based on the MO approach it has been possible [182] to establish propensity rules for photoabsorption and autoionization processes. With these rules the different oscillator strengths and widths for different Rydberg series within a N manifold have been explained .

5.6.2   Complex rotation method

For a direct numerical calculation of energies and widths of doubly excited states various methods have been applied. In particular, we mention here the Feshbach projection operator formalism [183, 184], the close-coupling approximation [185], the hyperspherical coordinate method [186], and the multiconfiguration Hartree-Fock method [187].

The most accurate resonance computations are nowadays performed using the complex coordinate rotation method [188, 189, 190, 191, 192, 186, 193, 194, 195, 196, 197]. The idea of this approach is based on the dilatation analytic continuation [198, 199, 200]. The advantage of the complex rotation method is that it transforms the continuum resonance wave function to a square-integrable (localized) wave function typical for a bound state. This brings about considerable simplifications, for the resonance wave function can then be calculated using existing bound-state codes which are mostly based on the variational ansatz for a complex Hamiltonian[201], i.e. bound and resonant states are treated on equal footing. This method has yielded accurate results for few-body resonant states. To illustrate the basic elements of the complex rotation method let us consider the Hamiltonian of a two-electron atomic system which, in Rydberg units, reads

$$H = T + V = -\Delta_1 - \Delta_2 - \frac{2Z}{r_1} - \frac{2Z}{r_2} + \frac{2}{r_{12}},$$

where Z is the charge of the residual ion. The complex-rotation method consists in transforming formally the radial coordinate r_j according to the rule

$$r_j \to r_j \exp(i\theta),$$

where θ is referred to as the rotational angle. The Hamiltonian H transforms as

$$H \rightarrow \tilde{H}(\theta) = T \exp(-2i\theta) + V \exp(-i\theta).$$

The eigenvalues of the transformed Hamiltonian are obtained by solving the complex-eigenvalue problem

$$E(\theta) = \frac{\langle \Psi | \tilde{H}(\theta) | \Psi \rangle}{\langle \Psi | \Psi \rangle},$$

where the wave function Ψ is usually expanded in a basis set functions.

To find the resonances one search for eigenvalues E_{res} of the operator $\tilde{H}(\theta)$ that do not depend on the rotational angle θ. The real part of the complex eigenvalue is interpreted as the resonance position E_r, whereas the imaginary part as the half-width of the resonance $\Gamma/2$ (i.e. $E_{res} = E_r + i\Gamma/2$). Due to the finite-size of the employed basis, however, in practice the resonance eigenvalues do depend weakly on θ. Resonances are then identified by a minimal value of $\partial E/\partial \theta$. Furthermore, the photoabsorption cross section $\sigma(\omega)$ is obtained by a sum over all (resonant and continuum) eigenstates of $\tilde{H}(\theta)$, more precisely [202]

$$\sigma(\omega) \propto \omega \Im \left[\sum_i \frac{\langle \Psi_0 | \sum_j r_j \exp(i\theta) | \Psi_i \rangle \langle \Psi_i | \sum_j r_j \exp(i\theta) | \Psi_0 \rangle}{E_i - E_0 - \hbar\omega} \right],$$

where the index i runs over the states of all the electrons j and the index 0 labels the ground state.

6 Two-electrons systems at the complete fragmentation threshold: Wannier theory

The treatment of many-body systems within an effective single-particle or a mean-field approach (e.g., as done by DFT in section 4.6 or by HF in section 4.2) implies that the interaction of a single particle with any other particular particle in the system is in general weaker than the interaction with the surrounding rest of the system. This picture losses its validity when two or more particles are highly excited, for the excited particles can access a large manifold of degenerate orbitals and will adjust their motion as dictated by their mutual interaction in the presence of the mean field. As a direct demonstration of this statement we discussed in the previous section the properties of doubly excited states of helium. From this example one expects that inter-electronic correlation plays a major role when the electrons are excited to states with energies close to the double ionization threshold I^{++}. Therefore, considerable attention (e.g. [90, 131, 132, 133, 134, 135, 136, 137, 138] and references therein) has been devoted to the question of how the double ionization cross section σ^{++} at I^{++} is influenced by the correlated motion of the electrons. In a seminal work Wannier [90] addressed this question using classical arguments and derived in 1953 a threshold law for the double electron escape. Meanwhile, the conclusions of his work have been, to a large extent, confirmed experimentally and using a variety of theoretical approaches (cf. e. g. Refs. [250, 91, 138, 134] and further references therein). Here we adhere to Wannier's original work and sketch the main steps of Wannier's arguments and results. For details of other theoretical and experimental approaches we refer to Refs. [90, 131, 132, 133, 134, 135, 136, 137, 138].

6.1   Classical mechanics of two excited electrons at the double escape threshold

Let us consider two electrons with vanishing total orbital angular momentum ($L = 0$) and ask the cross section $\sigma^{2+}(E)$ behaves near $E = I^{++}$, where E is the total energy of the electron pair. Since $L = 0$ the motion takes place in one plane depicted in Fig. 6.1. Only three variables are sufficient for description of the two electron trajectories:

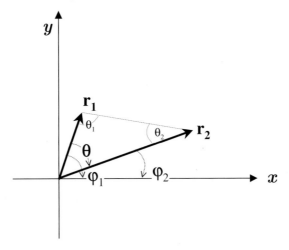

Figure 6.1: The coordinate system for two electrons moving in a plane.

$$R \; = \; \sqrt{r_1 + r_2}, \quad \theta = \arccos(\hat{\mathbf{r}}_1 \cdot \hat{\mathbf{r}}_2), \quad \alpha = \arctan(\frac{r_1}{r_2}) \tag{6.1}$$

$$\Leftrightarrow \mathbf{r}_1 \; = \; \begin{pmatrix} R & \sin\alpha & \cos\varphi_1 \\ R & \sin\alpha & \sin\varphi_1 \end{pmatrix}, \tag{6.2}$$

$$\mathbf{r}_2 \; = \; \begin{pmatrix} R & \cos\alpha & \cos\varphi_2 \\ R & \cos\alpha & \sin\varphi_2 \end{pmatrix}, \quad \theta = \varphi_1 - \varphi_2. \tag{6.3}$$

The coordinates R, θ and α are called the hyperspherical coordinates [203, 204, 205, 164, 206, 165, 129]. The hyperradius R quantifies the size of the triangle formed by the two vectors $\mathbf{r}_1$ and $\mathbf{r}_2$ whereas the mock angles θ and α describe the shape of this triangle. The kinetic energy $T = \frac{1}{2}\left(\dot{\mathbf{r}}_1^2 + \dot{\mathbf{r}}_2^2\right)$ casts in the coordinate system (6.1)

$$T = \frac{1}{2}\dot{R}^2 + \frac{1}{2}R^2\left[\dot{\alpha}^2 + \sin^2\alpha\,\dot{\varphi}_1^2 + \cos^2\alpha\,\dot{\varphi}_2^2\right].$$

The classical orbital momentum is $\mathbf{L} = L_z$, where $L_z = R^2 \sin^2 \alpha \, \dot{\varphi}_1 + R^2 \cos^2 \alpha \, \dot{\varphi}_2$. From $L_z = 0$ and $\dot{\theta} = \dot{\varphi}_1 - \dot{\varphi}_2$ we deduce that $\dot{\varphi}_1 = \dot{\theta} \cos^2 \alpha$ and $\dot{\varphi}_2 = -\dot{\theta} \sin^2 \alpha$, so that the kinetic energy is written as

$$T = \frac{1}{2}\dot{R}^2 + \frac{1}{2}R^2\dot{\alpha}^2 + \frac{1}{8}R^2\dot{\theta}^2 \sin^2 2\alpha.$$

The potential energy for the two electrons moving in the field of a residual ion with a positive charge Z is

$$V = -\frac{Z}{r_1} - \frac{Z}{r_2} + \frac{1}{r_{12}} = \frac{C(\alpha, \theta)}{R},$$

where the mock-angle dependent part $C(\alpha, \theta)$ of the potential energy is given by

$$C(\alpha, \theta) = -\frac{Z}{\sin \alpha} - \frac{Z}{\cos \alpha} + \frac{1}{\sqrt{1 - \sin 2\alpha \cos \theta}}.$$

To write down the equations of motion we express the Lagrange function $\mathcal{L}$ in hyperspherical coordinates as

$$\mathcal{L} = \frac{1}{2}\dot{R}^2 + \frac{1}{2}R^2\dot{\alpha}^2 + \frac{1}{8}R^2\dot{\theta}^2 \sin^2 2\alpha - \frac{C(\alpha, \theta)}{R}.$$

$$(6.4)$$

Likewise, the total energy E can be written in the form

$$E = \frac{1}{2}\dot{R}^2 + \frac{1}{2}R^2\dot{\alpha}^2 + \frac{1}{8}R^2\dot{\theta}^2 \sin^2 2\alpha + \frac{C(\alpha, \theta)}{R}.$$

$$(6.5)$$

The equations of motion for the two electrons in hyperspherical coordinates are then given by the expressions

$$\ddot{R} = R\dot{\alpha}^2 + \frac{1}{4}R\dot{\theta}^2 \sin^2 2\alpha + \frac{C}{R^2} \qquad (6.6)$$

$$\frac{d}{dt}\left(R^2\dot{\alpha}\right) = \frac{1}{4}R^2\dot{\theta}^2 \sin 4\alpha - \frac{1}{R}\frac{\partial C}{\partial \alpha} \qquad (6.7)$$

$$\frac{d}{dt}\left(\frac{1}{4}R^2\dot{\theta} \sin^2 2\alpha\right) = -\frac{1}{R}\frac{\partial C}{\partial \theta}. \qquad (6.8)$$

From a key importance is the similarity principle which states the following. When the system is expanding the trajectories remain topologically invariant. This is due to scaling transformation

$$R \quad \rightarrow \quad R' = \beta R,$$

$$\alpha \quad \rightarrow \quad \alpha' = \alpha,$$

$$\theta \quad \rightarrow \quad \theta' = \theta,$$

$$t \quad \rightarrow \quad t' = \beta^{3/2} t,$$

$$E \quad \rightarrow \quad E' = \frac{E}{\beta}, \tag{6.9}$$

where β is an arbitrary real positive constant. Using this scaling feature one deduces, that for $E \geq 0$, and along each trajectory the function $R(t)$ has, if at all, one minimum and no maximum. The proof is straightforward. From Eq. (6.6) we deduce that $R\ddot{R} = R^2 \dot{\alpha}^2 + \frac{1}{4} R^2 \dot{\theta}^2 \sin^2 2\alpha + \frac{C}{R} = R^2 \dot{\alpha}^2 + \frac{1}{4} R^2 \dot{\theta}^2 \sin^2 2\alpha + E - T$, meaning that

$$R\ddot{R} = E - \frac{1}{2}\dot{R}^2 + \frac{1}{2}R^2\dot{\alpha}^2 + \frac{1}{8}R^2\dot{\theta}^2 \sin^2 2\alpha.$$

Therefore, if R is stationary (i.e. if $\dot{R} = 0$) we obtain $\ddot{R} > 0$, in other words if $R(t)$ has an extremum it can not be more than one minimum (two minima means at least one maximum has to be in between them, in which case $\ddot{R} < 0$ must apply)[1].

Another important statement is the following: Statistically almost all trajectories with $E = 0$ belong to the single ionization channel, i.e. asymptotically only one electron escapes and the other remains bound. The proof is straightforward. Let us label the individual electrons' energies by E_1 and E_2. The constraint $E = 0 = E_1 + E_2$ is satisfied by an infinity of trajectories for which one electron has a positive energy $E_1 > 0$ (unbound) and the other has a negative (bound) energy $E_2 = -Z^2/(2n^2)$. In contrast if we require $E_1 > 0$ and $E_2 > 0$ the condition $E = E_1 + E_2 = 0$ has only one solution. Thus, from a statistical point of view the double ionization is extremely unlikely compared to the single ionization. Nevertheless, one observes experimentally a considerable amount of double ionization events. Therefore, the time evolution of the electrons' distances from the residual ion (as described by $\alpha(t)$) can not be statistically distributed. Hence, the key point is to find out the (dynamical)

---

[1] If $E = 0$ and all velocities vanish ($\dot{R} = \dot{\alpha} = \dot{\theta} = 0$) then Eq. (6.5) requires $\ddot{R} = 0$. To investigate the stationarity of $R(t)$ in this case one has to inspect higher differentiations of $R(t)$. Doing so one concludes $\dddot{R} = 0$, however, for the four-fold time differentiation one calculates $R\ddddot{R} = R^2\ddot{\alpha}^2 + \frac{1}{4}R^2\ddot{\theta}^2 \sin^2 2\alpha$ which means $\ddddot{R} > 0$. Hence, the trajectory $R(t)$ shows for $E = 0$ only one minimum.

mechanism that stabilizes the two electrons' motion, in particular the motion along $\alpha(t)$ has to be stabilized at certain value α_0. Identifying such a stabilization process is equivalent to finding a non-statistical pathway for double ionization, for the scaling Eq. (6.9) implies that $\alpha = \alpha_0$ remains stabile when the system size, characterized by the hyperradius R, expands by an amount β.

To this end one inspects the variation of the potential with respect to α and θ and finds out that the conditions

$$
\frac{\partial C}{\partial \alpha} = \frac{Z \cos \alpha}{\sin^2 \alpha} - \frac{Z \sin \alpha}{\cos^2 \alpha} + \frac{\cos 2\alpha \cos \theta}{(1 - \sin 2\alpha \cos \theta)^{2/3}} = 0,
$$

$$
\frac{\partial C}{\partial \theta} = -\frac{\sin 2\alpha sin\theta}{2\sqrt{(1 - \sin 2\alpha \cos \theta)^{3/2}}} = 0, \tag{6.10}
$$

are fulfilled at

$$
\alpha_0 = \pi/2, \quad \text{and} \quad \theta_0 = \pi, \tag{6.11}
$$

i.e. when the two electrons are on opposite sides and at equal distances from the residual ion $(\mathbf{r}_1 = -\mathbf{r}_2)$.

For this reason it is appropriate to expand the potential $C(\alpha, \theta)$ around the extremal positions. This yields the following expansion coefficients

$$
C(\alpha, \theta) \approx -C_0 - \frac{1}{2} C_1 (\alpha - \alpha_0)^2 + \frac{1}{2} C_2 (\theta - \theta_0)^2, \tag{6.12}
$$

where

$$
C_0 = \frac{1}{\sqrt{2}} (4Z - 1), \; C_1 = \frac{1}{\sqrt{2}} (12Z - 1), \quad \text{and} \quad C_2 = \frac{1}{\sqrt{32}}. \tag{6.13}
$$

From Eq. (6.12) we deduce that the motion around the extremal (saddle) point is stable along θ but unstable along α (upon a small perturbations).

Using Eq. $(6.5)^2$ the equations of motion (6.8) can now be linearized around the saddle point to yield

$$
\ddot{R} = -\frac{C_0}{R^2}, \tag{6.14}
$$

$$
\frac{d}{dt} (R^2 \dot{\alpha}) = \frac{C_1}{R} (\alpha - \alpha_0), \tag{6.15}
$$

$$
\frac{d}{dt} (\frac{1}{4} R^2 \dot{\theta}) = -\frac{C_2}{R} (\theta - \theta_0). \tag{6.16}
$$

---

[2]Note that from Eq. (6.5) at the saddle point one deduces that $\dot{R} = \sqrt{2E + 2\frac{C_0}{R}}$.

Exchanging in these relations the variation in time into a variation with respect to the radial distance [3] R one finds for $E = 0$

$$\dot{R} = \sqrt{\frac{2C_0}{R}}, \tag{6.17}$$

$$R^2 \partial_R^2 \alpha + \frac{3R}{2} \partial_R \alpha - \frac{C_1}{2C_0}(\alpha - \alpha_0) = 0, \tag{6.18}$$

$$R^2 \partial_R^2 \theta + \frac{3R}{2} \partial_R \theta + \frac{2C_2}{C_0}(\theta - \theta_0) = 0. \tag{6.19}$$

For the solution of these equations we make the ansatz

$$\alpha(R) \approx \alpha_0 + R^n,$$

which yields for n

$$n = -\frac{1}{4} \pm \frac{1}{2}\mu, \tag{6.20}$$

$$\mu = \frac{1}{2}\sqrt{\frac{100Z - 9}{4Z - 1}}. \tag{6.21}$$

Thus, the general solution for $\alpha(R)$ can be written as

$$\alpha(R) = \alpha_0 + a\,R^{-\frac{1}{4}-\frac{1}{2}\mu} + b\,R^{-\frac{1}{4}+\frac{1}{2}\mu}, \tag{6.22}$$

where $a(E)$ and $b(E)$ are (energy dependent) integration constants. Likewise, we write $\theta(R)$ in the form

$$\theta(R) \approx \theta_0 + R^m \text{ and obtain } m = -\frac{1}{4} \pm \frac{i}{2}\bar{\mu},$$

where

$$\bar{\mu} = \frac{1}{2}\sqrt{\frac{9 - 4Z}{4Z - 1}}.$$

Thus, the general solution for $\theta(R)$ is

$$\theta(R) = \theta_0 + \bar{a}\,R^{-\frac{1}{4}} \cos\left(\frac{\bar{\mu}}{2} \ln R + \bar{b}\right), \tag{6.23}$$

where $\bar{a}(E)$ and $\bar{b}(E)$ are integration constants.

---

[3] This is valid for $t > t_0$ where t_0 is determined from the minimum of $R(t)$, i.e. from the condition $\dot{R}(t) = 0$.

6.1.1   Wannier threshold law: a classical approach

The (total) cross section for the double escape $\sigma^{2+}(E)$ is obtained from the time variation of
the (microcanonical) phase space volume Ω_E which is available for the two excited electrons
at a fixed energy E, i. e.

$$\sigma^{2+}(E) = \dot{\Omega}_E = \frac{d}{dt}\left[\int\int dRd\alpha d\theta \int dp_R dp_\alpha dp_\theta\, \delta\left(\frac{p_R^2}{2} - \frac{C_0}{R} - E\right)\right], \qquad (6.24)$$

where p_R, p_α and p_θ are the momenta conjugate to coordinates R, α and θ, respectively.
Noting that

$$\int dR \int p_R \delta\left(\frac{p_R^2}{2} - \frac{C_0}{R} - E\right) = \int dR\left(\frac{dR}{dt}\right)^{-1},$$

we can write Eq. (6.24) as

$$\sigma^{2+}(E) = \int d\alpha d\theta \int dp_\alpha dp_\theta. \qquad (6.25)$$

The momenta conjugate to α and θ can be calculated from the Lagrange function $\mathcal{L}$ (given by
Eq. (6.4)) according to the relations [4]

$$
\begin{aligned}
p_\alpha &= \partial_{\dot\alpha}\mathcal{L} = R^2\,\dot{R}\partial_R\alpha(R) = \sqrt{2C_0}R^{3/2}\,\partial_R\alpha(R), \quad\text{and} && (6.26)\\
p_\theta &= \partial_{\dot\theta}\mathcal{L} = \frac{1}{4}R^2\dot\theta\sin^2(2\alpha) \approx \frac{1}{4}R^2\,\dot{R}\partial_R\theta(R) = \frac{\sqrt{2C_0}}{4}R^{3/2}\,\partial_R\theta(R). && (6.27)
\end{aligned}
$$

Therefore, the cross section (6.25) can be written as

$$\sigma^{2+}(E) = \frac{C_0}{2}R^3\int d\alpha\, d(\partial_R\alpha)\, d\theta\, d(\partial_R\theta). \qquad (6.28)$$

For the calculation of the cross section (6.28) we need thus only the variation of the quantities
α, $\partial_R\alpha$ and θ, $\partial_R\theta$. The variation of these variables can be viewed as a variation of the
integration constants $a, b, \bar{a}, \bar{b}$, more precisely from Eqs. (6.23, 6.22) it follows that

$$\sigma^{++}(E) = \frac{\mu\bar\mu C_0}{4}\int(\bar{a})\, d\bar{a}\, da\, db\, d\bar{b}. \qquad (6.29)$$

Thus, the functional dependence of $a, b, \bar{a}, \bar{b}$ determines the energy behaviour of the cross
section. In particular the behaviour of b is decisive, as clear from the asymptotic behaviour of
α Eq. (6.22) at large distances R. Taking the scaling properties of the trajectories (Eq. (6.9))
into account and imposing the condition that the (divergent) trajectory has to stay bound,

---

[4]We recall that for $E = 0$ the relation applies $\dot{R} = \sqrt{\frac{2C_0}{R}}$.

when $E = 0$ is approach, we conclude that $b = E^{\mu/2 - 1/4}\tilde{b}$, where $\tilde{b}$ is energy independent. At $E = 0$ the constants $a, \bar{a}, \bar{b}$ become energy independent and the cross section Eq. (6.29) simplifies to the universal form

$$\sigma^{+2}(E) \propto E^{\mu/2 - 1/4}. \tag{6.30}$$

This universal energy dependence of $\sigma^{++}(E)$ is called the Wannier threshold law.

The classical mechanical derivation outlined above rests on the assumption of vanishing E, it gives no indications on the energy range of validity of (6.30). Furthermore, the above arguments make clear that the cross section (at $E = 0$) does not depend on E_1 and E_2 and hence all the energy-sharing possibilities are equally probable. This statement is called the ergodic theorem. The Wannier exponent $\mu/2 - 1/4$ depends on the charge Z through the dependence of μ (Eq. (6.21)) on Z: for $Z = 1$ one obtains $\mu/2 - 1/4 = 1.127$, whereas for $Z \to \infty$ which is the case for two independent particles one deduces $\mu/2 - 1/4 = 1$, i.e. the interaction between the particles changes qualitatively the behaviour of the ionization cross section.

6.1.2   Remarks on the classical treatment of two electrons at threshold

Summarizing this section we recall that the threshold law for double electron escape into the continuum of a residual positive ion can be derived in the framework of classical mechanics [90]. Inspection of the equations of motion indicates that the subspace relevant for double escape is the subset of the configuration space where both electrons are at equal distances from the ion and in opposite directions. The basic origin of this behaviour is the existence of a saddle point in the total potential surface for this (Wannier) configuration.

How the classical motion proceeds in the Wannier configuration is readily inferred from the following argumentation:

The classical motion for the two electrons is governed by the equations

$$\ddot{\mathbf{r}}_1 = -\frac{Z\mathbf{r}_1}{r_1^3} + \frac{\mathbf{r}_1 - \mathbf{r}_2}{|\mathbf{r}_1 - \mathbf{r}_2|^3}, \quad \ddot{\mathbf{r}}_2 = -\frac{Z\mathbf{r}_2}{r_2^3} + \frac{\mathbf{r}_2 - \mathbf{r}_1}{|\mathbf{r}_2 - \mathbf{r}_1|^3}. \tag{6.31}$$

As readily verified, the Wannier mode $\mathbf{r} = \mathbf{r}_1 = -\mathbf{r}_2$ solves for Eqs. (6.31). Both relations (6.31) collapse then to one equation

$$\ddot{\mathbf{r}} = -\frac{(Z - 1/4)\mathbf{r}}{r^3} \tag{6.32}$$

that describes a Kepler problem with an effective residual ion charge $\bar{Z} = Z - 1/4$, i.e. the electron-electron interaction is incorporated as a static screening of the ion's field. In this context we note that in the Fermi liquid theory of Landau [50, 84] the interacting electronic system is mapped onto a non-interacting one where the single particle (quasi-particles) properties (such as the particle's mass and charge) are re-normalized. The same approach can be followed here: Eq. (6.32) can be written in the form

$$\ddot{\mathbf{r}} = \frac{Z \, Z_e \mathbf{r}}{r^3}, \tag{6.33}$$

where the "quasi-particle" charge $Z_e = -1 + 1/(4Z)$ can be viewed as the re-normalized electron charge, i.e. the two electrons move independently in the field of the ion but each of them having the reduced charge Z_e. For a strong residual field $|Z| \gg 1$ we obtain $Z_e = -1$ a.u., i.e. the normalization due to the electron-electron interaction (and hence this interaction itself) can be neglected.

The details of the motion prescribed by Eq. (6.33) has been discussed at length in Sec. (1), both from a classical and quantum mechanical point of view. Let us recall the main findings again: Below the threshold I^{++} both electron move along elliptic orbits of equal size and are located in the same plane [cf. Fig. 1.1]. For all times t the Wannier condition $\mathbf{r}_1 = -\mathbf{r}_2$ applies. Hence, the motions within the two ellipses are strongly correlated in phase. As derived in Sec. (1), at threshold the ellipses degenerate into a straight line, which means that the electrons perform a symmetric stretch vibration without rotation. In case the ellipses degenerate into a circle we end up with the case of a linear rotor, whose constant length is determined by the condition that the electron-ion interaction compensates for the electron-electron interaction and the centrifugal force. The existence of a further rigid-body solution (a top) has been pointed out in Ref. [207]. A detailed review of the periodic orbit analysis of He can be found in Ref. [208].

7 Quantum mechanics of many-electron systems at the double escape threshold

7.1 Generalities of many-electron threshold escape

A number of quantum mechanical studies, e.g. [131, 132, 133, 134, 135] confirmed the Wannier threshold law (6.30) for the two electron escape. For three-electron escape only few works exist, e.g.[210, 211]. For more particles threshold emission little is known. Therefore, it is instructive to consider the generalities of many-electron escape at the fragmentation threshold.

7.1.1 Cross section dependence on the number of escaping particles

From the structure of the density of states one concludes the following: at threshold the probability for many-electron escape decreases rapidly with the number N of excited electrons. For $N \gg 1$, the threshold state for the N particle fragmentation, constitutes a set of measure zero, and hence it is extremely unlikely to be populated. This statement is evidenced by a comparison between the two and the three-electron threshold emission [Fig. 7.1].

As argued for the double emission case and demonstrated in Fig. 7.1 (a) the threshold constraint $E = 0 = E_1 + E_2$ is satisfied by an infinity of states for which one electron has a positive energy $E_1 > 0$ (unbound) and the other has a negative (bound) energy $E_2 = -Z^2/(2n^2)$, whereas only one state (I^{++}) satisfies $E_1 = 0$ and $E_2 = 0$. For triple electron escape (Fig. 7.1 (b)) the number of states that fulfill $0 = E_1 + E_2 + E_3 = E_1 + E$ increases, a subset is shown in Fig. 7.1 (b), where for clarity, the manifold of two electron states is combined into one axis $E_1 + E_2$: the case $E_3 > 0$ and $-E_3 = E = E_1 + E_2$ corresponds to the situation where electron "3" is unbound whereas the electron pair "1, 2" has a negative energy, i. e. either both electrons occupy a doubly excited bound state or one of them is bound and the other is unbound such that their energies combine to $-E_3$. Similar arguments apply when electron "3" is bound ($E_3 < 0$). This increase in the number of states with increasing N is to

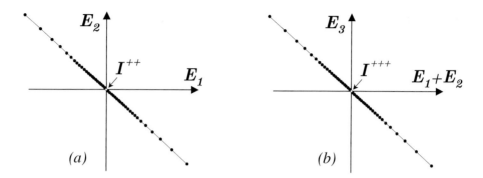

Figure 7.1: A schematic drawing for some of the available states for the total zero energy of two (a) and three (b) electron systems. E_1 and E_2 in (a) denote the energies of the two electrons ($E_1 + E_2 = 0$), whereas in (b) E_3 is the energy of the third electron ($E_1 + E_2 + E_3 = 0$). The double (triple) ionization threshold is indicated by I^{++} (I^{+++}).

be contrasted with the fact that the threshold is reached only if $E_j = 0$, $\forall j = 1 \cdots N$, which leads us to the conclusion that the threshold state is extremely unlikely to be populated with increasing number of particles. To put this statement in a mathematical language we recall that the total cross section at threshold is determined by phase-space arguments [212, 132]. Generally, the total cross section σ^{N+} for the emission of N particles has the structure [214]

$$\sigma^{N+} \quad \propto \quad \int |\langle f| O |i\rangle|^2 \, \delta(E_f - E_i) d^3\mathbf{k}_1 \cdots d^3\mathbf{k}_N, \tag{7.1}$$

$$\Rightarrow \sigma^{N+} \quad \propto \quad \int \left[\int |\langle f| O |i\rangle|^2 \, d\Omega_1 \cdots d\Omega_2 \right] (k_1 \ldots \ldots k_N) \, \delta(E_f - E_i) \, dE_1 \cdots dE_N, \tag{7.2}$$

where O is a (hermitian) operator that induces the transition from the initial state $|i\rangle$ (with energy E_i) to the final state $\langle f|$ (with energy E_f). The vectors $\mathbf{k}_j$, $j = 1 \cdots N$ are the momenta acquired by the escaping electrons in the asymptotic region. Hence, the parabolic dispersion relations $E_j = k_j^2/2$ apply. In Eq. (7.2) the wave vectors $\mathbf{k}_j$, $j = 1 \cdots N$ are characterized by the solid angles Ω_j and by their magnitudes k_j.

In a limited region around the threshold the state $(O |i\rangle)$ has only a very weak dependence on the total excess energy $E = E_1 + \cdots + E_N$. Hence, the energy dependence of $\sigma^{N+}(E)$ is determined by the factor

$$\rho(E) = \int (k_1 \cdots k_N) \, \delta(E - (E_1 + \cdots + E_N)) \, dE_1 \cdots dE_N = E^{3N/2-1} \tag{7.3}$$

and by the E-dependence of the normalization $\mathcal{N}$ of $|f\rangle$. If all electrons are considered free, i.e. if $f(\mathbf{r}_1, \cdots, \mathbf{r}_N) = (2\pi)^{-3N/2} \prod_{j=1}^{N} \exp(i\mathbf{k}_j \cdot \mathbf{r}_j)$, the function $\mathcal{N}$ is energy independent and we deduce

$$\sigma^{N+}(E) \stackrel{\text{PW}}{=} E^{3N/2-1} = e^{(3N/2-1)\ln E}. \tag{7.4}$$

This means, for $E \ll 1$ the cross section decreases rapidly with increasing number of particles N. For Coulomb systems, this N scaling behaviour is not expected to be changed qualitatively when including final-state interactions in $|f\rangle$. E.g. if all the N electrons are viewed as moving independently in the field of the ion we obtain for the final state $f(\mathbf{r}_1, \cdots, \mathbf{r}_N) = \mathcal{N}_{\text{ind}} \prod_{j=1}^{N} \exp(i\mathbf{k}_j \cdot \mathbf{r}_j)\varphi_j(\mathbf{r}_j, \mathbf{k}_j)$, where the distortion factor $\varphi_j(\mathbf{r}_j, \mathbf{k}_j)$ describes the influence of the residual ion field on the jth electron. As will be shown in detail below, the normalization factor $\mathcal{N}_{\text{ind}}$ in this case has the behaviour $\lim_{E_j \to 0} \mathcal{N}_{\text{ind}} \to \prod_{j=1}^{N} k_j^{-1} = \prod_{j=1}^{N} (2E_j)^{-1/2}$. Combining this threshold behaviour of $\mathcal{N}_{\text{ind}}$ with Eq. (7.3) we obtain

$$\sigma^{N+}(E) \stackrel{\text{ind}}{=} E^{N-1} = e^{(N-1)\ln E}. \tag{7.5}$$

This threshold law is valid when the strength of the ion Coulomb field dominates over all other interactions in the systems, e.g. for highly charged ions. Including the correlation between the electrons decreases the cross section, as shown for $N = 2$ (cf. Eqs. (6.30, 7.5)).

7.1.2   Structure of the total potential surface for N electron systems

A key ingredient of the Wannier theory for double escape is the existence of a saddle point in the total potential surface around which the total potential can be expanded. Thus, an important question to be answered for a many-electron system is how to determine the saddle points of the potential. Fig. 7.1 (b) gives a first hint that in many electron systems several (local) saddle points can exist, e.g. for $E_1 + E_3 = 0$ but for $E_3 < 0$, i.e. if electron "3" is bound to the core we obtain the saddle point known for the two electron escape. The saddle points we are interested in are singled out by the requirement that all excited particles recede from the residual ion into the asymptotic region [1] with a vanishing total energy.

---

[1] In the Wannier theory the R subspace is divided in three zones: the reaction zone $0 \leq R < a$, where a defines the region in which a description of the internal structure of the atom is important. In the Coulomb zone $a \leq R \leq R_0$ the internal structure of the residual ion is irrelevant for the motion of the two electrons, i.e. one can treat the ion as a point charge. In the free zone $R > R_0$ the kinetic energy is much larger than the potential energy. The constant hyperradius R_0 marks the border between the Coulomb zone and the free zone and is determined by the condition that the kinetic and the potential energies are of the same strength, i.e. $|E| = \frac{e^2}{R_0}$ where e is the charge of the electron and $E = (k_1^2 + k_2^2)/2$ is the kinetic energy. At threshold ($E \to 0$), the distance R_0 tends to infinity and therefore only the Coulomb zone is of interest for threshold fragmentation studies.

The total potential V can be divided into an attractive U_a and a repulsive part U_r

$$V = U_a + U_r, \quad U_a = -\sum_j^N \frac{Z}{r_j}, \quad U_r = \sum_{j>i}^N \frac{1}{|\mathbf{r}_i - \mathbf{r}_j|}. \tag{7.6}$$

As in the two-electron case, the potential V (7.6) can be transformed into hyperspherical coordinates. The hyperradius is defined as

$$R = \sqrt{\sum_{j=1}^N r_j^2}. \tag{7.7}$$

The total potential can be written as [213]

$$V = C(\omega)/R. \tag{7.8}$$

Here ω stands for a set of $3N - 1$ angles that are chosen appropriately [2]. Stationarity at a fixed R is then deduced from the condition

$$\nabla C(\omega) = 0.$$

Furthermore, the curvature of $C(\omega)$ is as well needed to pin down the nature of the extremal points, as demonstrated in the two-electron (Wannier) case.

The following analysis relies on the fact that U_a is always negative and U_r is always positive. As a consequence the total potential is minimal if U_a is maximal and U_r is minimal. Therefore, we investigate the stationarity of U_a and U_r separately. Let us first inspect the extremal positions of the attractive part U_a of the potential (7.6) at a given hyperradius R. For this purpose the function $g(\lambda) = U_a + \lambda(R^2 - r_1^2 \cdots - r_N^2)$ has to be analyzed where λ is a Lagrange multiplier. Due to the structure of the equation (7.8) for the total potential at a given R it is more convenient to inspect the function $\bar{C} = R\,g$. Introducing the dimensionless parameters

$$\alpha_j = r_j/R,$$

we find that $\bar{C}$ is stationary for

$$\alpha_{j0} = 1/\sqrt{N}.$$

---

[2] Due to rotational invariance only $3N-4$ angles enter dynamically into the solution of the problem, the remaining three (Euler) angles describe rigid-body space rotations.

Furthermore, $\bar{C}$ has the power expansion

$$\bar{C}(\alpha_j) = -C_{N0} - \frac{1}{2}C_{10}\sum_j(\alpha_j - \alpha_{j0})^2 + \cdots, \tag{7.9}$$

where the expansion coefficients are given by

$$C_{N0} = ZN^{3/2}, \quad C_{10} = 3ZN^{3/2}. \tag{7.10}$$

Eqs. (7.9, 7.10) are in complete analogy to Eqs. (6.12, 6.13) that result from the expansion of the total potential around the Wannier saddle points in the two-electron case. In fact it is straightforward to verify that Eqs. (7.9, 7.10) reduces to Eqs. (6.12, 6.13) for $N = 2$ and if the electron-electron interaction term is neglected. The stationarity condition $\alpha_j = 1/\sqrt{N}$ means that all electrons have the same distance from the residual ion. The actual geometrical arrangement of the electrons around the ion is deduced as follows: The gradient of the potential is the force $\mathcal{F}_r$ exerted by the ion on the electronic cloud. Since this force $\mathcal{F}_r$ does not change sign, the minimal value that can be acquired by $\mathcal{F}_r$ is zero. The minimum is reached when the center of the electronic charge coincides with the position of the ion. This is deduced from the following equation

$$\mathcal{F}_r = \frac{-ZN^{3/2}}{R^3}\left[\sum_j^N \mathbf{r}_j\right] = \frac{-ZN^{3/2}}{R^3}\mathbf{R}_{CM},$$
$$\mathcal{F}_r = 0 \quad \Rightarrow \quad \mathbf{R}_{CM} = 0, \tag{7.11}$$

where $\mathbf{R}_{CM}$ is the center of mass of the electronic cloud. The geometrical arrangement of the electrons in space at the stationary position of the total potential is further determined by minimizing the repulsive term U_r. The geometrical shape formed by arranging the electrons around the ion has to be invariant under rotation. This is because such a rotation is generated by a cyclic permutation of the electrons and such a permutation leaves the potential invariant. Furthermore, the geometrical shape has to be invariant under a reflection at any plane $\mathcal{P}$ that goes through the ion and bisecting any $\mathbf{r}_{ij}$, where $\mathbf{r}_{ij} \perp \mathcal{P}$. The reason for this is that this reflection operation is equivalent to an exchange of only i by j (and subsequent rotation if needed).

The arrangement of three electrons around the ion in the equilibrium configurations form an equilateral triangle, four form a regular tetrahedron and five form a symmetric bipyramid, these shapes are illustrated schematically in Fig. 7.2.

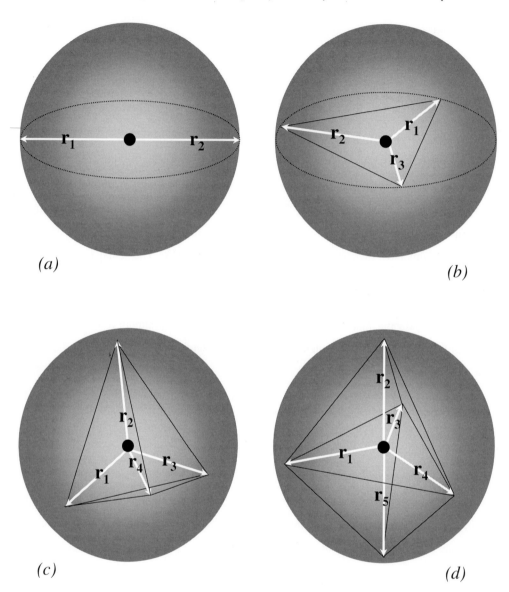

Figure 7.2: The Wannier configuration at the saddle point for two (a), three (b), four (c) and five (d) electrons.

7.1.3 Quantum mechanics of N electrons at low kinetic energies

The quantum mechanical state $\Psi_E(R, \omega)$, at the total energy E of N interacting, non-relativistic electrons is determined by the Schrödinger equation which in hyperspherical coordinates has the form [209]

$$\left[\Delta - 2V + 2E \right] \Psi_E(R, \omega) = 0,$$

$$\left\{ R^{1-3N} \partial_R R^{3N-1} \partial_R - \frac{\Lambda^2}{R^2} - \frac{2C(\omega)}{R} + 2E \right\} \Psi_E(R, \omega) = 0. \tag{7.12}$$

The second-order differential operator Λ acting only in the subspace spanned by the angles ω is the grand angular momentum, i.e. it is the quadratic Casimir operator of the $3N$-dimensional rotation group[3] $SO(3N)$. All Coulomb singularities, occurring when any two-particles distances tend to zero, reappear in the hyperspherical coordinates ω. The motion along R is singularity free for $N > 1$[4]. This statement is readily substantiated by considering the Poisson equation

$$\Delta V = -4\pi \sum_j Z\delta(r_j) + 4\pi \sum_{i \neq j} \delta(r_i - r_j). \tag{7.13}$$

From Eq. (7.12) we inferred the form of the hyperspherical Laplacian $\Delta = R^{1-3N} \partial_R R^{3N-1} \partial_R - \Lambda^2/R^2$ and insert it in (7.13) to obtain

$$\Delta V = \Delta C(\omega)/R = \begin{cases} -4\pi \, Z\delta(\mathbf{R}), & \text{for} \quad N = 1, \\ \frac{1}{R^3} \left[\Lambda^2 + 3(N-1) \right] C(\omega), & \text{for} \quad N \geq 2 \end{cases}. \tag{7.14}$$

Recalling that the volume element in hyperspherical coordinates[5] behaves as R^{3N-1} near the origin we conclude that for $N \geq 2$ the motion along the coordinate R is free from singularities. This means all two-particle Coulomb singularities must be contained in the coordinates ω. At the N particle coalescence point, i.e. for $R \to 0$, the wave function $\Psi_E(R, \omega)$ possesses a power series expansion containing logarithmic terms in R, the so-called Fock expansion, that will be briefly sketched in the next chapter.

As in the three-dimensional space, where Λ coincides with the orbital momentum operator, the eigenvalues and the eigenfunctions $F_K(\omega)$ of Λ^2 are determined from the relation[6]

$$\Lambda^2 F_K(\omega) = K \left(K + \frac{3N-1}{2} \right) F_K(\omega), \quad K = 0, 1, 2, \cdots. \tag{7.15}$$

---

[3]The grand angular momentum Λ^2 is given by $\Lambda^2 = \sum_{j>i} \Lambda_{ij}^2$, where $\Lambda_{ij} = x_i p_j - x_j p_i$ are the generators of space rotation in $\mathbb{R}_N$ (cf. appendix A.1). The position (x_i) and momentum (p_j) operators satisfy the canonical commutation relation $[x_i, p_j] = \delta_{ij}$ and the scalar operator Λ^2 commutes with all the elements Λ_{ij}, i.e. $[\Lambda^2, \Lambda_{ij}] = 0, \forall i, j$.

[4]This makes the radial coordiante R a possible suitable choice as an adiabatic coordinates as well as a reasonable measure for the size of the N particle system.

[5]The N dimensional volume element $dV_N = \prod_{j=1}^{N} d^3 x_j$ transforms in hyperspherical coordinates as $dV_N = dRdS = R^{N-1} dR \, d\Omega$, where dS is the element of the surface area. We note that $\int d\Omega = \frac{2\pi^{N/2}}{\Gamma(n/2)}$ and that the volume of an N dimensional sphere is $V_N = \frac{\pi^{N/2}}{\Gamma(N/2+1)} R^N$, where $\Gamma(x)$ is the Gamma function [99].

[6]The eigenfunctions of Λ are obtained by considering the harmonic homogeneous polynomials $\mathbf{Y}_\lambda(\mathbf{r})$ of degree λ. Homogeneity means that $\mathbf{Y}_\lambda(\mathbf{r}) = r^\lambda Y_\lambda(\omega)$. Acting with the hyperspherical Laplacian on $\mathbf{Y}_\lambda(\mathbf{r})$ and using the

The first order derivatives in Eq. (7.12) are transformed away upon the ansatz

$$\Psi_E(R, \omega) = (kR)^{-\frac{3N-1}{2}} \psi_E(R, \omega), \tag{7.16}$$

where $k = \sqrt{2E}$ is a hyperspherical wave vector. Since the whole system is invariant under overall rotation the angular and radial part of the wave function $\Psi_E(R, \omega)$ are separable, i. e.

$$\psi_E(R, \omega) = \chi(R) F_K(\omega). \tag{7.17}$$

For the region where around the saddle point the potential $C(\omega)$ varies slowly and therefore gradient terms of $C(\omega)$ can be neglected, i.e. one may approximate $C(\omega)$, to a first order, by the value $C_0 = C(\omega_0)$ at the equilibrium position ω_0. Taking Eqs. (7.15, 7.16) into account and inserting Eq. (7.17) into (7.12) we obtain a one dimensional determining equation for the $\chi(R)$

$$\left\{ \partial_R^2 - \frac{L(L+1)}{R^2} + \frac{2C_0}{R} + 2E \right\} \chi_L(R) = 0, \tag{7.18}$$

where

$$L(L+1) = K \left(K + \frac{3N-1}{2} \right) + (3N-1)(3N-3)/4.$$

Eq. (7.18) has the same structure as for one particle in an external potential that we treated before, [cf. Eq. (3.2)]. At $E = 0$ the solution of Eq. (7.18) is given in terms of Hankel functions

$$\chi_L(R) = \sqrt{R} \left\{ c_1 H^{(1)}_{2L+1}(\sqrt{-8C_0 R}) + c_2 H^{(2)}_{2L+1}(\sqrt{-8C_0 R}) \right\}, \tag{7.19}$$

where c_1 and c_2 are integration constants. Of interest for the fragmentation channel is the large R asymptotic of the function (7.19) which is

$$\chi(R) \underset{R \to 0}{\longrightarrow} R^{1/4} \left[c_1' e^{i\sqrt{8\bar{C}_0 R}} + c_2' e^{-i\sqrt{8\bar{C}_0 R}} \right], \tag{7.20}$$

where c_1 and c_2 are constants and $\bar{C}_0 = -C_0$.

Away from the equilibrium position $C(\omega) \neq C_0 = constant$, the problem is then to find a solution for the equation (at $E = 0$)

$$\left\{ \partial_R^2 - \left[\frac{\Lambda^2 + (3N-1)(3N-3)/4}{R^2} + \frac{2C(\omega)}{R} \right] \right\} \psi_E(R, \omega) = 0. \tag{7.21}$$

---

homogeneity and harmonicity properties of $\mathbf{Y}_\lambda(\mathbf{r})$ yields for the n dimensional operator Λ

$$\Lambda_n^2(\omega) Y_\lambda(\omega) = \lambda(\lambda + n - 2) Y_\lambda(\omega).$$

From all possible eigenfunctions $Y_\lambda(\omega)$ only those are then adopted which are finite, single-valued and continuous over ω. These are the so-called hyperspherical harmonics.

The solution (7.20) at the stationary point suggests the ansatz

$$\psi_E(R,\omega) = R^n \exp\left[i\sqrt{R}\,q(\omega)\right].$$ (7.22)

The function $q(\omega)$ is determined by inserting (7.22) into (7.21) which yields

$$
\begin{aligned}
\Big\{ -\frac{1}{R^2} \Big[\exp[-i\sqrt{R}\,q(\omega)]\,\Lambda^2\,\exp[i\sqrt{R}\,q(\omega)] \Big] \\
+ \frac{i(n-1/4)q}{R^{3/2}} + \frac{1}{R^2}\left[n(n-1) - (3N-1)(3N-3)/4\right] \\
- \frac{1}{R}\left[q^2/4 + 2C(\omega)\right] \Big\} R^n = 0.
\end{aligned}
$$ (7.23)

Expanding the exponential term in this equation in a power series we obtain

$$\exp(-i\varphi q)\,\Lambda^2\,\exp(i\varphi q)$$

$$= \Lambda^2 + i\varphi[\Lambda^2, q] - \frac{1}{2!}\varphi^2\left[[\Lambda^2, q], q\right] - \frac{i}{3!}\varphi^3\left[[[\Lambda^2, q], q], q\right] + \cdots.$$

In addition, one can show that the commutation relation

$$\left[[[\Lambda^2, q], q], q\right] = 0$$

applies which leads to the exact relation

$$\exp(-i\varphi q)\,\Lambda^2\,\exp(i\varphi q) = \Lambda^2 + i\varphi[\Lambda^2, q] - \frac{1}{2}\varphi^2\left[[\Lambda^2, q], q\right].$$

Inserting this relation in Eq. (7.23) leads to

$$
\begin{aligned}
\Big\{ \frac{i}{R^{3/2}} \left[(n-1/4)q - [\Lambda^2, q]\right] \\
+ \frac{1}{R^2}\left[n(n-1) - (3N-1)(3N-3)/4 - \Lambda^2\right] + \\
+ \frac{1}{R}\left[\frac{1}{2}[[\Lambda^2, q], q] - q^2/4 - 2C(\omega)\right] \Big\} R^n = 0.
\end{aligned}
$$ (7.24)

This equation can be further simplified by taking advantage of the relations $[\Lambda^2, q]g(\omega) = g\,\Lambda^2\,q - 2(\nabla g) \cdot (\nabla q)$ and $[[\Lambda^2, q], q]g(\omega) = -2g(\nabla q)^2,\ \forall g(\omega)$.

Since we are interested in the asymptotic (large R) behaviour we neglect the term that falls off as R^{-2} leading to the conclusion that Eq. (7.24) is valid identically if

$$q^2 + 4(\nabla q)^2 + 8C(\omega) = 0,$$ (7.25)

$$\frac{1}{4} + \left(\frac{\Lambda^2 q}{q}\right)_{\omega=\omega_0} = n.$$ (7.26)

The condition (7.26) is deduced from the fact that the gradient of the function R^n at the equilibrium position ω_0 has to vanish. Since Eq. (7.25) is quadratic in q we obtain two solutions Q_1 and Q_2 and accordingly, due to Eq. (7.26), two exponents n_1 and n_2. These two solutions merge together at ω_0 as shown above.

Equation (7.21) is invariant under the exchange

$$(kR, C/k) \leftrightharpoons (R, C).$$ (7.27)

This transformation combined with Eqs. (7.25, 7.26) leads to the constraint that the whole solution (7.22) should be invariant under the transformation

$$(kR, C/k, Q_1/\sqrt{k}, Q_2/\sqrt{k}, n_1.n_2) \leftrightharpoons (R, C, Q_1, Q_2, n_1, n_2).$$ (7.28)

The solution (7.22) has thus the form

$$\psi_E(kR, \omega) = (kR)^{1/4} \left\{ A \, (kR)^{n_1 - 1/4} \, \exp\left[i\sqrt{R} \, Q_1(\omega) \right] \right.$$

$$\left. + B \, (kR)^{n_2 - 1/4} \, \exp\left[i\sqrt{R} \, Q_2(\omega) \right] \right\},$$ (7.29)

where A and B are energy dependent constants. Thus, the total wave function (7.16) has the form

$$\Psi_E(R, \omega) = (kR)^{-\frac{3N-1}{2} + \frac{1}{4}} A \left\{ (kR)^{n_1 - 1/4} \, \exp\left[i\sqrt{k}R \, \frac{Q_1}{\sqrt{k}} \right] \right.$$

$$\left. + Dk^{n_1 - n_2} \, (kR)^{n_2 - 1/4} \, \exp\left[i\sqrt{k}R \, \frac{Q_2}{\sqrt{k}} \right] \right\}.$$ (7.30)

Here we made use of the condition $B = Dk^{n_1 - n_2} A$ where the constant D is energy independent. This constraint ensures that both parts of the solution (7.29) are of equal magnitudes near threshold.

7.1.4   Quantal calculations of the universal threshold behaviour

As discussed above using Eq. (7.2) the cross section for the escape of multiple electrons depends on the phase space factor $\rho(E)$, as given by Eq. (7.3), as well as on the excess-energy dependence of the transition matrix elements. The latter dependence is solely determined by the behaviour of the final state, i.e. of Eq. (7.30) as function of the energy: For $n_1 > n_2$ the first part of this wave function (7.30) (which behaves as $k^{-\frac{3N-1}{2}+\frac{1}{4}}$) dominates asymptotically and yields thus the normalization of the wave function. In the internal region, which the relevant region for the evaluation of the transition matrix elements and hence for the cross section Eq. (7.2), the second part of the wave function (7.30) is decisive. The energy dependence of this part is readily deduced from (7.30) as $k^{-\frac{3N-1}{2}+\frac{1}{4}+n_1-n_2}$. Combining this behaviour with Eq. (7.3) we deduce the universal threshold behaviour

$$\sigma^{N+} \quad \propto \quad E^{n_1-n_2-\frac{1}{4}}. \tag{7.31}$$

To obtain the value of the exponent for a certain number of particles N one has to solve the equations (7.25, 7.26). E.g. for two particles we obtain

$$n_1 = +\frac{\sqrt{\frac{100Z-9}{4Z-1}}-1}{8} + \frac{i}{4}\sqrt{\frac{9-4Z}{4Z-1}},$$

$$n_2 = -\frac{\sqrt{\frac{100Z-9}{4Z-1}}-1}{8} + \frac{i}{4}\sqrt{\frac{9-4Z}{4Z-1}}, \tag{7.32}$$

which determines the energy dependence of the cross section to be

$$\sigma^{2+} \propto E^{\frac{1}{4}\left(\sqrt{\frac{100Z-9}{4Z-1}}-1\right)}. \tag{7.33}$$

This relation is identical to the Wannier threshold law that we derived classically in section (6.1.1).

7.1.5   Incorporation of symmetry and spin in many-particle wave functions

The above treatment of the N electron wave function does not account for the spin variable. For non-relativistic systems, in which the spin and the spatial degrees of freedom are decoupled, the effect of the spin degrees of freedom can be considered separately. The spin part of electronic wave function dictates then the symmetry of the radial part in such a way that the antisymmetry of the total wave function is ensured. According to Knirk [213] the wave functions (7.16) can be classified by means of the following good quantum numbers: the parity (π), the total orbital momentum and its projections L, M_L, as well as by the total spin (S) and its components along a quantization axis (M_S). The wave function (7.16) is then written as

$$\Psi_{E,\pi L M_L S M_S}(R,\omega) = R^{-\frac{3N-1}{2}} \sum_j \psi_{j,\pi L M_L}(R,\omega)\Theta_{j,S M_S}, \tag{7.34}$$

where the functions $\psi_{j,\pi L M_L}(R,\omega)$ have the same radial structure as Eq. (7.22). Furthermore, the functions $\Theta_{j,S M_S}$ are spin joint eigenfunctions of S^2 and S_z. The index j runs over the degeneracies of the spin states of S^2. The Pauli principle imposes certain symmetry properties on $\psi_{j,\pi LM_L}(R,\omega)$ when two electrons are exchanged, in particular some of the functions $\psi_{j,\pi LM_L}(R,\omega)$ will have nodes at the equilibrium position. This situation can be accounted for by writing $\psi_{j,\pi LM_L}(R,\omega)$ in the form (cf. Eq. (7.22))

$$\psi_{j,\pi LM_L}(R,\omega) = f_{j,\pi LM_L}(\omega)\, R^n\, \exp[i\sqrt{R}\,q(\omega)].$$

Repeating the steps for this function as done for the ansatz (7.22) one derives the determining equation

$$\left\{ \frac{i}{R^{3/2}}\left[(n-1/4)q - [\Lambda^2, q] \right] + \frac{1}{R^2}\left[n(n-1) - (3N-1)(3N-3)/4 - \Lambda^2 \right] \right.$$
$$\left. + \frac{1}{R}\left[\frac{1}{2}[[\Lambda^2, q], q] - q^2/4 - 2C(\omega) \right] \right\} R^n\, f_{j,\pi LM_L}(\omega) = 0. \quad (7.35)$$

Comparing this relation with Eq. (7.23) one can repeat the steps leading from Eq. (7.23) to the determining relations for $q(\omega)$ and (n) (7.25, 7.26) and obtain the same relation (7.25) $q(\omega)$, i.e. $q(\omega)$ is independent of the nodal structure of the wave functions. The exponents n depend however on f, i.e. on the nodes of the wave functions near the stationary point:

$$n = \frac{1}{4} + \left(\frac{\Lambda^2 q}{q} \right)_{\omega=\omega_0} - 2\left(\frac{(\nabla q)\cdot(\nabla f_j)}{q\, f_j} \right)_{\omega=\omega_0}. \quad (7.36)$$

From Eq. (7.31) it is clear that this change in n will be reflected in a modified threshold law when the wave function possesses nodes at the equilibrium[7].

---

[7] If the wave function has no nodes at the saddle point the function f_j is unity.

8 Highly excited states of many-body systems

In the preceding chapters we have seen that the description of quantum mechanical two-particle systems can, in general, be reduced to the treatment of a one particle problem for the relative motion, that can in general be handled exactly by theory. Ground state properties of many-body systems can as well in most cases be treated theoretically by utilizing variational techniques to find the minimal (ground-state) energy of the system. For excited systems the situation is more complex. Already for three particles one observes the formation of resonant states that are strongly affected by electronic correlation. Semi-quantitative and exact diagonalization methods provide a satisfactory description of these autoionizing states. With an increasing number of electrons and/or higher excitations such methods become however intractable. Under certain special conditions (e.g. $E = 0$), quantities (such as the low-energy total cross section) which are determined by the available N particle phase-space can be derived from a general consideration of the structure of the N particle wave function. Apart from these special cases, we are faced with the problem of how to construct approximate many-body wave functions. It is the aim of this chapter to address this question for a system consisting of N interacting excited electrons.

8.1 General remarks on the structure of the N particle Schrödinger equation

Before addressing the solution techniques of the many-body Schrödinger equation it is instructive to recall some general aspects of the many-body wave functions.

8.1.1 The Fock expansion

As discussed in previous chapters, the treatment of the two-body problem involves in general the solution of a one-dimensional, second-order ordinary differential equation. Solutions of such equations can be expressed in terms of a power series in the relevant variables [1]. The existence and convergence of such solutions are guaranteed by well-established theorems in the field of ordinary differential equations [215]. In essence, a power series expansion for regular solutions exist if, and only if, the potential has a power series expansion including at most a weak singularity.

The treatment of non-separable, many-body problems requires the solution of higher dimensional partial differential equations. To connect to the theory of ordinary differential equation one may convert the N particle Schrödinger equation into an equivalent system of second-order ordinary differential equations, e.g. as done in the hyperspherical treatment. On the other hand, one may attempt to solve approximately the N particle Schrödinger equation directly. For both routes, there exist, in general, no power series solutions analogous to the one-dimensional case. The main reason behind this difference between the two and many-body problems is that the Schrödinger equation for N particle systems contains a multi-center potential for $N > 2$, and a non-central interaction generates in the solution logarithmic terms, in addition to powers in the relevant particle variables.

A prominent example of this situation is the fact that the ground state of helium can not be expanded in an analytic series of the interparticle coordinates [216]. Bartlett proved the existence of a formal expansion including logarithmic functions of the interparticle coordinates. A direct practical demonstration of this (anomaly) has been demonstrated by Fock [204], who used the hyperspherical coordinates to show that the exact three-body wave function $\Psi(R, \omega)$ has, in the neighborhood of the three-body coalescence point ($R = 0$) the expansion

$$\Psi(R, \omega) = \sum_{k=0}^{\infty} R^k \sum_{p=0}^{[k/2]} (\ln R)^p \; \psi_{kp}(\omega) \Bigg|_{|R| \ll 1} , \tag{8.1}$$

where the upper limit on the second summation $[k/2]$ denotes the largest integer that does not exceed $k/2$. Substituting this expansion into the Schrödinger equation Fock obtained a recurrence relation, involving differential operators, for the angular functions $\psi_{kp}(\omega)$ for $k = 0$ and $k = 1$, and he obtained solutions which are in agreement with those of Bartlett. Since the original work of Bartlett and Fock a considerable amount of studies on the Fock expansion have been carried out (e.g. [217, 218, 219, 220, 221, 222, 223, 224]). In particular it has been shown [220, 221] that the Fock expansion can be extended to an arbitrary system of charged particles and to states of any symmetry. In fact Leray [225] proved that every solution of the N electron Schrödinger equation can be written in the form of the Fock expansion. However, an explicit numerical investigation of the expansion coefficients for an arbitrary number of electrons is still outstanding.

8.1.2   The Kato cusp conditions

The Coulomb interaction between two charged particles diverges at their coalescence point. On the other hand the normalization condition of the wave function Ψ implies that $|\Psi|$ has to have an upper bound everywhere. In fact Kato [140] has shown that all eigenfunctions of a many-particle Coulombic system are continuous everywhere and hence finite, even at the Coulomb singularities. This means the divergences of the potentials at the two-particle collision points have to be compensated for by equivalent diverging terms in the kinetic energy. Kato has shown that this behaviour is reflected in certain properties of the exact wave function around the two-body coalescence points. Namely, he proved that if two particles of masses m_i and m_j and charges z_i and z_j approach each other ($r_{ij} \to 0$) and all other interparticle distances remain finite then the many-body wave function Ψ has to satisfy (the Kato cusp conditions) [140]

$$\left. \frac{\partial \tilde{\Psi}}{\partial r_{ij}} \right|_{r_{ij}=0} = z_i z_j \, \mu_{ji} \, \Psi(r_{ij}=0), \quad \forall \; i, j \in [1, N], \tag{8.2}$$

where $\tilde{\Psi}$ is the wave function averaged over a sphere of small radius $r_\epsilon \ll 1$ around the singularity $r_{ij} = 0$ and μ_{ij} is the reduced mass of the particles i and j.

Later on Kato's result has been extended [226] by inspecting the solution in the vicinity of the coalescence points and requiring all the terms that diverge in the limit of $r_{ij} \to 0$ to cancel each other. This yields an expansion for the wave function Ψ in the form ($r_{ij} \equiv \mathbf{r}$)

$$\Psi = \sum_{l=0}^{\infty} \sum_{m=-l}^{m=1} r^l \, f_{lm}(r) \, Y_{lm}(\theta, \phi), \tag{8.3}$$

where

$$f_{lm}(r) = f_{lm}^0(r) \left[1 + \frac{z_i z_j \mu_{ij}}{l+1} r + \mathcal{O}(r^2) \right]. \tag{8.4}$$

The function f_{lm}^0 is the first term in the series expansion [1] in powers of r

$$f_{lm}(r) = \sum_{j=0}^{\infty} f_{lm}^j r^j.$$

---

[1] The correct power expansion of the wave function around the two-body collision point is best illustrated for the

8.1.3   Boundary conditions for the N-body problem

In addition to the regularity conditions (8.1, 8.2) certain asymptotic boundary conditions can be imposed on the solution of the Schrödinger equation. As illustrated in the case of two particles (page 14), according to the prescribed boundary conditions, appropriate solutions for standing (i.e. bound) waves (2.54), outgoing waves (2.55), or incoming waves (2.56) have been singled out. When carrying over this procedure to many-body systems one encounters two problems: (1) For $N > 2$ a variety of (mixed) boundary conditions (or channels) occurs, e.g. one may consider a system of N (indistinguishable) electrons, where few electrons are bound and few others are in the continuum, as detailed in section 9.8. (2) For Coulomb systems the boundary conditions, as such are difficult to derive. This is because the Coulomb interaction has an infinite range and the particles are strictly speaking never free.

Since we know the exact form of the Coulomb two-body wave functions (2.54, 2.55, 2.56), the asymptotic behaviour is readily deduced in this case. This same behaviour is as well easily obtained by considering the Schrödinger equation for the relative motion of two charged particles (with charges z_1 and z_2 having a reduced mass μ)

$$\left[\Delta - \frac{2\mu z_1 z_2}{r} + k^2 \right] \Psi_{\mathbf{k}}(\mathbf{r}) = 0. \tag{8.5}$$

The position vector $\mathbf{r}$ is the two-particle relative coordinte and $\mathbf{k}$ is the momentum conjugate to $\mathbf{r}$. The distortion of the plane wave motion due to the presence of the potential is exposed by the ansatz

$$\Psi_{\mathbf{k}}(\mathbf{r}) = N_k e^{i\mathbf{k}\cdot\mathbf{r}} \bar{\Psi}_{\mathbf{k}}(\mathbf{r}), \tag{8.6}$$

where N_k is a normalization factor. The large relative distance (asymptotic) behaviour of (8.5) is obtained by neglecting terms that fall off faster than the Coulomb potential. This leads to

$$\left[i\mathbf{k} \cdot \nabla - \frac{\mu z_1 z_2}{r} \right] \bar{\Psi}_{\mathbf{k}}(\mathbf{r}) = 0. \tag{8.7}$$

---

case of a one-electron hydrogenic positive ion with a charge Z. In this case the wave function behaves as

$$\Psi \propto r^l \left[1 - \frac{Z}{l+1} r + \mathcal{O}(r^2) \right] Y_{lm}(\theta, \phi) \ \Rightarrow \ \partial_r \Psi \propto \left[l r^{l-1} - Z r^l + \mathcal{O}(r^{l+1}) \right] Y_{lm}(\theta, \phi).$$

Thus, for $l = 0$ we readily conclude

$$\partial_r \tilde{\Psi} \Big|_{r=0} = -Z \Psi(r = 0).$$

This equation admits a solution of the form $\bar{\Psi} = \exp(i\phi)$, where

$$\phi_{\mathbf{k}}^{\pm}(\mathbf{r}) = \pm\frac{z_1 z_2 \mu}{k} \ln k(r \mp \hat{\mathbf{k}} \cdot \mathbf{r}). \qquad (8.8)$$

Thus, the asymptotic wave function reads

$$\Psi_{\mathbf{k}}(\mathbf{r}) = N_k e^{i\mathbf{k}\cdot\mathbf{r}} e^{\pm i\alpha_k \ln k(r \mp \hat{\mathbf{k}}\cdot\mathbf{r})}, \qquad (8.9)$$

which explicitly displays the Coulomb potential-induced modification of the asymptotic plane-wave motion. The factor $z_1 z_2 \mu/k = z_1 z_2/v$ (v is the relative velocity) is the Sommerfeld parameter that we introduced in Eq. (2.57).

The result (8.9) coincides with the asymptotic form of the wave function (2.56), i.e. in the two-particle case the procedure of solving the asymptotic Schrödinger equation is redundant. This solution strategy becomes however useful when considering many-particle systems, where the exact wave function is unknown.

In fact it has been suggested [71] that in the limit of large interparticle separations, the wave function $\Psi(\mathbf{r}_1, \cdots, \mathbf{r}_N)$ for N charged particles (with charges Z_j) moving in the continuum of a massive residual ion of charge Z takes on a generalized form of Eq. (8.9), namely

$$\lim_{\substack{r_{lm}\to\infty \\ r_n\to\infty}} \Psi(\mathbf{r}_1, \cdots, \mathbf{r}_N) \;\to\; (2\pi)^{-3N/2} \prod_{s=1}^{N} \xi_s(\mathbf{r}_s)\psi_s(\mathbf{r}_s) \prod_{\substack{i,j=1 \\ j>i}}^{N} \psi_{ij}(\mathbf{r}_{ij}), \qquad (8.10)$$

$$\forall \; l,m,n \in [1,N]; m > l.$$

The functions $\xi_j(\mathbf{r}_j)$, $\psi_j(\mathbf{r}_j)$, $\psi_{ij}(\mathbf{r}_{ij})$ are given by

$$\xi_j(\mathbf{r}_j) \;=\; \exp(i\mathbf{k}_j \cdot \mathbf{r}_j), \qquad (8.11)$$

$$\psi_j(\mathbf{r}_j) \;=\; \exp\left[\mp i\alpha_j \ln(k_j\, r_j \pm \mathbf{k}_j \cdot \mathbf{r}_j)\right], \qquad (8.12)$$

$$\psi_{ij}(\mathbf{r}_{ij}) \;=\; \exp\left[\mp i\alpha_{ij} \ln(k_{ij}\, r_{ij} \pm \mathbf{k}_{ij} \cdot \mathbf{r}_{ij})\right]. \qquad (8.13)$$

The vectors $\mathbf{k}_{ij}$ are the momenta conjugate to $\mathbf{r}_{ij}$, i. e. $\mathbf{k}_{ij} := (\mathbf{k}_i - \mathbf{k}_j)/2$. The Sommerfeld parameters α_j, α_{ij} are given by

$$\alpha_{ij} = \frac{Z_i Z_j}{v_{ij}}, \qquad \alpha_j = \frac{Z Z_j}{v_j}. \qquad (8.14)$$

In Eq. (8.14) v_j denotes the velocity of particle j relative to the residual charge whereas $\mathbf{v}_{ij} := \mathbf{v}_i - \mathbf{v}_j$. While the functional form of Eq. (8.10) is plausible accounting for each pair of interaction by a corresponding Coulomb-phase distortion, the actual derivation of Eq. (8.10) is not as straightforward as in the two-body case. This is due to the many-body correlation between the particles, even in the asymptotic region.

9 The three-body Coulomb system

The classical three-body problem, e. g. the study of the dynamics of the system moon-earth-sun, has been the subject of research since the early days of modern physics, yet there are still some unanswered questions to be addressed [86, 87]. The quantum mechanical atomic analog, namely the Coulomb three-body problem is as well one of the first "many-body problems" to be investigated quantum mechanically [88] and is still the subject of lively research, as detailed below. From a formal point of view, the three-body problem is generally not solvable exactly in the sense, that the number of integrals of motion is less than the number of degrees of freedom. On the other hand, it has been demonstrated using several methods that, e.g. for a system consisting of two-electrons and a positive ion, physical observables can be calculated numerically and are in an impressively good agreement with experimental findings [93, 94, 95, 92, 96, 97] (see also [321, 322, 317] and references therein for different types of three-body systems). This kind of numerical studies, not covered in this work, is extremely important for predicting reliably and/or comparing with experiments. On the other hand, it is highly desirable to uncover analytically the features pertinent to the three-body motion and trace their footprints when considering systems with a larger number or containing different types of particles. In this context it should be mentioned that, from a conceptual and a practical point of view, methods and tools that are developed for the three-body problem are not a priori relevant for may-body (thermodynamic) systems. This is because, in contrast to few-body systems, in an extended medium the fluctuations in the values of the single-particle quantities (such as energies and momenta) are generally of a less importance for the mean-field values of the respective quantities. This observation can be utilized for the description of phenomena inherent to thermodynamic many-body systems, such as their collective response. On the other hand, however, with increasing number of particles in a "small" system, a crossover behaviour is expected to emerge that marks the transition from a small (atomic-like) to an extended, thermodynamic behaviour, and vice versa (see Refs. [105, 21, 368] for concrete examples). It is one of the aims of this and of the forthcoming chapters to address these issues using purely analytical methods.

This chapter is devoted to the analytical treatment of the three-body Coulomb scattering problem, which has received recently much of attention [71, 72, 73, 74, 75, 76, 77, 79, 80, 81, 82]. While, in contrast to the two-body problem, exact three-body solutions are not available, under certain, asymptotic conditions analytical solutions can be obtained. These solutions are found to carry some of the general features of the two-body scattering, such as the characteristic asymptotic phases. In section 7 we encountered another special situation, where the solution can be expanded in a power series around the saddle point of the potential with the expansion coefficients being determined by the properties of the total potential surface. Apart from such special cases, the treatment of the Coulomb three-body problem is complicated, mainly because of the infinite range of Coulomb forces which forbids free asymptotic states of charged particles. This excludes the straightforward use of standard tools of scattering theory, such as the standard perturbation expansion (more precise details are given in chapter 11).

On the other hand, one can always attempt to solve the Schrödinger equation directly. An important prototype of three-body systems, where such an attempt has been undertaken, consists of two electrons moving in the field of a massive nucleus. Such systems are realized as the final state achieved in the electron impact ionization (the so-called (e,2e) process) and in the double photoionization of atomic systems (this process is referred to as $(\gamma, 2e)$ process). One of the traditional method to solve the Schrödinger equation in this case is to reduce the three-body system approximatively to two two-body subsystems which are uncoupled in the configuration space. The correlation between these two subsystems is then accounted for parametrically, e.g. by the use of momentum-dependent effective product charges [122, 123, 124]. This results in a six-dimensional wave function of the three-body continuum state, which is expressed in the coordinates $\mathbf{r}_a, \mathbf{r}_b$ of two electrons a, b with respect to the nucleus. No explicit dependence appears in the solution on the electron-electron relative coordinate $\mathbf{r}_{ab} = \mathbf{r}_a - \mathbf{r}_b$. This means that wave functions provided by such methods do not satisfy the Kato cusp conditions at the electron-electron collision point (the derivative with respect to $\mathbf{r}_{ab}$ vanishes, cf. Eq. (8.2)). Furthermore, asymptotically these solutions do not go over into the known asymptotic form. Therefore, it is necessary to address the full dependence of the Schrödinger equation on the coordinate $\mathbf{r}_{ab}$ that describes the electronic correlation.

Mathematically, the three-body Schrödinger equation constitutes an elliptical partial differential equation in six variables with a non-denumerable infinity of solutions. Therefore, to single out the physically meaningful solution, appropriate boundary conditions are needed which are prescribed on an asymptotic five-dimensional closed manifold $\mathcal{M}$. Unfortunately, even the specification of such asymptotic states is an involved task. Redmond [71] and others [125, 126, 72] proposed asymptotic scattering states valid in a subspace of $\mathcal{M}$ in which all interparticle distances tend to infinity. Alt and Mukhamedzhanov [75] argued that a correct description of the whole asymptotic region $\mathcal{M}$ requires the introduction of local relative momenta. It should be remarked here, that asymptotic states are needed as boundary conditions to select acceptable solutions of the Schrödinger equation. From a practical point of view, however, the asymptotic wave functions are of limited value, for the evaluation of transition amplitudes involves an integration over regions in the configuration space which are outside the asymptotic domain $\mathcal{M}$.

The next sections outline a strategy to construct three-body continuum states which are, to a leading order, exact asymptotic solutions on the manifold $\mathcal{M}$. The Kato-cusp conditions are shown to be satisfied at all three two-body collision points. The finite-distance behaviour is also studied and the analytical structure of the wave function at the three-body dissociation threshold is investigated.

9.1   Appropriate coordinate systems

As well-known from classical mechanics the appropriate choice of generalized coordinates is an essential step in the solution of the equation of motion. Thus, it is useful to inspect the possible suitable choices of coordinate systems for the formulation of few-body Coulomb problems. Special emphasis is put on the coordinate systems that will be utilized in the next chapters.

9.1.1   Separation of internal and external coordinates

In the center of mass system, six coordinates are required for the description of the three-body problem. Since the total potential is rotationally invariant one usually decomposes these six coordinates into three internal coordinates describing the size and the shape of the triangle formed by the nucleus and by the two electrons (cf. Fig. 6.1). Further three external coordinates are needed to specify the orientation of the principal axes of inertia of the three particles with respect to a space-fixed coordinate frame [these coordinates are usually chosen to be the Eulerian angles $(\alpha_e, \beta_e, \gamma_e)$].

9.1.2   Spherical polar coordinates

The total potential depends only on the internal coordinates. Therefore, one can use the spherical polar coordinates r_1, r_2 and θ, that are displayed in Fig. 6.1 (page 80), and write the wave function in the separable form [227, 143]

$$\Psi_{LM}(\mathbf{r}_1, \mathbf{r}_2) = \sum_{k=-L}^{L} \psi_{Lk}(r_1, r_2, \theta) d^L_{Mk}(\alpha_e, \beta_e, \gamma_e). \tag{9.1}$$

The finite rotation matrices $d^L_{Mk}(\alpha_e, \beta_e, \gamma_e)$ are eigenfunctions of the total angular momentum operator[1]. The advantage of this separable form of the wave function is most clear when considering S three-body states, in which case there is no need to consider the external coordinates[2].

9.1.3   Hyperspherical coordinates

Hyperspherical coordinates (R, ω), as previously introduced in Eq. (7.12), are a prototypical example of accounting for the isotropy of space, i.e. for the invariance of the system under overall rotations. The advantage of this coordinate system is that it displays explicitly the good quantum numbers resulting from this symmetry, namely the orbital quantum numbers. Disadvantage is the complicated form of the potential energy $C(\omega)/R$. The hyperspherical

---

[1] See Ref. [228] for the complications that arise for $N \geq 4$ and appendix A.1 for further details.

[2] For S states the kinetic and the potential energy terms have a transparent structure in the coordinate system (r_1, r_2, θ), e.g. the Laplacian reads [229]

$$\Delta_{breit} = r_1^{-2}\partial_{r_1} r_1^2 \partial_{r_1} + r_2^{-2}\partial_{r_2} r_2^2 \partial_{r_2} + \left[r_1^{-2} + r_2^{-2}\right](\sin\theta)^{-1}\partial_\theta(\sin\theta)\partial_\theta. \tag{9.2}$$

coordinate system is very well suited for the study of collective phenomena that occur when the system expands or condenses (recall Eq. (7.13) on page 95 and the discussion thereafter). The Wannier threshold behaviour that we discussed in the preceding chapter is just one of numerous examples.

9.1.4 Relative coordinates

An obvious coordinate system in which the potential energy is diagonal, is the relative coordinate system [230]

$$r_1,\ r_2,\ r_{12}.\tag{9.3}$$

These quantities are scalar and hence can be regarded as the internal (body-fixed) variables. The price for the potential being diagonal is that the kinetic energy term is not separable, e.g. for S states the Laplacian reads

$$
\begin{aligned}
\Delta_{rel} &= r_1^{-2}\partial_{r_1}r_1^2\partial_{r_1} + r_2^{-2}\partial_{r_2}r_2^2\partial_{r_2} + 2r_{12}^{-2}\partial_{r_{12}}r_{12}^2\partial_{r_{12}} + \\
&\quad + \frac{r_1^2 + r_{12}^2 - r_2^2}{r_1 r_{12}}\partial_{r_1}\partial_{r_{12}} + \frac{r_2^2 + r_{12}^2 - r_1^2}{r_2 r_{12}}\partial_{r_2}\partial_{r_{12}}.
\end{aligned}\tag{9.4}
$$

The weights of the mixed derivatives have a simple meaning, namely

$$\cos\theta_j = \frac{r_i^2 + r_{12}^2 - r_j^2}{2 r_i r_{12}};\ i \neq j = 1, 2.$$

The angle θ is $\theta = \arccos(\hat{\mathbf{r}}_1 \cdot \hat{\mathbf{r}}_2)$, whereas the angles θ_1, θ_2 are displayed in Fig. 6.1. These angles are the internal angles of the triangle formed by the three-particles. The rotation of this triangle in space may be described by Euler angles.

In contrast to the hyperspherical coordinates that emphasize the collective aspects of many-body systems, the relative coordinate system underlines the two-body nature of the Coulomb interactions. Thus, this system is useful when considering situations in which the two-body dynamics is dominant, e. g. in the asymptotic region. A major shortcoming of the relative coordinate system is its restriction to the three-body problem in which case the number of internal variables coincides with the number of particles. In contrast, for N-body system $N(N - 1)/2$ relative coordinates are present. Therefore, the definition (9.3) is unique only for the three-body system.

9.1.5   Elliptic coordinates

To account for the fact that a three-body electronic system has a definite symmetry under particle interchange one may use the elliptic coordinates, defined as

$$s = r_1 + r_2, \quad t = r_1 - r_2, \quad u = r_{12}. \tag{9.5}$$

As clear from this definition, t (s) is odd (even) under particle exchange. The angular correlation is described by the coordinate u. As readily verified, the potential energy has a simple structure in these coordinates on the expense of having non-separable kinetic energy terms, e. g. for S states the Laplacian has the form [231]

$$\Delta_{ec} = 2 \bigg\{ \partial_{s^2}^2 + \partial_{t^2}^2 + u^{-2} \partial_u u^2 \partial_u + \frac{4}{(s^2 - t^2)} \left(s \partial_s - t \partial_t \right)$$

$$+ \frac{2}{u(s^2 - t^2)} \Big[s(u^2 - t^2) \partial_u \partial_s - t(u^2 - s^2) \partial_u \partial_t \Big] \bigg\}. \tag{9.6}$$

For the three-body problem the elliptic coordinates turned out to be useful to exhibit the symmetry properties of the wave functions. For the N-particle case however we encounter the same problem as in the relative coordinate case, because according to Eq. (9.5) one operates with pairs of particles and the number of these pairs grows quadratically with N.

9.2   Coordinate systems for continuum problems

As discussed in details in Chapter 2 for a two-body Coulomb system there is a decisive difference between the generic symmetry operations associated with the bound ($O(4)$ rotations) and continuum states (rotation plus translation). This is clearly exhibited by the structure of these states: While the bound states (2.26) explicitly indicate the isotropy of space (and hence the use of spherical coordinates is appropriate), the analytical structure of continuum states (2.55, 2.56) underlines the existence of a preferential direction (the polar vector $\mathbf{k}$). Thus, the appropriate symmetry operations are translation along and rotation around the quantum number $\mathbf{k}$. The overall rotation symmetry is still conserved but in the sense that the position and the direction $\mathbf{k}$ is rotated. As a consequence of this difference between bound and continuum states we have seen that the most appropriate coordinate for the continuum (Stark) states are the parabolic coordinate set (2.49).

For continuum N particle systems additional complications arise: The potential energy depends on (few) internal coordinates only, and hence, from a dynamical point of view, it seems reasonable to employ an internal body-fixed coordinate system, such as (9.3). On the other hand, the structure of the two-body Coulomb continuum wave functions (2.55, 2.56) indicates the necessity to account for the space fixed direction $\mathbf{k}$. In addition, since N body continuum states are labelled by a number of wave vectors $\{\mathbf{k}_j\}$, $j = 1, \cdots, N$, the translational and cylindrical symmetry around one specific direction, say $\mathbf{k}_1$, is generally lost. Therefore, for the N-body continuum case there is not a uniquely preferential choice for the most appropriate coordinate system, as in the two-body case (2.49).

9.2.1   Jacobi coordinates

A coordinate system which is most suitable to describe the kinetic energy terms and to treat all particles (of different masses) democratically is the Jacobi coordinate system [232]. In contrast to the relative coordinate system (9.4) where the potential energy has a diagonal term on the expense of having cross terms (even for S states), in the Jacobi coordinate system, the kinetic energy is diagonal whereas potential energy has a complicated form. Since we are going to employ this reference frame intensively later on, we examine it in some details for the case of three-particles. For the general case of N particles the reader is referred to Ref. [233] and references therein.

Having separated out the uniform center-of-mass motion, a system of three particles, with masses m_i and charges Z_i $i \in 1, 2, 3$, can be described by one set of the three Jacobi coordinates sets $(\mathbf{r}_{ij}, \mathbf{R}_k)$; $i, j, k \in \{1, 2, 3\}$; $\epsilon_{ijk} \neq 0$; $j > i$. An illustration of this coordinate system is shown in Fig. 9.1. A priori, no preference is given to any of the three sets of Jacobi coordinates, however physically the different sets correspond to different grouping of the three-body system, e.g. if two particles are bound the third is in the continuum it is advantageous to use the set in which one of the Jacobi coordinates is the relative position between the bound particles. The Jacobi sets of coordinates are linked to each others via the transformation

$$\begin{pmatrix} \mathbf{r}_{13} \\ \mathbf{R}_2 \end{pmatrix} = \mathbf{D}_3 \begin{pmatrix} \mathbf{r}_{12} \\ \mathbf{R}_3 \end{pmatrix} \quad \text{and} \quad \begin{pmatrix} \mathbf{r}_{23} \\ \mathbf{R}_1 \end{pmatrix} = \mathbf{D}_2 \begin{pmatrix} \mathbf{r}_{12} \\ \mathbf{R}_3 \end{pmatrix}, \qquad (9.7)$$

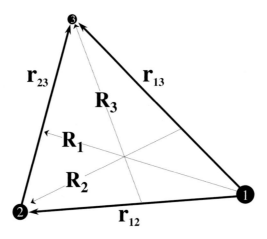

Figure 9.1: The different sets of the Jacobi coordinates $(\mathbf{r}_{ij}, \mathbf{R}_k)$; $i, j, k \in \{1, 2, 3\}$; $\epsilon_{ijk} \neq 0$; $j > i$. The coordinate $\mathbf{r}_{ij}$ is the relative position between particle i and particle j whereas $\mathbf{R}_k$ is the relative position of particle k with respect to the center of mass of the two particles i and j.

where the dimensionless rotation matrices are given by the relations

$$\mathbf{D}_3 = \begin{pmatrix} \mu_{12}/m_1 & 1 \\ 1 - \mu_{13} \cdot \mu_{12}/m_1^2 & -\mu_{13}/m_1 \end{pmatrix}, \quad \text{and}$$

$$\mathbf{D}_2 = \begin{pmatrix} -\mu_{12}/m_2 & 1 \\ -1 + \mu_{23} \cdot \mu_{12}/m_2^2 & -\mu_{23}/m_2 \end{pmatrix}. \tag{9.8}$$

The reduced masses are defined as

$$\mu_{ij} = m_i m_j/(m_i + m_j) \, ; \; i, j \in \{1, 2, 3\}; \; j > i.$$

In what follows we denote the momenta conjugate to $(\mathbf{r}_{ij}, \mathbf{R}_k)$ by $(\mathbf{k}_{ij}, \mathbf{K}_k)$. These momenta are related to each others in a similar manner as in the case of the positions vectors, namely

$$\begin{pmatrix} \mathbf{k}_{12} \\ \mathbf{K}_3 \end{pmatrix} = \mathbf{D}_2^t \begin{pmatrix} \mathbf{k}_{23} \\ \mathbf{K}_1 \end{pmatrix} = \mathbf{D}_3^t \begin{pmatrix} \mathbf{k}_{13} \\ \mathbf{K}_2 \end{pmatrix}, \tag{9.9}$$

where $\mathbf{D}_3^t$ and $\mathbf{D}_2^t$ are respectively the transposed matrices of $\mathbf{D}_1$ and $\mathbf{D}_2$. With these inter-relations between the three sets of the Jacobi coordinates it is straightforward to show that the scalar product $(\mathbf{r}_{ij}, \mathbf{R}_k) \cdot \begin{pmatrix} \mathbf{k}_{ij} \\ \mathbf{K}_k \end{pmatrix}$ has the same value for all three sets of the Jacobi coordinates.

The special feature of the Jacobi coordinates is that the kinetic energy operator H_0 is represented in a diagonal form, i. e.[3]

---

[3]For brevity, here and unless otherwise stated, no distinction is made in the notation between an operator O and its matrix representation $\langle \, \alpha|O|\alpha' \, \rangle$ with respect to the basis $\{\alpha\}$.

$$H_0 = -\frac{1}{2\mu_{ij}}\Delta_{\mathbf{r}_{ij}} - \frac{1}{2\mu_k}\Delta_{\mathbf{R}_k}, \qquad \forall\ (\mathbf{r}_{ij}, \mathbf{R}_k)\,, \tag{9.10}$$

where the reduced mass μ_k is given by

$$\mu_k = m_k(m_i + m_j)/(m_1 + m_2 + m_3).$$

The eigensolution of the free Hamiltonian (9.10) are plane waves expressed in Jacobi coordinates with an eigenenergy

$$E_0 = \frac{\mathbf{k}_{ij}^2}{2\mu_{ij}} + \frac{\mathbf{K}_k^2}{2\mu_k}, \qquad \forall\ (\mathbf{k}_{ij}, \mathbf{K}_k)\,. \tag{9.11}$$

The full three-body time-independent Schrödinger equation

$$\left[H_0 + \sum_{\substack{i,j \\ j>i}}^{3} \frac{Z_{ij}}{r_{ij}} - E \right] \langle \mathbf{r}_{kl}, \mathbf{R}_m | \Psi \rangle = 0 \tag{9.12}$$

has however a complicated form in the Jacobi coordinates, because the relative coordinates r_{ij} occurring in the potential part have to be expressed in terms of the chosen Jacobi set $(\mathbf{r}_{kl}, \mathbf{R}_m)$. This done by utilizing Eq. (9.7). In Eq. (9.12) product charges have been introduced as $Z_{ij} = Z_i Z_j; j > i \in \{1, 2, 3\}$.

9.2.2   Parabolic coordinates

As discussed in the case of two particle scattering the appropriate coordinates for continuum two-body states are the parabolic coordinates (2.49). Thus, it is suggestive to extend the definition of the coordinate system (2.49) to the three-body case by introducing the curvilinear coordinates [74]

$$
\begin{aligned}
\xi_1 &= r_{23} + \hat{\mathbf{k}}_{23}\cdot\mathbf{r}_{23}, & \eta_1 &= r_{23} - \hat{\mathbf{k}}_{23}\cdot\mathbf{r}_{23}, \\
\xi_2 &= r_{13} + \hat{\mathbf{k}}_{13}\cdot\mathbf{r}_{13}, & \eta_2 &= r_{13} - \hat{\mathbf{k}}_{13}\cdot\mathbf{r}_{13}, \\
\xi_3 &= r_{12} + \hat{\mathbf{k}}_{12}\cdot\mathbf{r}_{12}, & \eta_3 &= r_{12} - \hat{\mathbf{k}}_{12}\cdot\mathbf{r}_{12}.
\end{aligned}
\tag{9.13}
$$

The conditions for the uniqueness of this transformations are provided by the Jacobi determinant

$$\frac{d\xi_1 \wedge d\xi_2 \wedge d\xi_3 \wedge d\eta_1 \wedge d\eta_2 \wedge d\eta_3}{d^3\mathbf{r}_{ij} \wedge d^3\mathbf{R}_k}$$

$$= \left\{ \left[(\hat{\mathbf{k}}_{12} \times \hat{\mathbf{k}}_{23}) \cdot \mathbf{r}_{23} \right] \left[\hat{\mathbf{k}}_{13} \cdot (\mathbf{r}_{13} \times \mathbf{r}_{23}) \right] \right.$$

$$\left. + \left[(\hat{\mathbf{k}}_{13} \times \hat{\mathbf{k}}_{12}) \cdot \mathbf{r}_{13} \right] \left[\hat{\mathbf{k}}_{23} \cdot (\mathbf{r}_{13} \times \mathbf{r}_{23}) \right] \right\} \frac{8}{r_{23}\, r_{13}\, r_{12}}. \qquad (9.14)$$

This means the transformation (9.13) is unique if the vectors $\hat{\mathbf{k}}_{12}$ and $\hat{\mathbf{k}}_{23}$ are linearly independent. Note that in our case the vectors $\hat{\mathbf{k}}_{12}$, $\hat{\mathbf{k}}_{23}$ and $\hat{\mathbf{k}}_{13}$ satisfy a triangular relation and hence it suffices to ensure that $\hat{\mathbf{k}}_{12}$ and $\hat{\mathbf{k}}_{23}$ are linearly independent. The definition of the coordinate set (9.13) does not include the variable φ, which appears in the two-particle case in Eq. (2.49). That is a consequence of the loss of the cylindrical symmetry around one specific wave vector $\mathbf{k}_j$, due to the presence of the other wave vectors $\mathbf{k}_i$, $i \neq j$, which are physically relevant.

9.2.3   Parabolic-relative coordinates

The derivation of the two-body Coulomb bound states (2.54) in the parabolic coordinates (2.49), as well as the analytical continuation of these states to the continuum (2.55, 2.56) lead to an important observation: A plane wave state subjected to a Coulomb potential is modified in a characteristic way. The Coulomb distortion effects are described by one coordinate, either ξ or η, depending on whether incoming or outgoing wave boundary conditions are imposed. For the formation of standing waves a (coherent) combination of incoming and outgoing waves is required and hence the need for both ξ and η for the description of the bound states (2.54).

For the treatment of continuum states with well-specified boundary conditions, there is no need to account, in the definition of the coordinates, for both incoming and outgoing wave boundary conditions, as done in Eq. (9.13). Only η_j or ξ_j, $j = 1, 2, 3$ are sufficient to account for the Stark-like behaviour of the two-body Coulomb continuum states. This leaves us with three other coordinates to choose.

A suitable choice is made upon recalling that the coordinates η_j or ξ_j, $j = 1, 2, 3$ are pertinent to separate two-body systems. To account for a collective behaviour of the three-body system one has to introduce coordinates similar to, e.g. the hyperspherical, relative or elliptical coordinates. Thus, we introduce the curvilinear coordinates [77]

$$\xi_1^\pm = r_{23} \pm \hat{\mathbf{k}}_{23} \cdot \mathbf{r}_{23},$$

$$\xi_2^\pm = r_{13} \pm \hat{\mathbf{k}}_{13} \cdot \mathbf{r}_{13},$$

$$\xi_3^\pm = r_{12} \pm \hat{\mathbf{k}}_{12} \cdot \mathbf{r}_{12},$$

$$\xi_4 = r_{23},$$ \hspace{4cm} (9.15)

$$\xi_5 = r_{13},$$

$$\xi_6 = r_{12} .$$

The coordinates (ξ_4, ξ_5, ξ_6) parameterize the shape and the size of the triangle spanned by the three particles. The space-fixed dynamics is described by (ξ_1, ξ_2, ξ_3). As in the case of (9.13), the uniqueness of the transformations (9.15) is inferred from the Jacobi determinant

$$\frac{d\xi_1^+ \wedge d\xi_2^+ \wedge d\xi_3^+ \wedge d\xi_4 \wedge d\xi_5 \wedge d\xi_6}{d^3\mathbf{r}_{ij} \wedge d^3\mathbf{R}_k}$$

$$\propto \left\{ (\hat{\mathbf{k}}_{12} \times \hat{\mathbf{k}}_{23}) \cdot \mathbf{r}_{23} \left[\hat{\mathbf{k}}_{13} \cdot (\mathbf{r}_{13} \times \mathbf{r}_{23})\right] \right.$$

$$\left. + (\hat{\mathbf{k}}_{13} \times \hat{\mathbf{k}}_{12}) \cdot \mathbf{r}_{13} \left[\hat{\mathbf{k}}_{23} \cdot (\mathbf{r}_{13} \times \mathbf{r}_{23})\right] \right\} \frac{1}{r_{23}\, r_{13}\, r_{12}} . \quad (9.16)$$

From this relation it is evident that the transformation (9.15) is unique if, e. g. $| \hat{\mathbf{k}}_{13} \cdot \hat{\mathbf{k}}_{23} | \neq 1$.

The $\pm$ sign in the definition of the coordinates (9.15) indicate appropriate choices according to the type of the boundary conditions.

An essential point which, unfortunately has not been yet thoroughly investigated, derives from Eqs. (2.53, 2.60, 2.61) (page 14). Assume we are able to derive approximate expressions for the many-body continuum wave function Ψ with certain boundary conditions. Eqs. (2.53, 2.60) suggest that the bound states wave functions can be obtained by combining wave functions with incoming and outgoing wave boundary conditions. Continuing analytically the normalization N_Ψ of Ψ one may be able to identify the bound state spectrum from the poles of N_Ψ.

9.2.4   Parabolic-hyperspherical coordinates

A further important coordinate system, which makes use of the hyperspherical and the Jacobi coordinates frame is the scaled coordinate system, called the parabolic-hyperspherical coordinate system [78]

$$\xi_1^{\pm} = \frac{1}{r_{23}}(r_{23} \pm \hat{\mathbf{k}}_{23} \cdot \mathbf{r}_{23}),$$

$$\xi_2^{\pm} = \frac{1}{r_{13}}(r_{13} \pm \hat{\mathbf{k}}_{13} \cdot \mathbf{r}_{13}),$$

$$\xi_3^{\pm} = \frac{1}{r_{12}}(r_{12} \pm \hat{\mathbf{k}}_{12} \cdot \mathbf{r}_{12}),$$

$$\zeta = \arctan \frac{\overline{r}_{ij}}{\overline{R}_k},$$

$$\gamma = \hat{\overline{\mathbf{R}}}_k \cdot \hat{\overline{\mathbf{r}}}_{ij},$$

$$\rho = (\overline{R}_k^2 + \overline{r}_{ij}^2)^{1/2}.$$

(9.17)

As will be shown below, the form of the Schrödinger equation is considerably simplified by introducing the mass-dependent Jacobi coordinates

$$\overline{\mathbf{r}}_{ij} = \mu_{ij}^{1/2} \, \mathbf{r}_{ij}, \quad \overline{\mathbf{R}}_k = \mu_k^{1/2} \, \mathbf{R}_k.$$

(9.18)

Again the introduction of the $\pm$ signs in Eqs. (9.17) allows for the treatment of problems with different kinds of boundary conditions. In Eqs. (9.17) the coordinates ξ_i, $i \in \{1, 2, 3\}$ play the role of Euler angles in a hyperspherical treatment where the laboratory reference frame is specified by the directions of the relative momenta that are determined asymptotically. The body-fixed dynamics is described by the coordinates (ζ, γ, ρ).

9.3   Approximate three-body states and the parabolic-relative coordinates

Now let us inspect the structure of the Schrödinger equation in the parabolic-relative coordinates (9.15). The potential energy depends only on ξ_4, ξ_5, ξ_6 and is diagonal in these coordinates. Thus, it remains to clarify the form of the kinetic energy. To this end we operate in Jacobi coordinates and subsequently transform the Schrödinger equation to the parabolic-relative coordinates (9.15). As we are considering continuum solutions of the Schrödinger equation (9.12) at a fixed total energy (9.11) we make for the wave function the ansatz

$$\Psi(\mathbf{r}_{ij}, \mathbf{R}_k) = N_{\mathbf{k}_{ij}, \mathbf{K}_k} \exp(i \, \mathbf{r}_{ij} \cdot \mathbf{k}_{ij} + i \, \mathbf{R}_k \cdot \mathbf{K}_k) \, \overline{\Psi}(\mathbf{r}_{ij}, \mathbf{R}_k),$$

(9.19)

where $N_{\mathbf{k}_{ij}, \mathbf{R}_k}$ is a normalization factor. Upon inserting this ansatz in the Schrödinger Eq. (9.12) we obtain the following determining equation for the function $\overline{\Psi}(\mathbf{r}_{ij}, \mathbf{R}_k)$

$$
\begin{aligned}
\left[\frac{1}{\mu_{ij}} \Delta_{\mathbf{r}_{ij}} + \frac{1}{\mu_k} \Delta_{\mathbf{R}_k} + 2i \left(\frac{1}{\mu_{ij}} \mathbf{k}_{ij} \cdot \nabla_{\mathbf{r}_{ij}} + \frac{1}{\mu_k} \mathbf{K}_k \cdot \nabla_{\mathbf{R}_k} \right) \right. \\
\left. - 2 \sum_{\substack{m,n \\ n>m}}^{3} \frac{Z_{ij}}{r_{mn}} \right] \overline{\Psi}(\mathbf{r}_{ij}, \mathbf{R}_k) = 0.
\end{aligned}
\tag{9.20}
$$

The main task is now to transform this partial differential equation into the coordinate system (9.15). Doing so Eq. (9.20) casts

$$
[\, H_{\mathrm{par}} + H_{\mathrm{in}} + H_{\mathrm{mix}} \,] \overline{\Psi}(\xi_1, \dots, \xi_6) = 0 \,.
\tag{9.21}
$$

The key point is that the operator H_{par} is differential in parabolic (external) coordinates (ξ_1, ξ_2, ξ_3) only. It has namely the form

$$
\begin{aligned}
H_{\mathrm{par}} := \quad & \frac{2}{\mu_{23}\xi_4} \left[\partial_{\xi_1} \, \xi_1 \, \partial_{\xi_1} + ik_{23} \, \xi_1 \, \partial_{\xi_1} - \mu_{23} \, Z_{23} \right] \\
+ & \frac{2}{\mu_{13}\xi_5} \left[\partial_{\xi_2} \, \xi_2 \, \partial_{\xi_2} + ik_{13} \, \xi_2 \, \partial_{\xi_2} - \mu_{13} \, Z_{13} \right] \\
+ & \frac{2}{\mu_{12}\xi_6} \left[\partial_{\xi_3} \, \xi_3 \, \partial_{\xi_3} + ik_{12} \, \xi_3 \, \partial_{\xi_3} - \mu_{12} \, Z_{12} \right].
\end{aligned}
\tag{9.22}
$$

It is important to note that the potential enters these equations as the constant product charges Z_{ij}. This means the eigenfunctions of the operator (9.22) diagonalize exactly the total potential. Therefore, the terms H_{in} and H_{mix} in Eq. (9.21) must be due to parts of the kinetic energy operator.

Inspecting the equation (9.22) one readily concludes that H_{par} can be written as a sum of three commuting one-dimensional differential operators, i. e.

$$
H_{\mathrm{par}} = \frac{2}{\mu_{23}\xi_4} h_{\xi_1} + \frac{2}{\mu_{13}\xi_5} h_{\xi_2} + \frac{2}{\mu_{12}\xi_6} h_{\xi_3},
\tag{9.23}
$$

where the operators h_{ξ_i} satisfy the commutation relations

$$
[h_{\xi_i}, h_{\xi_j}] = 0, \quad i \neq j = 1, 2, 3.
\tag{9.24}
$$

The explicit form of the one-dimensional operators h_j is

$$
h_{\xi_j} = \partial_{\xi_j} \, \xi_j \, \partial_{\xi_j} + ik_{kl} \, \xi_j \, \partial_{\xi_j} - \mu_{kl} \, Z_{kl}, \quad \epsilon_{jkl} \neq 0,
\tag{9.25}
$$

which is readily deduced from Eqs. (9.23, 9.22).

Comparing Eq. (9.25) with the confluent hypergeometric differential equation (2.50) (on page 14) leads to the conclusion that the eigenfunctions of the operators h_{ξ_j} are the confluent hypergeometric functions with zero eigenvalues. In other words we can write

$$h_{\xi_j}\left[{}_1F_1\left(i\alpha_{kl}, 1, -ik_{kl}\,\xi_j\right)\right] = 0, \quad \epsilon_{jkl} \neq 0. \tag{9.26}$$

By virtue of Eqs. (9.24, 9.23) one concludes that the closed form eigenfunctions of H_{par} [Eq. (9.23)] with zero eigenvalues are

$$H_{\text{par}}\overline{\Psi}_{\text{par}} = 0, \tag{9.27}$$

where $\overline{\Psi}_{\text{par}}$ is a direct product of the eigenfunctions of the operators h_{ξ_j}, i. e.

$$\begin{aligned}\overline{\Psi}_{\text{par}}(\xi_1,\xi_2,\xi_3) \;=\; & {}_1F_1\left(i\alpha_{23}, 1, -ik_{23}\,\xi_1\right)\\ & {}_1F_1\left(i\alpha_{13}, 1, -ik_{13}\,\xi_2\right)\\ & {}_1F_1\left(i\alpha_{12}, 1, -ik_{12}\,\xi_3\right).\end{aligned} \tag{9.28}$$

Here $\alpha_{ij} = Z_{ij}\mu_{ij}/k_{ij}$ are the Sommerfeld parameters.

The entire solution (9.19) is expressible in terms of the coordinates (9.15) because the plane-wave arguments can be transformed into (9.15) and have then the form

$$\begin{aligned}\mathbf{k}_{ij}\cdot\mathbf{r}_{ij} + \mathbf{K}_k\cdot\mathbf{R}_k &= \sum_{j>i=1}^{3}\frac{m_i+m_j}{m_1+m_2+m_3}\mathbf{k}_{ij}\cdot\mathbf{r}_{ij},\\ &= \frac{\mu_1}{m_1}k_{23}\left(\xi_1-\xi_4\right) + \frac{\mu_2}{m_2}k_{13}\left(\xi_2-\xi_5\right) + \frac{\mu_3}{m_3}k_{12}\left(\xi_3-\xi_6\right)\end{aligned} \tag{9.29}$$

The solution (9.28) coincides with the so-called 3C approximation [130, 72, 73, 74]. The name refers to the fact that this wave function consists of three two-body parabolic wave functions. Due to Eq. (9.24) the normalization of the wave function (9.28) is readily obtained from the normalization of the hydrogenic wave function in parabolic coordinates (2.58), namely

$$N_{\mathbf{k}_{ij},\mathbf{K}_k} = (2\pi)^{-3}\prod_{i>j=1}^{3} e^{-\pi\alpha_{ij}/2}\,\Gamma(1 - i\alpha_{ij}). \tag{9.30}$$

The operator H_{in} in Eq. (9.21) is differential in internal (body fixed) coordinates only and depends parametrically on the coordinates $\xi_{1/2/3}$. It is explicit form is

$$\begin{aligned}H_{\text{in}} := \;& \frac{1}{\mu_{23}}\left[\frac{1}{\xi_4^2}\partial_{\xi_4}\xi_4^2\partial_{\xi_4} + 2\,i\,k_{23}\frac{\xi_1-\xi_4}{\xi_4}\partial_{\xi_4}\right]\\ &+ \frac{1}{\mu_{13}}\left[\frac{1}{\xi_5^2}\partial_{\xi_5}\xi_5^2\partial_{\xi_5} + 2\,i\,k_{13}\frac{\xi_2-\xi_5}{\xi_5}\partial_{\xi_5}\right]\\ &+ \frac{1}{\mu_{12}}\left[\frac{1}{\xi_6^2}\partial_{\xi_6}\xi_6^2\partial_{\xi_6} + 2\,i\,k_{12}\frac{\xi_3-\xi_6}{\xi_6}\partial_{\xi_6}\right].\end{aligned} \tag{9.31}$$

It is remarkable that this operator possesses a structure similar to H_{par}, namely H_{in} is the sum of three commuting operators operating in a one dimensional space:

$$H_{\text{in}} := h_{\xi_4} + h_{\xi_5} + h_{\xi_6}, \tag{9.32}$$

where $h_{\xi_j}, j = 4, 5, 6$ are respectively the three terms in Eq. (9.31) (for example $h_{\xi_4} = \frac{1}{\mu_{23}} \left[\frac{1}{\xi_4^2} \partial_{\xi_4} \xi_4^2 \partial_{\xi_4} + 2 i k_{23} \frac{\xi_1 - \xi_4}{\xi_4} \partial_{\xi_4} \right]$). From Eq. (9.31) it is readily deduced that the following commutation relations

$$[h_{\xi_i}, h_{\xi_j}] = 0, \quad i \neq j = 4, 5, 6, \tag{9.33}$$

apply. The eigenfunctions of the operators $h_{\xi_j}, j = 4, 5, 6$ can as well be found in closed form. Since these operators are parts of the kinetic energy, their eigenfunctions will not show the slow decay characteristic for the Coulomb potentials. In fact, if we inspect for example the operator h_{ξ_4} we arrive at the conclusion that the derivative f_{ξ_4} of its eigenfunction is determined by the differential equation

$$[\xi_4 \partial_{\xi_4} + 2(1 + i k_{23}\xi_1) - 2 i k_{23}\xi_4] f_{\xi_4} = 0. \tag{9.34}$$

This equation can be solved by the ansatz

$$f_{\xi_4} = N_{\xi_4} e^{\lambda_1 \xi_4 + \lambda_2 \ln(\xi_4)},$$

which after inserting in the differential equation (9.34) yields

$$f_{\xi_4} = N_{\xi_4} \left(\xi_4 \right)^{2 i k_{23}} e^{i k_{23} \xi_1 \xi_4} e^{-2\xi_4}. \tag{9.35}$$

This relation exhibits explicitly the fast decaying behaviour of the eigenfunctions of H_{in} with growing ξ_4, which underlines the unimportance of H_{in} for the asymptotic behaviour. At short distances however, the contribution of H_{in} to the total solution is generally not negligible.

The remainder term H_{mix} contains mixed derivatives resulting from off-diagonal elements of the metric tensor and couples internal to external motion[4]

$$H_{\text{mix}} := \sum_{u \neq v = 1}^{6} \left\{ (\nabla_{\mathbf{r}_{ij}} \xi_u) \cdot (\nabla_{\mathbf{r}_{ij}} \xi_v) + (\nabla_{\mathbf{R}_k} \xi_u) \cdot (\nabla_{\mathbf{R}_k} \xi_v) \right\} \partial_{\xi_u} \partial_{\xi_v}. \tag{9.36}$$

---

[4]Note that the operator H_{in} (9.31) contains as well coupling terms between the internal ($\xi_j, j = 4, 5, 6$) and the external ($\xi_j, j = 1, 2, 3$) coordinates. This is also obvious from Eq. (9.35). If one wishes to construct eigenfunctions that diagonalize H_{in} and H_{para} at the same time, it is more convenient to add the coupling terms, that appear in (9.31) and depend on ($\xi_j, j = 4, 5, 6$), to the mixing operator H_{mix}.

The structure and the role played by the mixing term H_{mix} is in analogy to what we discussed for the non-adiabatic mixing term (4.6) that appears in the context of the Born-Oppenheimer and the adiabatic approximations. The origin of this analogy is the parametric (adiabatic) dependence of H_{par} on the internal coordinates. It should be noted however, that this behaviour appears naturally from the structure of the Schrödinger equation and is not the result of a specific ansatz for the wave function, as done, for example by making the assumption (4.1) for the wave function when applying the BO approximation.

Recalling our discussion of the procedure that lead to the adiabatic solutions we conclude, that the simplest approximation is to neglect the action of mixing (non-adiabatic) terms H_{mix} on the wave function (9.28). If this doing is justified then the wave function (9.28) can be employed as a good approximate expression for the exact three-body continuum state. For this reason it is imperative to study the properties of the operators H_{in}, H_{mix} and their action on the wave function (9.28). At first it is instructive to explore the exact boundary conditions for the three-body Coulomb continuum. This we will perform first in Jacobi coordinates and then transform the results into the parabolic relative coordinate system.

9.4   Asymptotic properties of the three-body wave equation

The asymptotic behaviour of the three-body wave function (9.19) in the limit of large interparticle distances $\mathbf{r}_{ij}$ is obtained by specializing Eq. (8.10) to three-particles which yields in the Jacobi coordinate system [71, 125, 126, 72, 74]

$$
\lim_{\substack{r_{ij}\to\infty \\ R_k\to\infty}} \Psi(\mathbf{r}_{ij}, \mathbf{R}_k) \quad\longrightarrow\quad (2\pi)^{-3} \exp(i\,\mathbf{k}_{ij}\cdot\mathbf{r}_{ij} + i\,\mathbf{K}_k\cdot\mathbf{R}_k)
$$
$$
\times \prod_{\substack{i,j=1 \\ j>i}}^{3} \exp\left(\pm i\alpha_{ij}\ln(k_{ij}\,r_{ij} \pm \mathbf{k}_{ij}\cdot\mathbf{r}_{ij})\right), \quad \forall\ (\mathbf{r}_{ij}, \mathbf{R}_k).
$$

$$(9.37)$$

This expression is sometimes referred to as the *Redmond asymptotic*. The '+' and '−' signs refer to different boundary conditions. Note that the Sommerfeld parameters α_{ij} depend only on the relative velocities k_{ij}/μ_{ij} of the three pairs

$$
\alpha_{ij} = \frac{Z_{ij}\mu_{ij}}{k_{ij}}, \tag{9.38}
$$

i. e. these parameters have a pure two-body nature and do not carry any information on the three-body coupling strength. As already mentioned above, for N-particle systems different types of asymptotics can be defined. E. g., for a three-body system described using a certain Jacobi coordinate set $(\mathbf{r}_{ij}, \mathbf{R}_k)$, one can consider the case where one Jacobi coordinate tends to infinity whereas the other coordinate remains finite [75]. For this reason in a three-body system one defines the following asymptotic regions

$$\mathcal{L} \; := \; \left\{ \mathbf{r}_{ij}, \mathbf{R}_k \,;\, r_{ij} \to \infty \quad and \quad R_k \to \infty, \quad \forall \; (\mathbf{r}_{ij}, \mathbf{R}_k) \right\}, \tag{9.39}$$

$$\mathcal{L}_\alpha \; := \; \left\{ \mathbf{r}_{ij}, \mathbf{R}_\alpha \,;\, \frac{r_{ij}}{R_\alpha} \to 0 \quad and \quad R_\alpha \to \infty, \quad \forall \; (\mathbf{r}_{ij}, \mathbf{R}_\alpha) \right\}, \tag{9.40}$$

$$\mathcal{L}_{ij} \; := \; \left\{ \mathbf{r}_{ij}, \mathbf{R}_k \,;\, \frac{R_k}{r_{ij}} \to 0 \quad and \quad r_{ij} \to \infty, \quad \forall \; (\mathbf{r}_{ij}, \mathbf{R}_k) \right\}. \tag{9.41}$$

In the region $\mathcal{L}$ where the Redmond asymptotic is valid all particles are far apart and their position vectors are not linearly dependent. In the regime $\mathcal{L}_{ij}$ two particles (i and j) are far away from each other and the third particle resides at the center of mass of i and j. Thus, in this region all interparticle distances are large ($\mathcal{L}_{ij} \subset \mathcal{L}$ holds, but not the converse) and hence Eq. (9.37) applies. An important region contained in the domain $\mathcal{L}_{ij}$ is the Wannier configuration for two electrons moving in the field of a positive ion (the two electrons are the particles i and j). In the asymptotic region $\mathcal{L}_\alpha$ the particle α is far away from the compound formed by i and j. In this case Eq. (9.37) does not hold true [75], as explicitly demonstrated below.

The asymptotic form of the wave function in the region $\mathcal{L}_\alpha$ derives from the Schrödinger equation Eq. (9.12) which in $\mathcal{L}_\alpha$ has the leading order form

$$\left(H_0 \;+\; \frac{Z_{ij}}{r_{ij}} \;+\; \frac{Z_\alpha \, (Z_i + Z_j)}{R_\alpha} \;-\; E \right) \psi_\alpha^{as} = 0; \quad \forall \quad (\mathbf{r}_{ij}, \mathbf{R}_\alpha) \in \mathcal{L}_\alpha. \tag{9.42}$$

Since the kinetic energy operator is separable in any set of Jacobi coordinates it follows that Eq. (9.42) is separable in the coordinates $(\mathbf{r}_{ij}, \mathbf{R}_\alpha)$.

The solutions of (9.42) are simply a product of two Coulomb waves under the constraint that $R_\alpha \to \infty$, i.e.

$$
\begin{aligned}
\psi_\alpha^{as} \;=\; & (2\pi)^{-3/2} \exp(i \, \mathbf{k}_{ij} \cdot \mathbf{r}_{ij} + i \, \mathbf{K}_\alpha \cdot \mathbf{R}_\alpha) \\
& \times N_{ij} \; {}_1F_1 \left(i\alpha_{ij}, \, 1, \, -i \left[k_{ij} \, r_{ij} + \mathbf{k}_{ij} \cdot \mathbf{r}_{ij} \right] \right) \\
& \times \exp \left[i\bar{\gamma}_\alpha \ln(K_\alpha \, R_\alpha + \mathbf{K}_\alpha \cdot \mathbf{R}_\alpha) \right],
\end{aligned}
\tag{9.43}
$$

where the parameters $\overline{\gamma}_\alpha$ and N_{ij} are functions of the momenta, namely

$$\overline{\gamma}_\alpha = \frac{Z_\alpha(Z_i + Z_j)\mu_\alpha}{K_\alpha}, \qquad N_{ij} = (2\pi)^{-3/2}\,e^{-\pi\alpha_{ij}/2}\,\Gamma(1 - i\alpha_{ij}). \tag{9.44}$$

The existence of a global analytic asymptotic defined on $\mathcal{M} = \mathcal{L}_\alpha \bigcup \mathcal{L}_{ij}$ derives from the fact that the regions $\mathcal{L}_\alpha$ and $\mathcal{L}_{ij}$ are not disjoint. In order to find such an asymptotic form and to facilitate the investigation of the behaviour of the eigenfunctions of the operators H_{par}, H_{in} and H_{mix} at finite distances, we consider Eq. (9.28) in Jacobi coordinates which has the explicit form

$$\Psi(\mathbf{r}_{ij}, \mathbf{R}_k) \approx \Psi_{3C}(\mathbf{r}_{ij}, \mathbf{R}_k),$$
$$= (2\pi)^{3/2} \exp(i\,\mathbf{k}_{ij}\cdot\mathbf{r}_{ij} + i\,\mathbf{K}_k\cdot\mathbf{R}_k)$$
$$\prod_{\substack{m,n=1\\n>m}}^{3} N_{mn}\,{}_1F_1\big(i\alpha_{mn},\ 1,\ -i\,[k_{mn}\,r_{mn} + \mathbf{k}_{mn}\cdot\mathbf{r}_{mn}]\big). \tag{9.45}$$

To unravel the contributions of the (non-adiabatic mixing) operators H_{mix} and H_{in} we make, for the exact solution of (9.12), the general ansatz

$$\Psi(\mathbf{r}_{ij}, \mathbf{R}_k) = \Psi_{3C}(\mathbf{r}_{ij}, \mathbf{R}_k)(\,1 - f(\mathbf{r}_{ij}, \mathbf{R}_k)\,). \tag{9.46}$$

If H_{in} and H_{mix} can be neglected then $f \equiv 0$. In case only H_{mix} vanishes the function f can be expressed in terms of the eigenfunctions of H_{in} that can be derived in parabolic coordinates, as outlined above and then converted into Jacobi coordinates. Here we leave open the form of f and derive a general determining equation in Jacobi coordinates. Later on we will show how f can be incorporated effectively in the wave function (9.45).

To simplify notations let us choose the set $(\mathbf{r}_{13}, \mathbf{R}_2)$ and insert (9.46) in (9.12). This leads to the differential equation

$$\left[\frac{1}{2\mu_{13}}\Delta_{\mathbf{r}_{13}} + \frac{1}{\mu_{13}}(i\mathbf{k}_{13} + \alpha_{13}k_{13}\,\mathbf{F}_{13} + \alpha_{12}k_{12}\,\mathbf{F}_{12})\cdot\nabla_{\mathbf{r}_{13}}\right]\cdot f$$
$$+ \left[\frac{1}{2\mu_{23}}\Delta_{\mathbf{r}_{23}} + \frac{1}{\mu_{23}}(i\mathbf{k}_{23} + \alpha_{23}k_{23}\,\mathbf{F}_{23} - \alpha_{12}k_{12}\,\mathbf{F}_{12})\cdot\nabla_{\mathbf{r}_{23}}\right]\cdot f$$
$$- \alpha_{12}k_{12}\,\mathbf{F}_{12}\cdot(Z_{13}\,\mathbf{F}_{13} - Z_{23}\,\mathbf{F}_{23})(1 - f) = D_{pol}(f). \tag{9.47}$$

The terms $\mathbf{F}_{ij}$ are expressible in terms of confluent hypergeometric functions as

$$\mathbf{F}_{ij} = \frac{{}_1F_1(1 + i\alpha_{ij}, 2, -i[k_{ij}\,r_{ij} + \mathbf{k}_{ij}\cdot\mathbf{r}_{ij}])}{{}_1F_1(i\alpha_{ij}, 1, -i[k_{ij}\,r_{ij} + \mathbf{k}_{ij}\cdot\mathbf{r}_{ij}])}\,(\hat{\mathbf{k}}_{ij} + \hat{\mathbf{r}}_{ij}). \tag{9.48}$$

This derivation of (9.47) underlines the fact that the kinetic and the potential energy terms have different appropriate coordinates, i.e. while the generic coordinates for the Coulomb potential are the relative positions r_{ij}, the kinetic energy operator is diagonal in Jacobi coordinates. Therefore, cross terms are in general inevitable if one treats at the same time the kinetic and the potential energy terms. Here we transformed the Jacobi coordinates into the relative coordinates $(\mathbf{r}_{13}, \mathbf{r}_{23})$ (cf. Eq. (9.29)). This introduces the mass-polarization term D_{pol} in (9.47) which has the form

$$D_{pol}(f) = \frac{1}{m_3}(D_1 (f-1) + D_2 f), \tag{9.49}$$

where the differential operator D_2 has the explicit form

$$
\begin{aligned}
D_2 &= [\alpha_{13}k_{13}\, \mathbf{F}_{13} + \alpha_{12}k_{12}\, \mathbf{F}_{12}] \cdot \nabla_{\mathbf{r}_{23}} \\
&+ [\alpha_{23}k_{23}\, \mathbf{F}_{23} - \alpha_{12}k_{12}\, \mathbf{F}_{12}] \cdot \nabla_{\mathbf{r}_{13}} + \nabla_{\mathbf{r}_{23}} \cdot \nabla_{\mathbf{r}_{13}}.
\end{aligned}
\tag{9.50}
$$

The second operator D_1 that occurs in Eq. (9.49) is expressed as

$$
\begin{aligned}
D_1 &= (\alpha_{13}k_{13})(\alpha_{23}k_{23})\, \mathbf{F}_{13} \cdot \mathbf{F}_{23} \\
&- (\alpha_{13}k_{13})(\alpha_{12}k_{12})\, \mathbf{F}_{13} \cdot \mathbf{F}_{12} \\
&+ (\alpha_{23}k_{23})(\alpha_{12}k_{12})\, \mathbf{F}_{23} \cdot \mathbf{F}_{12} \\
&+ 2k_{12}(\alpha_{12}k_{12})(i - \alpha_{12})(1 + \hat{\mathbf{k}}_{12} \cdot \hat{\mathbf{r}}_{12}) \\
&\times \frac{{}_1F_1\, (2 + i\alpha_{12},\, 3,\, -i[k_{12}\, r_{12} + \mathbf{k}_{12} \cdot \mathbf{r}_{12}]\,)}{{}_1F_1\, (i\alpha_{12},\, 1,\, -i[k_{12}\, r_{12} + \mathbf{k}_{12} \cdot \mathbf{r}_{12}]\,)} \\
&+ (\alpha_{12}k_{12}) \frac{{}_1F_1\, (1 + i\alpha_{12},\, 2,\, -i[k_{12}\, r_{12} + \mathbf{k}_{12} \cdot \mathbf{r}_{12}]\,)}{{}_1F_1\, (i\alpha_{12},\, 1,\, -i[k_{12}\, r_{12} + \mathbf{k}_{12} \cdot \mathbf{r}_{12}]\,)}\, \frac{2}{r_{12}}.
\end{aligned}
\tag{9.51}
$$

The operators D_1 and D_2 are negligible in cases where the mass of one of the particles, say m_3 is much larger than the mass of the other two particles. This is, for example, the case for two electrons moving in the field of a massive ion. As clearly seen from Fig. 9.1 in this situation the relative and one set of the Jacobi coordinates coincide. This is also clear from the structure of the polarization term (9.49) for $(m_1/m_3) \to 0$.

Finding the solution of the Schrödinger equation (9.12) is equivalent to the task of solving for the partial differential equation (9.47). As a first attempt we may ask when the expression $\Psi_{3C}(\mathbf{r}_{ij}, \mathbf{R}_k)$ is a good approximate of the exact solution $\Psi(\mathbf{r}_{ij}, \mathbf{R}_k)$, or in other words when the function f is negligibly small ($\|f\| \ll 1$). Mathematically, the function $f = 0$ is a solution of Eq. (9.47), if the inhomogeneous term

$$\mathcal{R} = \mu_{12} Z_{12} \, \mathbf{F}_{12} \cdot (Z_{13} \, \mathbf{F}_{13} - Z_{23} \, \mathbf{F}_{23}) + \frac{1}{m_3} D_1 \qquad (9.52)$$

vanishes. If this were the case one obtains a manifold of solutions $f = constant$. The acceptable solution out of this manifold is then singled out by the normalization requirement.

If $\mathcal{R}$ is finite the solutions f, as defined by Eq. (9.47), is obtained upon solving the equation

$$f = f_0 - \int d^3 r'_{13} \int d^3 r'_{23} \, G(\mathbf{r}_{13} - \mathbf{r}'_{13}, \mathbf{r}_{23} - \mathbf{r}'_{23}) \, \mathcal{R}(\mathbf{r}'_{13}, \mathbf{r}'_{23}) \,. \qquad (9.53)$$

Here $G(\mathbf{r}_{13}, \mathbf{r}_{23})$ denotes the Green's function of a Schrödinger type equation, namely

$$\begin{aligned}
\Bigg[&\frac{1}{2\mu_{13}} \Delta_{\mathbf{r}_{13}} + \frac{1}{2\mu_{23}} \Delta_{\mathbf{r}_{23}} \\
&+ \frac{i}{\mu_{13}} \, \mathcal{K}_{13} \cdot \nabla_{\mathbf{r}_{13}} + \frac{i}{\mu_{23}} \mathcal{K}_{23} \cdot \nabla_{\mathbf{r}_{23}} \\
&- \frac{1}{m_3} \, D_2 \Bigg] G(\mathbf{r}_{13}, \mathbf{r}_{23}) = \delta^3(\mathbf{r}_{13}) \, \delta^3(\mathbf{r}_{23}).
\end{aligned}$$
$$(9.54)$$

The quantities $\mathcal{K}_{13}$ stands for complex effective vectors defined by the relations

$$\mathcal{K}_{13} = \mathbf{k}_{13} - i\alpha_{13} k_{13} \, \mathbf{F}_{13} - i\alpha_{12} k_{12} \, \mathbf{F}_{12},$$

$$\mathcal{K}_{23} = \mathbf{k}_{23} - i\alpha_{23} k_{23} \, \mathbf{F}_{23} + i\alpha_{12} k_{12} \, \mathbf{F}_{12}. \qquad (9.55)$$

To specify the boundary condition of this equation we have to inspect the asymptotic behaviour of the function $\mathcal{R}$.

Obviously the behaviour of the inhomogeneous term $\mathcal{R}$ in Eq. (9.52) is dictated by the generalized functions $\mathbf{F}_{ij}$ (9.48). Asymptotically, an expression is derived for $\mathbf{F}_{ij}$ by inspection of the asymptotic properties of the hypergeometric functions which leads to

$$\lim_{r_{ij} \to \infty} |\mathbf{F}_{ij}| \to \left| \frac{\hat{\mathbf{k}}_{ij} + \hat{\mathbf{r}}_{ij}}{\mathbf{k}_{ij} \cdot (\hat{\mathbf{k}}_{ij} + \hat{\mathbf{r}}_{ij}) \, r_{ij}} \right| + \mathcal{O}\left(|k_{ij} \, r_{ij} + \mathbf{k}_{ij} \cdot \mathbf{r}_{ij}|^{-2}\right) \,. \qquad (9.56)$$

From this asymptotic form of $\mathbf{F}_{ij}$ and from the dependence of expression $\mathcal{R}$ (9.52) on $\mathbf{F}_{ij}$ we conclude that the inhomogeneous term $\mathcal{R}$ decays asymptotically faster than the Coulomb potential, only in the case when two independent Jacobi coordinates tend to infinity. This limit can be expressed equivalently as $r_{ij} \to \infty \; \forall \, i, j \in \{1, 2, 3\}; \; j > i$. In this asymptotic regime we have

$$\lim_{\substack{r_{ij} \to \infty \\ R_k \to \infty}} \mathcal{R} \to \mathcal{O}\left(|k_{ij} \, r_{ij} + \mathbf{k}_{ij} \cdot \mathbf{r}_{ij}|^{-2}\right), \quad \forall \; \mathbf{r}_{ij}, \mathbf{R}_k \in \mathcal{L}, \qquad (9.57)$$

as is evident from (9.56). Since a negligible $\mathcal{R}$ is equivalent to $f = 0$ and $f = 0$ is equivalent to the conclusion that $\Psi \rightarrow \Psi_{3C}$ we infer that, in the subspaces $\mathcal{L}$ and $\mathcal{L}_{ij}$ (where $\mathcal{R} \rightarrow 0$), the wave function Ψ_{3C} is the leading order asymptotic eingenstate of the Schrödinger equation (9.12).

In the asymptotic subspace $\mathcal{L}_\alpha$ only the function $\mathbf{F}_{ij}$ takes on its asymptotic form (9.56). From Eq. (9.56) it is then clear that the expression $\mathcal{R}$ (Eq.9.52) is of the order of the Coulomb potential and hence can not be neglected. Accordingly, $f = 0$ does not solve (9.47) and Ψ_{3C} is not a global asymptotic solution of (9.12) in $\mathcal{M}$.

Thus, boundary conditions for equation (9.54) are as follows

$$f(\mathbf{r}_{ij}, \mathbf{R}_k) \ = \ 0, \qquad \forall \ \mathbf{r}_{ij}, \mathbf{R}_k \in \mathcal{L}, \tag{9.58}$$

$$f(\mathbf{r}_{ij}, \mathbf{R}_k) \ = \ 1 - \frac{\psi_\alpha^{as}}{\Psi_{3C}}, \quad \forall \ \mathbf{r}_{ij}, \mathbf{R}_\alpha \in \mathcal{L}_\alpha . \tag{9.59}$$

Due to the condition (9.58) the solution f_0 of the homogeneous equation is $f_0 = 0$.

An alternative (instead of solving for Eq. (9.54)) way for finding approximate wave function relies on the following observation [77, 75]: In the Schrödinger equation (9.12) the total energy and the total potentials occur as sums of single particle objects. As a matter of principle however, only the total energy and only the total potential are of relevance for the dynamics of the system. Any transformation leaving these quantities unchanged and not affecting the boundary conditions should not have an influence on the exact solution of the Schrödinger equation (9.12).

Therefore, one can write for the total potential the invariance relations

$$\sum_{\substack{i,j \\ j>i}}^{3} \frac{\bar{Z}_{ij}(\mathbf{r}_{ij})}{r_{ij}} \equiv \sum_{\substack{i,j \\ j>i}}^{3} \frac{Z_{ij}}{r_{ij}} . \tag{9.60}$$

Accordingly, the total energy E_0 (9.11) is invariant under any transformation that satisfies

$$\frac{\bar{\mathbf{k}}_{ij}^2(\mathbf{R}_k, \mathbf{r}_{ij})}{2\mu_{ij}} + \frac{\bar{\mathbf{K}}_k^2(\mathbf{R}_k, \mathbf{r}_{ij})}{2\mu_k} \equiv \frac{\mathbf{k}_{ij}^2}{2\mu_{ij}} + \frac{\mathbf{K}_k^2}{2\mu_k} = E_0, \qquad \forall \ (\mathbf{k}_{ij}, \mathbf{K}_k) . \tag{9.61}$$

These relations serve as a definition of the position-dependent product charges $\bar{Z}_{ij}$ and the position dependent magnitude of the wave vectors $\bar{K}_k$, $\bar{k}_{ij}$. The conditions (9.60, 9.61) ensure that the total potential and the total energies are conserved for any choice of the functional dependence of $\bar{Z}_{ij}$ and $\bar{\mathbf{K}}_k$, $\bar{\mathbf{k}}_{ij}$.

It should be emphasized that the splitting of the total potential and energy into two-body terms is unique only when the exact wave function separates in a product of three two-body wave functions. This is, for example, the case for the 3C approximation in its range of validity, i. e. in the domain $\mathcal{L}$. This puts on $\bar{Z}_{ij}$ and $\bar{\mathbf{K}}_k$, $\bar{\mathbf{k}}_{ij}$ the additional constraint

$$\overline{Z}_{ij}(\mathbf{r}_{ij}, \mathbf{R}_k) \quad \rightarrow \quad Z_{ij}, \quad \forall\; \mathbf{r}_{ij}, \mathbf{R}_\alpha \in \mathcal{L}_\alpha, \tag{9.62}$$

$$\bar{k}_{ij}(\mathbf{R}_k, \mathbf{r}_{ij}) \quad \rightarrow \quad k_{ij}, \quad \forall\; \mathbf{r}_{ij}, \mathbf{R}_\alpha \in \mathcal{L}_\alpha, \tag{9.63}$$

$$\bar{K}_k(\mathbf{R}_k, \mathbf{r}_{ij}) \quad \rightarrow \quad k_{ij}, \quad \forall\; \mathbf{r}_{ij}, \mathbf{R}_\alpha \in \mathcal{L}_\alpha. \tag{9.64}$$

At first sight it seems that Eqs. (9.60, 9.61) are just a formal manipulation of the total potential and the total energy and do not lead to any new insight into the solution of the Schrödinger equation (9.12). To clarify the advantage of Eqs. (9.60, 9.61) we note that there will be an infinity of solutions satisfying Eqs. (9.60, 9.61) and the boundary conditions imposed on the functions $\bar{Z}_{ij}$ and/or the functions $\bar{\mathbf{K}}_k$, $\bar{\mathbf{k}}_{ij}$.

This additional freedom can be used as follows. Given an approximate solution of the three-body problem one determines the parts R of (9.12) that are not described by this solution. In a second step we can use the freedom gained by introducing Eqs. (9.60, 9.61) and determine $\bar{Z}_{ij}$ and $\bar{\mathbf{K}}_k$, $\bar{\mathbf{k}}_{ij}$ in a way that the neglected terms R are minimized. This procedure is not unusual in theoretical physics. In fact it is at the heart of the Fermi liquid theory of Landau [50, 52]. There correlation effects, i. e. loosely speaking (correlation) parts of the Hamiltonian not incorporated in a single-particle picture, are subsumed as a modification of the single particle properties leading thus to the concept of quasi particles. We will elaborate on this point in chapter 14. The difference here is that we are treating all two-particle (pair) correlations to infinite orders. This we interpret as having three quasi particles, each of them is formed out of one of the possible pairs in the system. (3-body) Correlation effects that go beyond two-body correlations are then incorporated as modifications of the properties and the interactions of the quasi particles. In other words our quasi particles are in fact quasi particle pairs.

To formulate precisely the above statement we recall that in a three-body problem all (isolated) two-body interactions are described by the wave function Ψ_{3C}. Higher order correlation are isolated and assigned to the term $\mathcal{R}$, given by Eq. (9.52). Therefore, we require that the position-dependent product charges $\overline{Z}_{ij}$ are determined in such a way that the condition

(9.60) is fulfilled and the remainder term $\mathcal{R}$ is minimized. The next requirement concerns the asymptotic domain. As demonstrated above, the three-body system fragments in a set of three two-body systems only when all particles are well separated, which leads us to the conclusion that Ψ_{3C} is a correct asymptotic solution of the Schrödinger equation only in the region $\mathcal{L}$. To obtain asymptotic expressions in other asymptotic domains we require that

$$\mathcal{R}[\overline{Z}_{ij}(\mathbf{r}_{ij}, \mathbf{R}_{\alpha})] \quad \rightarrow \quad 0, \quad \forall\ \mathbf{r}_{ij}, \mathbf{R}_{\alpha} \in \mathcal{L}_{\alpha}, \tag{9.65}$$

$$\overline{Z}_{ij} \quad \text{is finite,} \quad \forall\ \mathbf{r}_{ij}, \mathbf{R}_{k} \in \mathcal{L}. \tag{9.66}$$

Leaving aside the question of whether or not we can satisfy in practice the above requirements imposed on $\overline{Z}_{ij}(\mathbf{r}_{ij}, \mathbf{R}_{\alpha})$, a serious problem arises in this concept. Once we have determined the functions $\overline{Z}_{ij}(\mathbf{r}_{ij}, \mathbf{R}_{\alpha})$ we have to insert them into the Schrödinger equation (9.12). How then can we solve the resulting differential equation? The key to resolve this question is provided by the observation that lead us to the introduction of the parabolic coordinates (9.15), namely that the interactions (governed by the total potential) can depend on body-fixed coordinates only. This restricts further the allowed functional dependence of $\overline{Z}_{ij}(\mathbf{r}_{ij}, \mathbf{R}_{\alpha})$ and call at the same time for the use of parabolic-relative coordinates (9.15) for the actual determination of $\overline{Z}_{ij}(\mathbf{r}_{ij}, \mathbf{R}_{\alpha})$.

9.5 Dynamical screening in few-body systems

To inspect how the position-dependent charges $\overline{Z}_{ij}(\mathbf{r}_{ij}, \mathbf{R}_{\alpha})$ will affect the structure of the Schrödinger equation let us recall our finding (9.21) (page 119) that the Schrödinger equation in the parabolic-relative coordinate system possesses an approximate (asymptotic) separability. Furthermore, the operator H_{par} (9.22) depends only *parametrically* on the internal coordinates (ξ_4, ξ_5, ξ_6). This means, given a set of 'parameters' (ξ_4, ξ_5, ξ_6) the eigenfunctions of H_{par} can be found exactly [cf. (9.45)]. This feature of H_{par} persists if the product charges Z_{ij} depend only on the internal coordinates (ξ_4, ξ_5, ξ_6), in which case the parametric dependence of H_{par} on (ξ_4, ξ_5, ξ_6) is maintained. In view of this situation, it is appropriate to make the ansatz

$$\bar{Z}_{ij} = \overline{Z}_{ij}(\xi_4, \xi_5, \xi_6). \tag{9.67}$$

With this functional dependence the regular exact eigenfunction of H_{par} with a zero eigen-value has the explicit form

$$
\overline{\Psi}_{\text{DS3C}}(\xi_1, \xi_2, \xi_3)|_{(\xi_4, \xi_5, \xi_6)} \;=\; {}_1F_1\left(i\beta_{23}(\xi_4, \xi_5, \xi_6), 1, -ik_{23}\,\xi_1\right)
$$
$$
{}_1F_1\left(i\beta_{13}(\xi_4, \xi_5, \xi_6), 1, -ik_{13}\,\xi_2\right)
$$
$$
{}_1F_1\left(i\beta_{12}(\xi_4, \xi_5, \xi_6), 1, -ik_{12}\,\xi_3\right). \tag{9.68}
$$

The functions β_{mn} play the role of position and momentum dependent Sommerfeld parameters and are given by

$$
\beta_{mn} = \frac{\overline{Z}_{mn}(\xi_4, \xi_5, \xi_6)\,\mu_{mn}}{k_{mn}}. \tag{9.69}
$$

The modifications of the Sommerfeld parameters β_{mn} can be interpreted as a dynamic screening of the interaction of the two particles m and n by the presence of the third one. Therefore, the wave function (9.68) has been termed the dynamically screened, three-body Coulomb wave function (DS3C) [77].

The structure of the wave function (9.68) illudes to a separation in two-body systems, one should note, however, that each of the hypergeometric functions occurring in (9.68) is gener-ally dependent on all coordinates and looses therefore the two-body character. From a physical point of view Eq. (9.69) states that the strength of the interaction between two particles i and j is no longer determined by their (constant) product charges $Z_i Z_j$, as is the case in two-body scattering or in Ψ_{3C} (9.46). It is rather described by a dynamical product-charge functions $\overline{Z}_{ij}(\xi_4, \xi_5, \xi_6)$ depending on the shape of the triangle formed by the three particles (regardless of its orientation in space). In searching for the explicit functional dependence we require a correct asymptotic behaviour everywhere and we seek a functional form of $\overline{Z}_{ij}(\xi_4, \xi_5, \xi_6)$ that minimizes the part not diagonalized by (9.28). To preserve the scaling properties of the Schrödinger Eq. (9.12) when introducing the functions $\overline{Z}_{ij}(\xi_4, \xi_5, \xi_6)$, we split the total po-tential in three terms that have the structure

$$
\overline{V}_{ij} = \overline{Z}_{ij}/r_{ij}, \quad j > i
$$

Each of these potentials is assumed to be the most general linear superposition of the three physical two-body potentials $V_{ij} := Z_i Z_j / r_{ij}$, with coefficients $\bar{a}_{ij}$ that depend on the internal coordinates. This is achieved by means of the linear expansion

$$
\begin{pmatrix} \overline{V}_{23} \\ \overline{V}_{13} \\ \overline{V}_{12} \end{pmatrix} = \overline{A} \begin{pmatrix} V_{23} \\ V_{13} \\ V_{12} \end{pmatrix} ,
\tag{9.70}
$$

where $\overline{A}(\xi_4, \xi_5, \xi_6)$ is a 3×3 matrix with elements $\overline{a}_{ij} = \overline{a}_{ij}(\xi_{4\cdots6})$:

$$
\overline{A} = \begin{pmatrix} \overline{a}_{11} & \overline{a}_{12} & \overline{a}_{13} \\ \overline{a}_{21} & \overline{a}_{22} & \overline{a}_{23} \\ \overline{a}_{31} & \overline{a}_{32} & \overline{a}_{33} \end{pmatrix} .
\tag{9.71}
$$

Since the relation $\overline{Z}_{ij} = \overline{V}_{ij}\, r_{ij}$ applies, the functions $\overline{Z}_{ij}(\xi_4, \xi_5, \xi_6)$ can be deduced from (9.70) once the matrix $\overline{A}$ is identified.

The requirement of the invariance of the total potential under the transformation (9.70) imposes the condition

$$
\sum_{i=1}^{3} \overline{a}_{ij} = 1 ; \quad j = 1, 2, 3 .
\tag{9.72}
$$

Any matrix $\overline{A}$ whose elements satisfy the three equations (9.72) leaves the total potential and hence the Schrödinger equation and its exact solutions invariant. To uniquely identify the coefficients $\overline{a}_{ij}$ six further determining equations are needed, in addition to (9.72). These conditions are chosen as to achieve certain desired properties of the resulting wave function. The simplest choice for the matrix $\overline{A}$, which is compatible with (9.72), is $\overline{A} = 1$. The resulting wave function Ψ_{DS3C} reduces in this case to Ψ_{3C} (9.28). From Eq. (9.70) we conclude that $\overline{A} = 1$ means that the coupling between any of the three two-body subsystems is disregarded. Hence, the wave function Ψ_{3C} contains only two-particle interactions.

9.5.1   Two electrons in the field of a positive ion

The dynamic of the three-body system depends decisively on the mass and on the charge state of the particles. Therefore, it is to be expected that the coupling matrix $\overline{A}$ is not universal but specific to the three-body problem under study. This feature, while understandable and unavoidable due to physical arguments, presents from a practical point of view a disadvantage of the DS3C approach, for in each specific three-body case, the matrix $\overline{A}$ has to be determined separately.

Presently, the properties of $\overline{A}$ have been studied for the case of two electrons moving in the field of a massive positive ion which has a charge Z. It is customary in the field of electron-atom collisions to designate the relative coordinates of the electrons with respect to the nucleus by $\mathbf{r}_a$ and $\mathbf{r}_b$. In our notation this corresponds to choosing m_3 as the mass of the residual ion. The coordinates $\mathbf{r}_{13} \equiv \xi_2$, $\mathbf{r}_{23} \equiv \xi_1$ and $\mathbf{r}_{12} \equiv \xi_3$ become then $\mathbf{r}_b$, $\mathbf{r}_a$ and $\mathbf{r}_{ba} = \mathbf{r}_b - \mathbf{r}_a$, respectively. To connect with traditional nomenclature we further rename correspondingly the conjugate momenta, product charges and Sommerfeld parameters.

The invariance condition of the Schrödinger equation under the introduction of the product-charge functions $\bar{z}_j(\xi_{4\dots 6})$ reads (here we use the notation $\bar{z}_b \equiv \bar{Z}_{13}$, $\bar{z}_a \equiv \bar{Z}_{23}$, $\bar{z}_{ba} \equiv \bar{Z}_{12}$)

$$\sum_j \frac{\bar{z}_j(\xi_{4\dots 6})}{r_j} \equiv \frac{-Z}{r_a} + \frac{-Z}{r_b} + \frac{1}{r_{ab}}, \quad j \in \{a, b, ab\}. \tag{9.73}$$

The wave functions containing $\bar{z}_j$ must be compatible with the three-body asymptotic boundary conditions. These are specified by the shape and by the size of the triangle formed by the three particles (two electrons and the ion). This means, the derived wave function must be, to a leading order, an asymptotic solution of the three-body Schrödinger equation when the aforementioned triangle tends to a line (two particles are close to each other and far away from the third particle) or in the case where, for an arbitrary shape, the size of this triangle becomes infinite. The latter limit implies that all interparticle coordinates $r_{a,b,ab}$ must grow with the same order, otherwise we eventually fall back to the limit of the three-particle triangle being reduced to a line [77], as described above. In addition we require the Wannier threshold law for double electron escape (given by Eq. (6.30)) to be reproduced when the derived wave functions are used for the evaluation of the matrix elements. The conditions specified above are sufficient to determine $\bar{z}_j$ and thus the wave function Ψ_{DS3C} (9.68).

The applicability of the wave function Ψ_{DS3C} to scattering reactions is hampered by the involved functional dependence leading to complications in the numerical determination of the normalization and of the scattering matrix elements.

The normalization of the wave function Ψ_{3C} derives directly from the normalization of the two-body Coulomb wave function. This is a consequence of the internal separability of the operator H_{par}. This normalization argument holds true if the product charges $\bar{z}_j$ were position independent or if they depend only parametrically on the internal coordinates. In the latter case one has still to tackle the problem of normalizing continuum functions of complicated functional form. This normalization problem is resolved upon making the approximation

$$\frac{r_i}{r_j} \propto \frac{v_i}{v_j}, \tag{9.74}$$

in which case the position dependence of $\bar{z}_j(r_a, r_b, r_{ab})$ is converted into velocity dependence. This assumption suffices for the conversion because the coordinate dependence of the product charges occurs, due to dimensionality arguments, as a ratio of positions. It should be emphasized that the approximation (9.74) is not a classical one, i. e. it is not assumed that the motion of the particles proceeds along classical trajectories. In fact the whole problem is still treated full quantum mechanically (by the wave function Ψ_{DS3C}). Eq. (9.74) merely means that the total potential is exactly diagonalized in the phase space where Eq. (9.74) is satisfied, as readily deduced from Eq. (9.73).

Eq. (9.74) renders possible the normalization of Ψ_{DS3C} since in this case we obtain $\bar{z}_j = \bar{z}_j(k_a, k_b, k_{ab})$ and hence we deduce for the normalization $N_{\Psi_{DS3C}}$ the expression

$$N_{\Psi_{DS3C}} = \prod_j N_j, \ j \in \{a, b, ba\},$$

$$N_j = \exp[-\beta_j(k_a, k_b, k_{ba})\pi/2]\, \Gamma[1 - i\beta_j(k_a, k_b, k_{ba})]. \tag{9.75}$$

It has been shown [89] that the velocity-dependent product charges for $Z = 1$ possess the form

$$\bar{z}_{ba}(\mathbf{v}_a, \mathbf{v}_b) = \left[1 - (f\, g)^2\, a^{b_1}\right] a^{b_2}, \tag{9.76}$$

$$\bar{z}_a(\mathbf{v}_a, \mathbf{v}_b) = -1 + (1 - \bar{z}_{ba}) \frac{v_a^{1+a}}{(v_a^a + v_b^a)v_{ab}},$$

$$\bar{z}_b(\mathbf{v}_a, \mathbf{v}_b) = -1 + (1 - \bar{z}_{ba}) \frac{v_b^{1+a}}{(v_a^a + v_b^a)v_{ab}}. \tag{9.77}$$

The functions occurring in Eqs. (9.76, 9.77) are defined as ($\mathbf{v}_a$, $\mathbf{v}_b$ are the electrons' velocities and $\mathbf{v}_{ab} = \mathbf{v}_a - \mathbf{v}_b$)

$$f := \frac{3 + \cos^2 4\alpha}{4}, \qquad \tan\alpha = \frac{v_a}{v_b}, \tag{9.78}$$

$$g := \frac{v_{ab}}{v_a + v_b}, \tag{9.79}$$

$$b_1 := \frac{2v_a v_b \cos(\theta_{ab}/2)}{v_a^2 + v_b^2}, \tag{9.80}$$

$$b_2 := g^2(-0.5 + \bar{\mu}), \tag{9.81}$$

$$a := \frac{E}{E + 0.5}, \tag{9.82}$$

where E is the total energy of the two continuum electrons, measured in atomic units. $\bar{\mu} = \mu/2 - 1/4$ is the Wannier index (cf. Eq. (6.30)), the value of $\bar{\mu}$ depends on the residual ion charge value, the numerical value of $\bar{\mu}$ for a unity charge of the residual ion is $\bar{\mu} = 1.127$). The interelectronic relative angle θ_{ab} is given by $\theta_{ab} = \arccos(\hat{v}_a \cdot \hat{v}_b)$. With increasing excess energies $(E \gg 1)$ one verifies that $a \to 1$ [Eq. (9.82)] and all dependencies of the product-charge functions (9.76-9.77) which are due to incorporating the Wannier threshold law, become irrelevant. The charges (9.76-9.77) reduce then to those given in Ref. [77] with Eq. (9.74) being applied. From the functional forms of the charges (9.76-9.77) it is clear that when two particles approach each other (in velocity space) they experience their full two-body Coulomb interactions, whereas the third one 'sees' a net charge equal to the sum of the charges of the two particles that are close to each other.

9.5.2   Dynamical screening via complex effective wave vectors

As documented by Eqs. (9.60, 9.61) three-body effects can be captured by theory upon the introduction of effective quantities, such as product charges and wave vectors. In the preceding section we discussed how the dynamical product-charge concept can be exploited to derive correlated three-body wave functions. Here we discuss the approach of local wave vectors.

The idea of dynamical effective wave vectors have been utilized in two cases, 1.) for the incorporation of some of the short-range three-body interactions, not included in H_{par} [83], and 2.) for accomplishing correct asymptotic behaviour in the entire asymptotic region [75]. In the latter case, and specializing to a system of two electrons in the field of an ion, local wave vectors are derived to be [75]

$$\bar{\mathbf{k}}_a = \mathbf{k}_a + \bar{f}(\alpha_{ab}, \mathbf{k}_{ba}, \mathbf{r}_b), \tag{9.83}$$

$$\bar{\mathbf{k}}_b = \mathbf{k}_b + \bar{f}(\alpha_{ab}, \mathbf{k}_{ab}, \mathbf{r}_a), \tag{9.84}$$

$$\bar{\mathbf{k}}_{ab} = \mathbf{k}_{ab} + \bar{f}(\alpha_a, \mathbf{k}_a, \mathbf{r}_{CM}) - \bar{f}(\alpha_b, \mathbf{k}_b, \mathbf{r}_{CM}), \tag{9.85}$$

where $\mathbf{r}_{CM} = (\mathbf{r}_a + \mathbf{r}_b)/2$ is the electron-pair center-of-mass coordinate, α_j, $(j = a, b, ab)$ are the Sommerfeld parameters and the distortion function $\bar{f}(\alpha, \mathbf{k}, \mathbf{r})$ is defined as

$$\bar{f}(\alpha, \mathbf{k}, \mathbf{r}) = \frac{r}{R} \left[\frac{{}_1F_1(1 + i\alpha, 2, -i(kr + \mathbf{k} \cdot \mathbf{r}))}{-i\,{}_1F_1(i\alpha, 1, -i(kr + \mathbf{k} \cdot \mathbf{r}))} \right] \left(\hat{\mathbf{k}} + \hat{\mathbf{r}} \right). \tag{9.86}$$

The distance $R = r_a + r_b + r_{ab}$ quantifies the "size of the triangle formed by the three particles. Using the position dependent (complex) momenta (9.83, 9.85) one can prove [75] that the wave function

$$\Psi_{\mathcal{M}}(\mathbf{r}_a, \mathbf{r}_b, \mathbf{r}_{ab}) \ = \ \bar{N} \exp(i\,\mathbf{k}_a \cdot \mathbf{r}_b + i\,\mathbf{k}_b \cdot \mathbf{r}_b)$$

$$_1F_1\left(i\alpha'_a,\ 1,\ -i\left[k'_a\,r_a + \mathbf{k}'_a \cdot \mathbf{r}_a\right]\right)$$

$$_1F_1\left(i\alpha'_a,\ 1,\ -i\left[k'_b\,r_b + \mathbf{k}'_b \cdot \mathbf{r}_b\right]\right)$$

$$_1F_1\left(i\alpha'_{ab},\ 1,\ -i\left[k'_{ab}\,r_{ab} + \mathbf{k}'_{ab} \cdot \mathbf{r}_{ab}\right]\right), \tag{9.87}$$

is (to leading order) correct in the entire asymptotic domain $\mathcal{M}$. The modified Sommerfeld parameters α'_j are obtained from α_j by replacing $\mathbf{k}_j$ by $\bar{\mathbf{k}}_j$. In Eq. (9.87) the constant $\bar{N}$ derives from the normalization of the wave function. Unfortunately, it has not been yet possible to evaluate this factor. When the manifold $\mathcal{M}$ is approached the complex wave vectors (9.85) turn real. This is deduced from the following behaviour of $\bar{f}$ in the domain $\mathcal{M}$

$$\bar{f}(\alpha, \mathbf{k}, \mathbf{r})\Big|_{\substack{r_j \in \mathcal{M} \\ j=a,b,ab}} \ \longrightarrow \ \frac{\hat{\mathbf{k}} + \hat{\mathbf{r}}}{kR(1 + \hat{\mathbf{k}} \cdot \hat{\mathbf{r}})}.$$

The difference between the DS3C approach and the wave function $\Psi_{\mathcal{M}}$ is the following. While in both cases the total potential is treated to all orders and parts of the kinetic energy are neglected, in the DS3C theory one tries to remedy this shortcoming by accounting effectively for the short-range dynamics. In contrast when constructing the wave function $\Psi_{\mathcal{M}}$ only asymptotic arguments are employed. The more important difference from a practical point of view is that the DS3C wave function can be normalized, whereas the wave function Eq. (9.87) has not yet been normalized. The importance of the normalization factor is most highlighted when considering the threshold behaviour for the double electron escape and contrasting it with the Wannier threshold law (6.30).

9.5.3   Threshold behaviour

In chapter 6 we showed that at very low energies the cross section for two-electron double escape exhibits a universal behaviour, the Wannier threshold law, as given by Eq. (6.24). Thus, it is of interest to establish the low-energy properties of the wave function $\Psi_{3C}(\mathbf{r}_a, \mathbf{r}_b)$ and $\Psi_{DS3C}(\mathbf{r}_a, \mathbf{r}_b)$ and to contrast the findings with the Wannier predictions. Such an analysis for the wave function (9.87) is not possible, for this wave function is not normalized. As can be anticipated from the derivation of the Wannier threshold law (see in particular Eq.(6.24)), the normalization, which encompasses the density of states available for the two electrons, is decisive for the value of the cross section.

From a conceptual point of view the key difference between the methods we employed to derive the wave functions $\Psi_{3C}(\mathbf{r}_a, \mathbf{r}_b)$ and $\Psi_{DS3C}(\mathbf{r}_a, \mathbf{r}_b)$ and those utilized to obtain the Wannier threshold law, is that in the Wannier case the kinetic energy is treated exactly, while the potential energy is expanded in a Taylor series around the saddle point and only the leading-order terms of this expansion are included in the theory. In contrast, in the 3C or DS3C wave function treatment one diagonalizes the potential exactly and neglects (cross) terms of the kinetic energy, i.e. the short-range dynamic is not properly treated. This shortcoming is reflected in a spurious threshold behaviour of the Ψ_{3C}.

The Ψ_{3C} leads to cross sections for the two-particle double escape that decreases exponentially with decreasing small excess energy [89, 98]. This is at variance with the Wannier theory and with experimental findings (cf. e. g. Refs. [250, 91, 138, 134] and further references therein). The physical reason for this behaviour is the following. Since the Ψ_{3C} regards the three-body system as three non-interacting two-body systems the three-particle density of states (DOS) generated by Ψ_{3C} is directly proportional to the density of states of isolated two-electron systems. The latter DOS decreases exponentially when the two electrons are close to each other in velocity space.

To formulate this qualitative arguments more precisely and to show how the coupling introduced by $\bar{z}_j$, $j = a, b, ba$ removes this deficiency we consider the cross section $\sigma^{2+}(E)$ for double escape, given by the formula

$$\sigma^{2+}(E) \propto \int |T|^2 \delta(E - E_i) d^3 k_a d^3 k_b, \tag{9.88}$$

where E_i is the total energy in the initial channel. The transition-matrix element occurring in Eq. (9.88) has the form

$$T = \langle \Psi_{\mathbf{k}_a, \mathbf{k}_b} | W_i | \Phi_i \rangle. \tag{9.89}$$

The transition operator and the initial state of the three-body system are denoted by W_i and $|\Phi_i\rangle$, respectively. As discussed in the context of the Wannier theory, near threshold the functions Φ_i and W_i are in general hardly dependent on the excess energy $E = E_a + E_b$ (both energies E_a and E_b of the electrons are positive and small). Therefore, Eq. (9.88) can be written in the form of an expectation value as follows

$$\sigma^{2+}(E) \propto \langle \Phi_i | W_i \, (DOS_v) \, W_i | \Phi_i \rangle. \tag{9.90}$$

In the context of the Green's function theory, which the subject of chapter 11, the quantity DOS_v is called the local density of states per unit volume and is given as

$$DOS_v(\epsilon, \mathbf{r}_a, \mathbf{r}_b) = \int d^3 k_a d^3 k_b \Psi_{\mathbf{k}_a, \mathbf{k}_b}(\mathbf{r}_a, \mathbf{r}_b) \delta(E_a + E_b - \epsilon) \Psi^*_{\mathbf{k}_a, \mathbf{k}_b}(\mathbf{r}_a, \mathbf{r}_b). \tag{9.91}$$

The density of states (DOS) is obtained from (9.91) as

$$DOS(\epsilon) = \int d\mathbf{r}_a d\mathbf{r}_b DOS_v(\epsilon, \mathbf{r}_a, \mathbf{r}_b). \tag{9.92}$$

Equations (9.88-9.92) make clear that the energy dependence of $\sigma^{2+}(E)$ is entirely determined by the energy behaviour of the two-particle density of states (Eq. (9.92)) that can be occupied by the two escaping electrons [5].

On the other hand, according to Eq. (9.91) the E dependence of the DOS is determined by the normalization factor $|N_\Psi|^2$ of the wave function Ψ, provided the radial part of the wave function Ψ_r is well behaved near threshold. To investigate how the energy dependence of $|N_\Psi|^2$ is reflected in the cross section behaviour we introduce the hyperspherical momenta

$$K := (k_a^2 + k_b^2)/2 = E, \ \tan \beta = \frac{k_a}{k_b}, \ \text{and} \ \cos \theta_k = \hat{\mathbf{k}}_a \cdot \hat{\mathbf{k}}_b. \tag{9.93}$$

As far as the DS3C wave function is concerned we note that the dynamical product charges are limited to the intervals, $\bar{z}_a, \bar{z}_b \in [-Z, 0]$; $\bar{z}_{ba} \in [0, 1]$, i.e. a two-body interaction can be screened by the presence of a third charged particle, but it does not change its sign. For small excess energies $E \to 0$, the wave function $\bar{\Psi}_{DS3C}(\mathbf{r}_a, \mathbf{r}_b)$ [see Eq. (9.68)] can be expanded in terms of Bessel functions [139, 99]. The leading order term with respect to the excess energy reads

$$\lim_{E \to 0} \Psi_{DS3C}(\xi_{1 \cdots 6}) = (2\pi)^{-3} N_{DS3C} J_0(2\sqrt{-\bar{z}_a \xi_1}) J_0(2\sqrt{-\bar{z}_b \xi_2}) I_0(\sqrt{2\bar{z}_{ab} \xi_3}), \tag{9.94}$$

where $J_0(x), I_0(x)$ are Bessel and modified Bessel functions, respectively. A similar equation applies to Ψ_{3C} upon the replacement $\bar{z}_a = -Z = \bar{z}_b, \bar{z}_{ab} = 1$. Expressing Eq. (9.88) in the

---

[5]Note that Eq. (9.90) can be written in the form

$$\sigma^{2+}(E) \propto \int d\mathbf{r}_a \, d\mathbf{r}_b \, \langle \Phi_i | W_i | \mathbf{r}_a \mathbf{r}_b \rangle DOS_{nl} \langle \mathbf{r}'_a \mathbf{r}'_b | W_i | \Phi_i \rangle.$$

The trace of the local part of the non-local density of state DOS_{nl} yields the DOS.

hyperspherical momentum coordinates (9.93), one performs the integration over the variable K. Furthermore, due to the overall rotational invariance of the system Eq. (9.88) reduces to

$$\sigma^{+2}(E) \propto E^2 \int \sin^2 2\beta \ |N_\Psi|^2 |T_r|^2 \ d\beta \ d(\cos\theta_k) \ d\varphi_{ab} \ d\varphi_a. \qquad (9.95)$$

N_Ψ is the normalization of the function Ψ, whereas φ_{ab} and φ_a are the azimuthal angles of $\mathbf{k}_{ab}$ and $\mathbf{k}_b$ with respect to an appropriately chosen axis. The quantity T_r is the transition matrix element T [Eq. (9.89)] with the normalization N_Ψ of the final-state wave function being factored out. As evident from Eq. (9.95) the excess-energy dependence of $\sigma(E)$ is directly related to the excess-energy behaviour of $|N_\Psi|^2$, if for $E \ll 1$ the variation of T_r as function of E is insignificantly slow. On the other hand, the function N_Ψ is given, as a matter of definition, by the integral behaviour of the radial part of the wave function. This interrelation is best demonstrated by the threshold behaviour of the cross section σ^{+2} that is obtained by various wave functions having the threshold expansion (9.94). E.g. the Bessel function $J_0(x)$ has an oscillatory bound asymptotic behaviour, whereas the modified Bessel function $I_0(x)$, corresponding to the electron-electron interaction, is unbound for large arguments x. Therefore, the normalization $|N_{ab}|^2$ of the electron-electron Coulomb wave decreases exponentially with vanishing excess energy. Specifically, one derives for the normalization factors the following expressions

$$|N_j|^2 = 2\pi\beta_j \left(e^{2\pi\beta_j} - 1\right)^{-1}, \ j = a, b, ab. \qquad (9.96)$$

For $E \ll 1$ the factors N_j behaves as

$$|N_b|^2 \ = \ -2\pi\beta_b = \frac{-2\pi\bar{z}_b}{\sqrt{2E}\cos\beta}, \quad \forall \bar{z}_b < 0, \qquad (9.97)$$

$$|N_a|^2 \ = \ -2\pi\beta_a = \frac{-2\pi\bar{z}_a}{\sqrt{2E}\sin\beta}, \quad \forall \bar{z}_a < 0, \qquad (9.98)$$

$$|N_{ab}|^2 \ = \ \frac{2\pi\bar{z}_{ab}}{\sqrt{2E}f(\theta_k,\beta)} \exp\left(-2\pi\frac{\bar{z}_{ab}}{\sqrt{2E}f(\theta_k,\beta)}\right), \quad \forall \bar{z}_{ab} > 0, \qquad (9.99)$$

where the function f is defined as

$$f(\theta_k,\beta) = \sqrt{1 - \sin 2\beta \cos\theta_k} \ . \qquad (9.100)$$

For $\bar{z}_{ab} > 0$ the behaviour (9.99) of $|N_{ab}|$ results in an exponential decline of $\sigma^{2+}(E)$ when the excess energy is lowered (note that to obtain Ψ_{3C} we employ $\bar{z}_{ab} = 1$). Cross sections calculated with Ψ_{DS3C} do not exhibit this spurious behaviour as the term $I_0(\sqrt{2\bar{z}_{ab}\xi_3})$ contribute to the wave function only in a limited region of the Hilbert space (because $\bar{z}_{ab} \to 0$ for

$\xi_3 \rightarrow \infty$). From this argument however, it is not clear whether the Wannier threshold law is reproduced or not by Ψ_{DS3C}.

Depending on the employed dynamical mode the cross section $\sigma^{2+}(E)$ behaves as follows. In a free particle model, i.e. when the motion of the two electrons is described by plane waves we have $N_{PW} = 2\pi^{-3} = constant$) and one deduces from Eq. (9.95) the threshold law $\sigma(E) \propto E^2$. In an independent Coulomb particle model, i.e. if $\bar{z}_{ab} \equiv 0$, $\bar{z}_b = -Z = \bar{z}_a$, we obtain $N_{ab} = 1$ and thus $\sigma(E) \propto E$. This is deduced upon substitution of Eqs. (9.97, 9.98) into Eq. (9.95). In the first Born approximation, i.e. $\bar{z}_b \equiv 0 \equiv \bar{z}_{ab}$, $\bar{z}_a = -Z$, Eqs. (9.95, 9.98) yield $\sigma(E) \propto E^{1.5}$.

To address the question of the threshold behaviour of the cross section calculated within the DS3C model we insert Eqs. (9.97-9.99) into Eq. (9.95) and obtain

$$\sigma(E) \propto E \int \frac{\bar{z}_b \bar{z}_a \bar{z}_{ab}}{\sqrt{2E}} \exp\left(-2\pi \frac{\bar{z}_{ab}}{\sqrt{2E}} f^{-1}(\theta_k, \beta)\right)$$
$$\left[\sin 2\beta \ f^{-1}(\theta_k, \beta) \ |T_r(\theta_k, \beta, \varphi_b, \varphi_a)|^2\right] d\beta d\varphi_b d\varphi_a \ d(\cos\theta_k). \quad (9.101)$$

Eq. (9.101) is valid for $\bar{Z}_{ij} \neq 0$. Due to the structure of Eq. (9.101) it is convenient to write the function $\bar{z}_{ab}$ in the form

$$\bar{z}_{ab} = (1 - \eta \, a^{b_1}) a^{b_2}, \quad (9.102)$$

where $a = E/E_i$ is a dimensionless parameter. Within the DS3C theory the functions b_1 and b_2 are determined to be

$$b_2 \propto n \qquad \text{where} \qquad n = 0.5 + \frac{\mu}{2} - \frac{1}{4} - 1; \quad b_1 = \frac{2v_a v_b \cos(\theta_{ab}/2)}{v_a^2 + v_b^2}. \quad (9.103)$$

One verifies that with this functional dependence of $\bar{z}_{ab}$ the Wannier threshold law is reproduced and the integrand in Eq. (9.101) is a slowly varying function of E (note that $T_{r,DS3C}(\mathbf{k}_b, \mathbf{k}_a)$ varies insignificantly slow with E), i. e.

$$\sigma^{2+}(E) = E^{\mu/2 - 1/4} \int T_{r,DS3C}(\mathbf{k}_b, \mathbf{k}_a) \ d\varphi_b \ d\varphi_a \ d(\cos\theta_k) \ d\beta. \quad (9.104)$$

9.5.4   Kato cusp conditions

In the preceding sections we have seen that the properties of the three-body wave functions are well defined in the asymptotic region. For short distances, the structure of the wave function can as well be deduced. For example, if only two particles are close to each other, the wave function satisfies certain constraints known as the Kato cusp conditions [140, 141] (this is valid in case $\Psi(\mathbf{r}_{ij}, \mathbf{R}_k)$ does not vanish at the two-body coalescence points). The Kato cusp conditions have been introduced in section 8.1.2 (page 103). Here we inspect the behaviour of the approximate three-body functions $\Psi_{3C/DS3C}(\mathbf{r}_{ij}, \mathbf{R}_k)$ around the two-body coalescence points, i.e. the question is whether the relation

$$\left(\frac{\partial \tilde{\Psi}(\mathbf{r}_{ij}, \mathbf{R}_k)}{\partial r_{ij}}\right)_{r_{ij}=0} = Z_{ij}\,\mu_{ji}\,\Psi(r_{ij}=0, \mathbf{R}_k)\,, \quad \forall\ (\mathbf{r}_{ij}, \mathbf{R}_k) \tag{9.105}$$

is satisfied by $\Psi_{3C/DS3C}$, where $\tilde{\Psi}(\mathbf{r}_{ij}, \mathbf{R}_k)$ is the wave function averaged over a sphere of small radius $r_\epsilon \ll 1$ around the singularity $r_{ij} = 0$. To investigate the behaviour of $\tilde{\Psi}_{DS3C}(\mathbf{r}_a, \mathbf{r}_b)$ at, e.g. the collision point ($r_b = 0, r_a/r_b \to \infty$), we linearize $\Psi_{DS3C}(\mathbf{r}_a, \mathbf{r}_b)$ around $r_b = 0$ and find the leading order terms to be

$$\tilde{\Psi}_{DS3C}(\mathbf{r}_b, \mathbf{r}_a) = \mathcal{N}\exp(i\mathbf{k}_a \cdot \mathbf{r}_a)\ _1F_1\left(i\beta_a,\ 1,\ -i\left[k_a\,r_a + \mathbf{k}_a \cdot \mathbf{r}_a\right]\right)$$
$$\times\ _1F_1\left(i\beta_{ba},\ 1,\ -i\left[k_{ba}\,r_{ba} + \mathbf{k}_{ba} \cdot \mathbf{r}_{ba}\right]\right) D(\mathbf{r}_b)\,, \tag{9.106}$$

where $\mathcal{N}$ is a normalization constant and the function $D(\mathbf{r}_b)$ has the form

$$D(\mathbf{r}_b) = \frac{2\pi}{4\pi r_\epsilon^2}\int_{-1}^{1} r_\epsilon^2\, d\cos\theta\left[1 + ik_b\cos\theta + \alpha_b k_b\, r_b(1 + \cos\theta)\right],$$
$$= 1 + \alpha_b\, k_b\, r_b\,. \tag{9.107}$$

To derive Eq. (9.107) one chooses the z axes to be along the direction $\mathbf{k}_b$ and defines $\cos\theta = \hat{\mathbf{k}}_b \cdot \hat{\mathbf{r}}_b$. In the limit ($r_b \to 0$; $r_a/r_b \to \infty$) the effective Sommerfeld parameter β_b tends to the conventional parameter α_b. Therefore, we obtain

$$\left(\frac{\partial \tilde{\Psi}_{DS3C}(\mathbf{r}_b, \mathbf{r}_a)}{\partial r_b}\right)_{r_b=0}$$
$$= z_b\mathcal{N}\exp(i\mathbf{k}_a \cdot \mathbf{r}_a)\ _1F_1\left(i\beta_a,\ 1,\ -i\left[k_a\,r_a + \mathbf{k}_a \cdot \mathbf{r}_a\right]\right)$$
$$\times\ _1F_1\left(i\beta_{ba},\ 1,\ -i\left[k_{ba}\,r_{ba} + \mathbf{k}_{ba} \cdot \mathbf{r}_{ba}\right]\right), \tag{9.108}$$
$$= z_b\,\Psi_{DS3C}(r_b = 0, \mathbf{r}_a)\,.$$

It is straightforward to repeat the above steps and show that the Kato cusp conditions at ($r_a = 0$, $r_b/r_a \to \infty$) and ($r_{ba} = 0$, $r_b \gg 1$, $r_a \gg 1$) are fulfilled. As far as the wave function $\Psi_{3C}(\mathbf{r}_b, \mathbf{r}_a)$ is concerned, one can use the procedure employed for $\Psi_{DS3C}(\mathbf{r}_b, \mathbf{r}_a)$ and show that $\Psi_{3C}(\mathbf{r}_b, \mathbf{r}_a)$ satisfies the Kato cusp conditions and hence possesses a regular behaviour at all two-body collision points. This finding is not surprising since the 3C model treats all two-body interactions to all orders [73, 82, 142].

9.5.5   Compatibility with the Fock expansion

Both functions $\Psi_{DS3C}(\mathbf{r}_b, \mathbf{r}_a)$ and $\Psi_{3C}(\mathbf{r}_b, \mathbf{r}_a)$ are not compatible with the Fock expansion (8.1), i.e. these functions do not exhibit correct behaviour at the three-body collision point. The conclusion is evident from the fact that at the three-body coalescence point all distances are small. Hence, the confluent hypergeometric functions occurring in $\Psi_{DS3C}(\mathbf{r}_b, \mathbf{r}_a)$ and $\Psi_{3C}(\mathbf{r}_b, \mathbf{r}_a)$ can be expanded in a power series (2.51) in the variable r_a, r_b and r_{ab}. The resulting expansions of the wave functions $\Psi_{DS3C}(\mathbf{r}_b, \mathbf{r}_a)$ and $\Psi_{3C}(\mathbf{r}_b, \mathbf{r}_a)$ do not contain any logarithmic terms and hence are at variance with the Fock condition (8.1). The reason why the wave function $\Psi_{DS3C}(\mathbf{r}_b, \mathbf{r}_a)$ and $\Psi_{3C}(\mathbf{r}_b, \mathbf{r}_a)$ do not satisfy the Fock expansion is that at the triple collision point the collective behaviour is dominant rather then the successive two-body collisions. To incorporate collective features of the system it is more convenient to formulate the Schrödinger equation in an appropriate coordinate system, such as the parabolic hyperspherical coordinates.

9.6   Parabolic-hyperspherical approach

In the preceding section the formulation of the three-body problem within the parabolic-relative coordinate approach uncovered the approximate separable structure of the Schrödinger equation (9.12). In particular, we were able to derive exact eigensolutions $\overline{\Psi}_{par}(\xi_1, \xi_2.\xi_3)$ (cf. Eq. (9.28)) of the operator (9.22) which is differential in the coordinates $\xi_{1,2,3}$. The remaining coordinates $\xi_{4,5,6}$ enter this solution parametrically, i. e.

$$\overline{\Psi}(\xi_1, \ldots, \xi_6) \approx \overline{\Psi}_{par}(\xi_1, \xi_2, \xi_3)|_{(\xi_4, \xi_5, \xi_6)}. \tag{9.109}$$

This interrelation between the six degrees of freedom resembles similar situations encountered in adiabatic treatments where some degrees of freedom are varied parametrically, or even 'frozen' (cf. section 4.1 on the Born-Oppenheimer and the adiabatic approximation). Thus, it seems appropriate to treat the coordinates $\xi_{4,5,6}$ adiabatically. The problem which arises when trying to realize this suggestion is that the potential as function of $\xi_{4,5,6}$ becomes singular at certain points, these are the two-body collision points. This rules out a treatment of these coordinates as a slow varying variable in the whole Hilbert space. On the other hand we have shown explicitly upon analyzing Eq. (7.14) (page 95) that when the problem is formulated in hyperspherical coordinates, the motion along the hyperradius ρ is smooth and does not exhibit any a singular behaviour. Therefore, for an adiabatic treatment one should switch to hyperspherical coordinates. On the other hand we have seen that it is of great advantage to include in the definition of the coordinate system the generic coordinates for Coulomb scattering, namely the parabolic coordinates $\xi_{1,2,3}$. For this reason we consider the three-body problem formulated in the parabolic-hyperspherical coordinate system (9.17).

9.6.1   Parabolic-hyperspherical Schrödinger equation

Transforming the three-body Schrödinger equation (9.12) into the scaled Jacobi coordinates (9.18) and making the ansatz (9.19) leads to the expression

$$
\left[\Delta_{\overline{\mathbf{r}}_{ij}} + \Delta_{\overline{\mathbf{R}}_k} + 2i \left(\frac{1}{\sqrt{\mu_{ij}}}\, \mathbf{k}_{ij} \cdot \nabla_{\overline{\mathbf{r}}_{ij}} + \frac{1}{\sqrt{\mu_k}}\, \mathbf{K}_k \cdot \nabla_{\overline{\mathbf{R}}_k} \right) \right.
$$
$$
\left. - 2 \sum_{\substack{m,n \\ n>m}}^{3} \frac{q_{mn}}{\overline{r}_{mn}} \right] \overline{\Psi}(\overline{\mathbf{r}}_{ij}, \overline{\mathbf{R}}_k) = 0, \quad (9.110)
$$

where

$$
q_{mn} = \mu_{ij}^{1/2}\, Z_{ij}.
$$

The task is now to transform Eq. (9.110) into the parabolic hyperspherical coordinates (9.17). The Jacobi determinant for this transformation scales as $\rho^5 \sin^2 2\zeta / 4$ and does not vanish except for cases where a pair of the three vectors $\mathbf{k}_{ij}$ or $\overline{\mathbf{r}}_{ij}$ and $\overline{\mathbf{R}}_k$ are linearly dependent.

Similar to the N-particle hyperspherical formulation (Eq. (7.12)), in the curvilinear coordinates (9.17) the six-dimensional Laplacian $\Delta := \Delta_{\overline{\mathbf{r}}_{ij}} + \Delta_{\overline{\mathbf{R}}_k}$ is the sum of a hyperradial kinetic energy term and a centrifugal term, i.e.

$$
\Delta_{\overline{\mathbf{r}}_{ij}} + \Delta_{\overline{\mathbf{R}}_k} = \rho^{-5}\, \partial_\rho\, \rho^5\, \partial_\rho - \frac{\Lambda^2}{\rho^2}\,. \tag{9.111}
$$

The differential operator Λ^2 is a self-adjoint scalar operator defined in the Hilbert space $L_2(\omega, d\omega)$, on the domain $\omega = [0,2] \times [0,2] \times [0,2] \times [0,\pi/2] \times [-1,1]$, where $\omega \equiv \{\xi_1, \xi_2, \xi_3, \zeta, \gamma\}$. Since this domain is compact the operator Λ^2 has a discrete spectrum and is associated with the grand angular momentum, i. e. the Casimir operator of the $O(6)$ group. The differential operator Λ^2 can be decomposed further as

$$\Lambda^2 = \Lambda_{\text{in}}^2 + \Lambda_{\text{ext}}^2 + \Lambda_{\text{mix}}^2 \, . \tag{9.112}$$

The operator Λ_{in}^2 is differential in the internal angles $\{\zeta, \gamma\}$ only, whereas Λ_{ext}^2 operates only in the subspace spanned by $\{\xi_i; \ i = 1, 2, 3\}$. The operator Λ_{mix}^2 plays the role of a rotational coupling term in the conventional hyperspherical approach. It contains the mixed derivatives resulting from off-diagonal elements of the metric tensor and couples internal to external motion. The explicit expressions for Λ_{in}^2 is

$$\Lambda_{\text{in}}^2 = -\frac{4}{\sin^2 2\zeta} \left[-2\gamma\, \partial_\gamma + (1-\gamma)\, \partial_\gamma^2 + \sin 2\zeta\, \cos 2\zeta\, \partial_\zeta + \frac{1}{4}\sin^2 2\zeta\, \partial_\zeta^2 \right],$$

$$\Lambda_{\text{in}}^2 = -\frac{4}{\sin^2 2\zeta} \left[\partial_{2\zeta} \sin^2 2\zeta\, \partial_{2\zeta} - \hat{\mathbf{L}}_\gamma^2 \right]. \tag{9.113}$$

For a given Jacobi coordinate set $(\overline{\mathbf{r}}_{ij}, \overline{\mathbf{R}}_k)$ the quantity $\hat{\mathbf{L}}_\gamma^2$ is the operator of the squared orbital angular momentum of the particle 'k' with respect to the centre-of-mass of the pair 'ij'. This is concluded by expressing Λ_{in}^2 in terms of the angle $\theta := \arccos \gamma$, which yields the following expression for $\hat{\mathbf{L}}_\gamma^2$

$$\hat{\mathbf{L}}_\gamma^2 = -\sin^{-1}\theta\, \partial_\theta\, \sin\theta\, \partial_\theta \, . \tag{9.114}$$

The differential operator Λ_{ext}^2 has the form

$$\Lambda_{\text{ext}}^2 := -\rho^2 \left\{ \frac{1}{r_{23}^2\, \mu_{23}} \left[2\partial_{\xi_1}\, \xi_1\, \partial_{\xi_1} - \partial_{\xi_1}\, \xi_1^2\, \partial_{\xi_1} \right] \right.$$
$$\frac{1}{r_{13}^2\, \mu_{13}} \left[2\partial_{\xi_2}\, \xi_2\, \partial_{\xi_2} - \partial_{\xi_2}\, \xi_2^2\, \partial_{\xi_2} \right]$$
$$\left. \frac{1}{r_{12}^2\, \mu_{12}} \left[2\partial_{\xi_3}\, \xi_1\, \partial_{\xi_3} - \partial_{\xi_3}\, \xi_3^2\, \partial_{\xi_3} \right] \right\}, \tag{9.115}$$

whereas the coupling term Λ_{mix}^2 has the structure

$$\Lambda_{\text{mix}}^2 := -\rho^2 \sum_{u \neq v} \left\{ (\nabla_{\overline{\mathbf{r}}_{ij}} u) \cdot (\nabla_{\overline{\mathbf{r}}_{ij}} v) + (\nabla_{\overline{\mathbf{R}}_k} u) \cdot (\nabla_{\overline{\mathbf{R}}_k} v) \right\} \partial_u \partial_v;$$
$$u, v \in \{\xi_1, \xi_2, \xi_3, \zeta, \gamma, \rho\} \, . \tag{9.116}$$

A key property of the operator Λ_{ext}^2 (9.115) is its *parametric* dependence on the internal coordinates, and in particular on the hyperradius ρ. This is readily seen by expressing the function r_{ij} that occurs in (9.115) in terms of (ζ, γ, ρ), i.e.

$$r_{13} = \rho \sin \zeta,$$

$$r_{23} = \rho \left[\cos^2 \zeta + (\frac{m_1}{m_1 + m_3})^2 \sin^2 \zeta + \gamma \sin 2\zeta \right]^{1/2},$$

$$r_{12} = \rho \left[\cos^2 \zeta + (\frac{m_3}{m_1 + m_3})^2 \sin^2 \zeta - \gamma \sin 2\zeta \right]^{1/2}. \tag{9.117}$$

The gradient terms occurring in the Schrödinger equation (9.110) have to be expressed in the parabolic-hyperspherical coordinates. Having done that one finds the following expressions

$$2i \left(\frac{1}{\sqrt{\mu_{ij}}} \, \mathbf{k}_{ij} \cdot \nabla_{\bar{\mathbf{r}}_{ij}} + \frac{1}{\sqrt{\mu_k}} \, \mathbf{K}_k \cdot \nabla_{\bar{\mathbf{R}}_k} \right) = D_{\text{ext}} + D_{\text{in}}, \tag{9.118}$$

where the function D_{ext} has the form

$$
\begin{aligned}
D_{\text{ext}} := \quad & 2i \left(\frac{k_{23}}{r_{23}\mu_{23}} (2\xi_1 - \xi_1^2) \partial_{\xi_1} \right. \\
& + \frac{k_{13}}{r_{13}\mu_{13}} (2\xi_2 - \xi_2^2) \partial_{\xi_2} \\
& \left. + \frac{k_{12}}{r_{12}\mu_{12}} (2\xi_3 - \xi_3^2) \partial_{\xi_3} \right).
\end{aligned}
\tag{9.119}
$$

The differential operator D_{in} depends only on the gradient of the internal coordinates

$$
\begin{aligned}
D_{\text{in}} \quad := \quad & 2i \left[\frac{1}{\sqrt{\mu_{ij}}} \, \mathbf{k}_{ij} \cdot \nabla_{\bar{\mathbf{r}}_{ij}} u + \frac{1}{\sqrt{\mu_k}} \, \mathbf{K}_k \cdot \nabla_{\bar{\mathbf{R}}_k} u \right] \partial_u, \\
& u \in \{\zeta, \gamma, \rho\} \, .
\end{aligned}
\tag{9.120}
$$

Using the above relations for the kinetic energy operators the Schrödinger equation (9.12) can now be written as the sum of internal (body-fixed) and external (laboratory-fixed) differential operators with an additional (rotational) mixing term

$$\left[H_{\text{in}} + H_{\text{ext}} - \frac{\Lambda_{\text{mix}}^2}{\rho^2} \right] \overline{\Psi}(\xi_1, \xi_2, \xi_3, \zeta, \gamma, \rho) = 0. \tag{9.121}$$

The differential operator H_{in} depends on body-fixed degrees of freedom and has the form

$$H_{\text{in}} = \rho^{-5} \partial_\rho \, \rho^5 \, \partial_\rho - \frac{\Lambda_{\text{in}}^2}{\rho^2} + D_{\text{in}}, \tag{9.122}$$

whereas the external differential operator takes on the form

$$
\begin{aligned}
H_{\text{ext}} \;=\;& -\frac{\Lambda_{\text{ext}}^2}{\rho^2} + D_{\text{ext}} - 2 \sum_{\substack{m,n \\ n>m}}^{3} \frac{q_{mn}}{\bar{r}_{mn}}, \\[2mm]
=\;& \frac{2}{r_{23}^2\,\mu_{23}}\left[\partial_{\xi_1}\,\xi_1\,\partial_{\xi_1} - \partial_{\xi_1}\,\xi_1^2\,\partial_{\xi_1} + ik_{23}r_{23}(2\xi_1 - \xi_1^2)\partial_{\xi_1} - \mu_{23}r_{23}q_{23}\right] \\[2mm]
& + \frac{2}{r_{13}^2\,\mu_{13}}\left[\partial_{\xi_2}\,\xi_2\,\partial_{\xi_2} - \partial_{\xi_2}\,\xi_2^2\,\partial_{\xi_2} + ik_{13}r_{13}(2\xi_2 - \xi_2^2)\partial_{\xi_2} + \mu_{13}r_{13}q_{13}\right] \\[2mm]
& + \frac{2}{r_{12}^2\,\mu_{12}}\left[\partial_{\xi_3}\,\xi_1\,\partial_{\xi_3} - \partial_{\xi_3}\,\xi_3^2\,\partial_{\xi_3} + ik_{12}r_{12}(2\xi_3 - \xi_3^2)\partial_{\xi_3} + \mu_{12}r_{12}q_{12}\right].
\end{aligned}
$$

$$(9.123)$$

The various terms in this equation can be grouped as follows

$$
\begin{aligned}
H_{\text{ext}} \;=\;& \frac{2}{r_{23}^2\,\mu_{23}}\left[\partial_{\xi_1}\,\xi_1\,\partial_{\xi_1} + ik_{23}r_{23}\xi_1\partial_{\xi_1} - \mu_{23}r_{23}q_{23}\right] \\[2mm]
& + \frac{2}{r_{13}^2\,\mu_{13}}\left[\partial_{\xi_2}\,\xi_2\,\partial_{\xi_2} + ik_{13}r_{13}\xi_2\partial_{\xi_2} + \mu_{13}r_{13}q_{13}\right] \\[2mm]
& + \frac{2}{r_{12}^2\,\mu_{12}}\left[\partial_{\xi_3}\,\xi_1\,\partial_{\xi_3} + ik_{12}r_{12}\xi_3\partial_{\xi_3} + \mu_{12}r_{12}q_{12}\right] + H_{\text{rm}}.
\end{aligned}
$$

$$(9.124)$$

The differential operator H_{rm} depends on ξ_1, ξ_2, ξ_3 only. It has the explicit form

$$
\begin{aligned}
H_{\text{rm}} \;=\;& \frac{2}{r_{23}^2\,\mu_{23}}\left[-\partial_{\xi_1}\,\xi_1^2\,\partial_{\xi_1} + ik_{23}r_{23}(\xi_1 - \xi_1^2)\partial_{\xi_1}\right] \\[2mm]
& + \frac{2}{r_{13}^2\,\mu_{13}}\left[-\partial_{\xi_2}\,\xi_2^2\,\partial_{\xi_2} + ik_{13}r_{13}(\xi_2 - \xi_2^2)\partial_{\xi_2}\right] \\[2mm]
& + \frac{2}{r_{12}^2\,\mu_{12}}\left[-\partial_{\xi_3}\,\xi_3^2\,\partial_{\xi_3} + ik_{12}r_{12}(\xi_3 - \xi_3^2)\partial_{\xi_3}\right].
\end{aligned}
$$

$$(9.125)$$

The eigenfunctions of the operator $H_{\text{ext}} - H_{\text{rm}}$ have already been treated, these functions are namely (cf. Eq. (9.22))

$$
\overbrace{(H_{\text{ext}} - H_{\text{rm}})}^{H_{\text{DS3C}}}\overline{\Psi}_{\text{DS3C}} = 0.
$$

The explicit form of $\overline{\Psi}_{\text{DS3C}}$ in the coordinates Eq. (9.17) is

$$
\begin{aligned}
\overline{\Psi}_{\text{DS3C}}(\xi_1,\xi_2,\xi_3,\zeta,\gamma,\rho) \;=\;& {}_1F_1\left(i\beta_{23}(\zeta,\gamma),\,1,\,-i\left[k_{23}\,r_{23}\xi_1\right]\right) \\[2mm]
& {}_1F_1\left(i\beta_{13}(\zeta,\gamma),\,1,\,-i\left[k_{13}\,r_{13}\xi_2\right]\right) \\[2mm]
& {}_1F_1\left(i\beta_{12}(\zeta,\gamma),\,1,\,-i\left[k_{12}\,r_{12}\xi_3\right]\right).
\end{aligned}
$$

$$(9.126)$$

The relative coordinates r_{ij} have to be expressed in terms of parabolic-hyperspherical coordinates using (9.117). The plane-wave part can as well be written in terms of (9.17) since the relation applies

$$
\begin{aligned}
\mathbf{k}_{ij} \cdot \mathbf{r}_{ij} + \mathbf{K}_k \cdot \mathbf{R}_k &= \sum_{j>i=1}^{3} \frac{m_i + m_j}{m_1 + m_2 + m_3} \mathbf{k}_{ij} \cdot \mathbf{r}_{ij}, \\
&= \frac{m_2 + m_3}{m_1 + m_2 + m_3} (\xi_1 - 1) k_{23}\, r_{23} \\
&\quad + \frac{m_1 + m_3}{m_1 + m_2 + m_3} (\xi_2 - 1) k_{13}\, r_{13} \\
&\quad + \frac{m_1 + m_2}{m_1 + m_2 + m_3} (\xi_3 - 1) k_{12}\, r_{12} \, .
\end{aligned}
\tag{9.127}
$$

The Sommerfeld parameters appearing in Eq. (9.126) are expressed in terms of product charge functions $\overline{Z}_{ij}$

$$
\beta_{ij}(\zeta, \gamma) := \frac{\overline{Z}_{ij} \mu_{ij}}{k_{ij}}.
\tag{9.128}
$$

The effective product charge functions $\overline{Z}_{ij}$ are dependent on the internal coordinates. Their specific functional dependence for a given system can be determined as done in the previous section.

9.7   Parabolic-hyperspherical adiabatic expansion

Recalling the splitting $H_{\text{ext}} = H_{\text{DS3C}} + H_{\text{rm}}$ we rearrange equation (9.121) in the form

$$
\left[-\frac{1}{2} \rho^{-5} \partial_\rho \rho^5 \partial_\rho + \frac{\Lambda_{\text{in}}^2}{2\rho^2} - \frac{1}{2} H_{\text{DS3C}} - \frac{1}{2} \mathcal{F}_{\text{rm}} \right] \overline{\Psi}(\xi_1, \xi_2, \xi_3, \zeta, \gamma, \rho) = 0.
\tag{9.129}
$$

Here we have separated the function $\mathcal{F}_{\text{rm}}$ which is given by the formula

$$
\mathcal{F}_{\text{rm}} = H_{\text{rm}} - \frac{\Lambda_{\text{mix}}^2}{\rho^2} + D_{\text{in}}.
$$

In this section we outline how the ideas developed in the context of hyperspherical treatments can be utilized to deal with the dynamics in the internal coordinates. To this end we discard at first the mixing term $\mathcal{F}_{\text{rm}}$ and introduce the hyperspherical adiabatic Hamiltonian as

$$
\mathcal{U}(\rho) = \frac{1}{2} \left[\Lambda_{\text{in}}^2 - \rho^2 H_{\text{DS3C}}(\rho, \omega_p) \right],
\tag{9.130}
$$

where the collective symbol ω_p stands for all coordinates but ρ. Now let us consider the eigenvalue problem

$$
\left[\mathcal{U}(\rho) - U_\nu \right] \Phi_\nu(\rho, \omega_p) = 0.
\tag{9.131}
$$

The eigenvalues and eigenfunctions of $\mathcal{U}(\rho)$ depend parametrically on ρ. They are usually called the hyperspherical adiabatic eigenvalues and channel functions, respectively. For a given ρ the functions $\Phi_\nu(\rho, \omega_p)$ form a complete orthonormal basis

$$\langle \Phi_\nu(\rho, \omega_p) | \Phi_{\nu'}(\rho, \omega_p) \rangle = \delta_{\nu\nu'}.$$

Using this basis the solutions of Eq. (9.129) are expanded as

$$\overline{\Psi}(\omega_p, \rho) = \frac{1}{\rho^{5/2}} \sum_\nu F_\nu(\rho) \Phi_\nu(\rho, \omega_p). \tag{9.132}$$

The substitution of this expansion in Eq. (9.129) yields a set of defining equations for the radial functions $F_n(\rho)$

$$\left[-\frac{1}{2}\partial_\rho^2 + W_\nu(\rho) \right] F_\nu(\rho) = \sum_{\nu'} \left[P_{\nu\nu'}(\rho)\partial_\rho + \frac{1}{2}Q_{\nu\nu'}(\rho) \right] F_{\nu'}(\rho), \tag{9.133}$$

where the quantity $W_\nu(\rho)$ constitutes the so-called hyperspherical potentials and is given by the equation

$$W_\nu(\rho) = \frac{8U_\nu(\rho) + 15}{8\rho^2}. \tag{9.134}$$

The non-adiabatic coupling terms that appear in Eq. (9.133) are given by the matrices

$$\begin{aligned} P_{\nu\nu'}(\rho) &= \langle \Phi_\nu(\rho, \omega_p) | \partial_\rho | \Phi_{\nu'}(\rho, \omega_p) \rangle, \\ Q_{\nu\nu'}(\rho) &= \langle \Phi_\nu(\rho, \omega_p) | \partial_\rho^2 | \Phi_{\nu'}(\rho, \omega_p) \rangle. \end{aligned} \tag{9.135}$$

The numerical task is then to include a sufficient number of expansion terms in (9.132) and to solve Eq. (9.134) taking account of the appropriate boundary conditions. This task is simplified greatly by truncating the expansion (9.132) after the first term which yields the known hyperspherical adiabatic approximation (this amounts to operate within the spirit of the adiabatic Born-Oppenheimer approximation that we discussed in chapter 4.1). The wave function assumes then the form

$$\overline{\Psi}_{\nu,HSA}(\omega_p, \rho) = \frac{1}{\rho^{5/2}} F_\nu(\rho) \Phi_\nu(\rho, \omega_p). \tag{9.136}$$

The radial function in this equation satisfies the one-dimensional differential equation

$$\left[-\frac{1}{2}\partial_\rho^2 + W_\nu(\rho) - \frac{1}{2}Q_{\nu\nu}(\rho) \right] F_\nu(\rho) = 0. \tag{9.137}$$

For the present choice of the operator $\mathcal{U}$ the eigenfunctions $\Phi_\nu(\rho, \omega_p)$ are expressible in terms of the known eigenfunctions of H_{DS3C} and Λ_{in}^2.

9.8   Three-body wave functions with mixed boundary conditions

In section 2.3.3 (page 13) we considered the behaviour of the two-body Coulomb wave functions when the energy varies from negative to positive values crossing the ionization threshold. In particular, we discussed how the bound and the continuum energies and wave functions are interrelated. For the three-body problem we considered separately the wave function below (doubly excited states) and above the complete fragmentation threshold. The question which will be addressed in this section is how the structure of the three-body wave function changes when some of the particles are bound and the others are in the continuum. Clearly this situation is akin to many-body systems. The asymptotic boundary conditions are then mixed in the sense that they consist of both, decaying and oscillating parts. As evident from Eq. (9.37), for the case of three continuum particles, the presence of Coulomb potentials forbids a simple asymptotic behaviour, such as plane waves in case of short-range potentials. Therefore, it is not clear from the outset how the bound state is polarized by the presence of a continuum particle and at the same time how the motion of the continuum electron is distorted due to the structured residual ion.

For the sake of clarity we investigate systems consisting of one continuum electron moving in the field of a electron ion. This means, the latter consists of a structureless core with charge Z and one bound electron. In this case the laboratory frame coincides (to a very good approximation) with the center of mass system. The wave function ψ describing the system at the energy E derives as the solution of the time-independent Schrödinger equation

$$\left[\Delta_a + \Delta_b + \frac{2Z}{r_a} + \frac{2Z}{r_b} - \frac{2}{r_{ab}} + 2E\right]\psi(\mathbf{r}_a, \mathbf{r}_b) = 0 . \tag{9.138}$$

As previously mentioned, the positions of the two electrons with respect to the residual ion are denoted traditionally by $\mathbf{r}_a$ and $\mathbf{r}_b$. The total energy E of the system is the sum of the energy of the continuum electron $E_a = k_a^2/2$ and that of the bound electron which resides in a state specified by the principle, orbital and magnetic quantum numbers n, ℓ, m, respectively, i. e.

$$E = -\frac{Z^2}{2\,n^2} + \frac{k_a^2}{2} . \tag{9.139}$$

As discussed in some detail in Section (2.3.3) the unperturbed (hydrogenic) state of the bound electron 'b' has the structure [158, 161]

$$\zeta_{n,\ell,m}^b(\mathbf{r}_b) = r_b^{-1}\, \chi_{n,\ell}(r_b)\, \exp(-\frac{Z\,r_b}{n})\, \mathbf{Y}_{\ell m}(\hat{\mathbf{r}}_b) , \tag{9.140}$$

where $\chi_{n,\ell}(r_b)$, $Y_{\ell m}(\hat{\mathbf{r}}_b)$ are, respectively, the radial wave functions and the spherical harmonics. In view of the structure of Eq. (9.140) and of the general dependence of the continuum wave functions, it is appropriate to write the solution of the full equation (9.138) as the ansatz

$$\Psi_{\mathrm{bf}}(\mathbf{r}_a,\mathbf{r}_b) \;=\; \bar{\psi}\, r_b^{-1}\, e^{-Z\, r_b/n} \sum_{k=0}^{\infty} \xi_k. \tag{9.141}$$

Here the distorted motion of the continuum electron is described by the (continuum) part of the ansatz

$$\bar{\psi}(\mathbf{r}_a,\mathbf{r}_b) = \exp(i\mathbf{k}_a \cdot \mathbf{r}_a + i\phi)\,, \tag{9.142}$$

where the distortion factor ϕ is generally complex and needs to be determined. On the other hand, the expansion (coefficient) functions ξ_k are assumed to have the form

$$\xi_k = \frac{1}{r_b^k}\chi_{n,\ell}(r_b)Y_{\ell,m}(\hat{\mathbf{r}}_b)\,. \tag{9.143}$$

Inserting the ansatz (9.141) in the Schrödinger equation (9.138) yields the relation

$$\left[-\left(\frac{2\,Z}{n}+\frac{2}{r_b}\right)\partial_{r_b} + \Delta_b + 2i\,\mathbf{k}_a\cdot\boldsymbol{\nabla}_a \right.$$
$$\left. +\,\Delta_a + \frac{2\,Z}{r_b}+\frac{2\,Z}{r_a}-\frac{2}{r_{ab}}\right]\left(e^{i\phi}\sum_k \xi_k\right) = 0 \quad (9.144)$$

Carrying out the calculations for the differential operators acting on $e^{i\phi}\sum_k \xi_k$ we obtain the following determining differential equation for the functions ξ_k

$$\left[\Delta_a + \Delta_b + i(\Delta_a + \Delta_b)\phi - (\boldsymbol{\nabla}_a\phi)^2 - (\boldsymbol{\nabla}_b\phi)^2 \right.$$
$$+\,2i(\boldsymbol{\nabla}_b\phi\cdot\boldsymbol{\nabla}_b + \boldsymbol{\nabla}_a\phi\cdot\boldsymbol{\nabla}_a) - 2i\left(\frac{Z}{n}+\frac{1}{r_b}\right)\left(\frac{\partial\phi}{\partial r_b} - i\frac{\partial}{\partial r_b}\right)$$
$$\left. -\,2\,\mathbf{k}_a\cdot\boldsymbol{\nabla}_a\phi + 2i\,\mathbf{k}_a\cdot\boldsymbol{\nabla}_a + \frac{2\,Z}{r_b}+\frac{2\,Z}{r_a}-\frac{2}{r_{ab}}\right]\sum_k \xi_k = 0\,. \quad (9.145)$$

Let us now inspect the asymptotic behaviour of this equation. First we note that the leading order terms in the asymptotic regime are those which fall off faster than the Coulomb potential,

i. e. asymptotically Eq. (9.145) reduces to

$$
\left[-2i \frac{Z}{n} \left(\frac{\partial \phi}{\partial r_b} - i \frac{\partial}{\partial r_b} \right) - 2 \, \mathbf{k}_a \cdot \boldsymbol{\nabla}_a \phi \right.
$$

$$
\left. + 2i \, \mathbf{k}_a \cdot \boldsymbol{\nabla}_a + \frac{2Z}{r_b} + \frac{2Z}{r_a} - \frac{2}{r_{ab}} \right] \sum \xi_k = 0. \quad (9.146)
$$

Since the functions ξ_k and ϕ are independent and the equation (9.146) is valid irrespective of the values of ξ_k and ϕ we conclude that the function ϕ must satisfy

$$
\frac{i Z}{n} \frac{\partial \phi}{\partial r_b} + \mathbf{k}_a \cdot \boldsymbol{\nabla}_a \phi - \frac{Z}{r_a} - \frac{Z}{r_b} + \frac{1}{r_{ab}} = 0 . \quad (9.147)
$$

For the solution of this equation we make the ansatz

$$
\phi^\pm = -in \ln(r_b) + \Phi^\pm, \quad (9.148)
$$

$$
\Phi^\pm = \mp \frac{Z}{k_a} \ln(k_a \, r_a \mp \mathbf{k}_a \cdot \mathbf{r}_a) + \tilde{\phi}^\pm . \quad (9.149)
$$

The complex function $\tilde{\phi}^\pm$ is arbitrary and will be determined below. Since ϕ describes the distortion of the continuum electron motion, a distinction has been made between outgoing ($+$ sign) and incoming ($-$ sign) wave boundary conditions. The term $-in \ln(r_b)$ in Eq. (9.148) results in a real exponential factor in Eq. (9.141). This exponentially decaying function is then included in the functions ξ_k, as introduced by Eq. (9.143). The term $\mp \frac{Z}{k_a} \ln(k_a \, r_a \mp \mathbf{k}_a \cdot \mathbf{r}_a)$ in Eq. (9.149) describes the phase distortion of the continuum electron due to the presence of a residual charge (the core) with charge Z. This is basically, the Coulomb phase distortion if the ion were structureless. The fact that the ion has a bound electron is reflected by the presence of a second distorting factor in Eq. (9.149) namely the phase $\tilde{\phi}^\pm$. Substitution of of Eq. (9.148) in (9.147) results in the relation (only incoming wave boundary conditions are considered for brevity)

$$
\frac{i Z}{n} \frac{\partial \tilde{\phi}^+}{\partial r_b} + \mathbf{k}_a \cdot \boldsymbol{\nabla}_a \tilde{\phi}^+ + \frac{1}{r_{ab}} = 0 . \quad (9.150)
$$

The solution of this equation has the form

$$
\tilde{\phi}^+ = \frac{1}{\lambda} \ln(\lambda \, r_{ab} + \mathbf{c} \cdot \mathbf{r}_{ab}) , \quad (9.151)
$$

where the independent *complex* quantities λ and $\mathbf{c}$ are determined upon substituting (9.151) in (9.150) which yields

$$
\mathbf{c} = -\mathbf{k}_a + i \frac{Z}{n} \hat{\mathbf{r}}_b,
$$

$$
\lambda^2 = \left(\mathbf{k}_a - i \frac{Z}{n} \hat{\mathbf{r}}_b \right)^2 . \quad (9.152)
$$

For outgoing-wave boundary conditions we can repeat the steps outlined above and arrive at the final result

$$\widetilde{\phi}^- = -\frac{1}{\lambda} \ln(\lambda\, r_{ab} - \mathbf{c} \cdot \mathbf{r}_{ab}) \,. \tag{9.153}$$

From Eqs. (9.152) it follows that

$$\text{if} \quad r_a \gg r_b, \; k_a \gg Z/n \quad \overset{\text{Eq. (9.149)}}{\Longrightarrow} \quad \Phi^\pm = \mp \frac{Z-1}{k_a} \ln(k_a\, r_a \mp \mathbf{k}_a \cdot \mathbf{r}_a). \tag{9.154}$$

This means if the bound system consists of a neutral atom, i. e. an electron and a singly charged positive ion (so that $Z - 1 = 0$), and if the continuum electron energy is much larger than the bound state energy then the distortion of the continuum electron motion can be neglected.

The effect of the polarization of the bound state and the phase distortion of the continuum electron motion is best illustrated by considering the real and the imaginary part of the complex function $\exp(i\widetilde{\phi}^-)$. This can be done by rewriting λ in the form

$$\lambda = x + iy \,, \tag{9.155}$$

where x and y are real functions and hence the function $\widetilde{\phi}^-$ takes on the form

$$\widetilde{\phi}^- = \frac{\lambda^*}{|\lambda|^2} \ln\left[v^2 + u^2\right]^{1/2} + i \arctan\left(\frac{u}{v}\right) \,. \tag{9.156}$$

The real functions v and u are given by the equations

$$
\begin{aligned}
v &= x r_{ab} - \mathbf{k}_a \cdot \mathbf{r}_{ab}, \\
u &= y r_{ab} + \frac{Z}{n} \hat{\mathbf{r}}_b \cdot \mathbf{r}_{ab} \,.
\end{aligned}
\tag{9.157}
$$

Furthermore, we can express the complex function $\exp(i\widetilde{\phi}^-)$ in terms of the real phase function ζ and the amplitude function A by writing

$$\exp(i\widetilde{\phi}^-) = A \exp(i\zeta). \tag{9.158}$$

The amplitude A is a measure for the polarization degree of the initial state due to correlation effects between the bound and the continuum electrons. On the other hand ζ quantifies the distortion of the motion of the continuum electrons. The explicit expression for the polarization function A is

$$
\begin{aligned}
A &= \exp(-\Im\widetilde{\phi}^-), \\
&= \left[v^2 + u^2\right]^{y/2|\lambda|^2} \exp\left[-\frac{x}{|\lambda|^2} \arctan\frac{u}{v}\right] \,.
\end{aligned}
\tag{9.159}
$$

Following the same steps we derive for the phase distortion ζ the expression

$$
\begin{aligned}
\zeta &= \Re\,\widetilde{\phi}^-, \\
&= \ln\left[v^2 + u^2\right]^{x/2|\lambda|^2} + \left[\frac{y}{|\lambda|^2}\arctan\frac{u}{v}\right].
\end{aligned}
\tag{9.160}
$$

The expressions for x and y are then obtained by substituting (9.156) into (9.150). This yields two coupled differential equations that can be solved and the solution is given in a closed analytical form as

$$
x = \left\{\left[\frac{1}{4}\left(k_a^2 - \frac{Z^2}{n^2}\right)^2 + \frac{Z^2}{n^2}(\hat{\mathbf{r}}_b\cdot\mathbf{k}_a)^2\right]^{1/2} + \frac{1}{2}\left(k_a^2 - \frac{Z^2}{n^2}\right)\right\}^{1/2}, \tag{9.161}
$$

$$
y = -\frac{Z(\hat{\mathbf{r}}_b\cdot\mathbf{k}_a)}{x\,n}. \tag{9.162}
$$

This result concludes our analysis of the bound-continuum mixed asymptotic behaviour. At finite distances one can proceed along the same lines followed for the derivation of the asymptotic behaviour and derive for the term $\bar{\psi}$ in Eq. (9.141) the closed analytical expression

$$
\bar{\psi}^{\mp} = \mathcal{N}_{norm}\exp(i\mathbf{k}_a\cdot\mathbf{r}_a) \quad {}_1F_1\left[\pm i\alpha_a,\,1,\,\mp i k_a\,(r_a \pm \hat{\mathbf{k}}_a\cdot\mathbf{r}_a)\right]
$$

$$
{}_1F_1\left[\pm i\alpha_\lambda,\,1,\,\mp i(\lambda r_{ab}\mp\mathbf{c}\cdot\mathbf{r}_{ab}))\right]. \tag{9.163}
$$

The complex vector $\mathbf{c}$ and λ are determined by Eqs. (9.152), $\mathcal{N}_{norm}$ is a normalization constant and $\alpha_\lambda = 1/\lambda$. Physical processes involving the wave function (9.163) are numerous. E. g., in the case of excitation of neutral atoms by electron impact a final state is achieved that consists of one continuum electron and one excited electron. The latter electron is still bound to the positive ion core. In fact it is for this case where the wave function (9.163) has been successfully employed. The calculations have been performed [160] for the angular correlation parameters of the $1s \to 2p$ transition in atomic hydrogen.

9.8.1   Asymptotic states of two bound electrons

In the case that both electrons are bound one can as well find asymptotic solutions $\psi^{asy}(\mathbf{r}_a,\mathbf{r}_b)$ using similar method [234, 235], as those outlined above. It should be noted however, that, as explained in section (9.5.5), methods relying on separating the three-body system into non-interacting three two-body subsystems will not be able to reproduce the correct behaviour at the condensation point where the Fock expansion (8.1) has to be satisfied. For bound states, where the electrons are confined to a small region in space, the Fock expansion is more relevant than for continuum states. From a physical point of view the problem becomes clear when recalling that the centrifugal potential arises as a part of the kinetic energy and hence will always falls off faster than the Coulomb potential. Thus, when seeking asymptotic solutions of the wave equation only to the first order in the Coulomb potentials the effect of the centrifugal terms (and further terms) is completely neglected. This is a decisive difference to the treatment of the bound-continuum states in the preceding section, because the centrifugal force on the bound electron is exactly taken into account, as seen from the ansatz Eq. (9.140), when the continuum electron is well separated from the residual ion.

The treatment of two electrons bound to a structureless core with a charge Z employs the same arguments of the preceding section. In brief, in analogy to Eq. (9.142), one makes for the wave function $\psi^{asy}(\mathbf{r}_a, \mathbf{r}_b)$ of two electrons the ansatz

$$\psi^{asy}(\mathbf{r}_a, \mathbf{r}_b) = e^{-\lambda_a r_a - \lambda_b r_b + \varphi(\mathbf{r}_a, \mathbf{r}_b)}. \tag{9.164}$$

Inserting this ansatz in the Schrödinger equation and neglecting terms that fall off asymptotically faster than the Coulomb potentials one obtains determining equations for $\lambda_{a/b}$ and $\varphi(\mathbf{r}_a, \mathbf{r}_b)$. The resulting form of the wave function $\psi^{asy}(\mathbf{r}_a, \mathbf{r}_b)$ is

$$\psi^{asy}(\mathbf{r}_a, \mathbf{r}_b) = N_{asy} \mathcal{A}(\hat{\mathbf{r}}_a, \hat{\mathbf{r}}_b) \, r_a^{Z/\lambda_a - 1} \, r_b^{Z/\lambda_b - 1}$$
$$\left\{ \left[\mathbf{r}_{ab} - F\left(\lambda_a \hat{\mathbf{r}}_a - \lambda_b \hat{\mathbf{r}}_b \right) \right] \cdot (\mathbf{r}_a - \mathbf{r}_b) \right\}^F e^{-\lambda_a r_a - \lambda_b r_b}, \tag{9.165}$$

where N_{asy} is a normalization factor and

$$F^{-1} = |\lambda_a \hat{\mathbf{r}}_a - \lambda_b \hat{\mathbf{r}}_b|. \tag{9.166}$$

The angular function $\mathcal{A}(\hat{\mathbf{r}}_a, \hat{\mathbf{r}}_b)$ remains undetermined. At a given total energy E, the relation $E = -\lambda_b^2/2 - \lambda_a^2/2$ applies.

9.9   Partial-wave decomposition of three-body wave functions

Expanding wave functions in terms of partial waves is one of the most widely used methods in quantum mechanics. In particular, for systems with a rotational symmetry an expansion in terms of spherical harmonics is very effective. The method and its features are best illustrated for the one-particle case.

9.9.1   Expansion of Coulomb wave functions in spherical harmonics

Let us consider the scattering of one particle with charge Z_1 from the Coulomb field of a second charge Z_2. In spherical coordinates the relative motion along the radial variable r is governed by the Schrödinger equation (cf. 7.12)

$$\left[-\frac{1}{2\mu_m} \left(r^{-2} \partial_r r^2 \partial_r - r^{-2} L^2 \right) + \frac{Z_1 Z_2}{r} - E \right] \psi = 0, \tag{9.167}$$

where μ_m is the reduced mass and L^2 is the orbital angular momentum. Expanding ψ in the form

$$\begin{aligned}
\psi(\mathbf{r}, \mathbf{k}) &= N_\psi r^{-1} \sum_{l=0}^{\infty} F_l(k, r) \left(\frac{4\pi}{2l+1} \right) \sum_{m=-l}^{l} Y_{lm}^*(\hat{\mathbf{k}}) Y_{lm}(\hat{\mathbf{r}}), \\
&= N_\psi r^{-1} \sum_{l=0}^{\infty} F_l(k, r) P_l(\hat{\mathbf{k}} \cdot \hat{\mathbf{r}}), \tag{9.168}
\end{aligned}$$

where N_ψ is a normalization constant and $\mathbf{k}$ is the wave vector conjugate to $\mathbf{r}$. The expansion coefficients $F_l(k, r)$ are determined by inserting the ansatz (9.168) into the Schrödinger equation (9.168) which yields

$$\left[\partial_r^2 + k^2 - l(l+1)\, r^{-2} - \frac{2\alpha_c k}{r} \right] F_l(k, r) = 0, \tag{9.169}$$

where $\alpha_c = \mu_m Z_1 Z_2 / k$ is the Sommerfeld parameter. Making the ansatz

$$F_l(k, r) = e^{ikr} (kr)^{l+1} \bar{F}_l(\xi), \tag{9.170}$$

where $\xi = -2ikr$ and inserting it in (9.169) we obtain the Kummer-Laplace differential equation (2.50) (page 14)

$$\left[\xi \partial_\xi^2 + (2l + 2 - \xi) \partial_\xi - (l + 1 + i\alpha_c) \right] \bar{F}_l(\xi) = 0. \tag{9.171}$$

Therefore, the solution which is regular at $r = 0$ reads

$$\bar{F}_l = n_l \, {}_1F_1(l + 1 + i\alpha_c, 2l + 2, \xi), \tag{9.172}$$

where n_l is an integration constant. Inserting this equation in (9.170) we conclude that the regular spherical Coulomb wave function has the (large r) asymptotic behaviour

$$F_l(k, r) \overset{r \to \infty}{\longrightarrow} n_l \frac{(2l+1)!}{2^l} \frac{e^{\pi \alpha_c/2 + i\sigma_l}}{\Gamma(l + 1 + i\alpha_c)} \sin\left(kr - l\pi/2 - \alpha_c \ln(2kr) + \sigma_l \right). \tag{9.173}$$

The quantity

$$\sigma_l = \arg \Gamma(l + 1 + i\alpha_c)$$

is the Coulomb phase shift and hence the Coulomb scattering (S-)matrix element $S_l(k)$ is given by

$$S_l(k) = e^{2i\sigma_l} = \frac{\Gamma(l + 1 + i\alpha_c)}{\Gamma(l + 1 - i\alpha_c)}.$$

Using Eqs. (9.173, 9.170, 9.168) we obtain the Coulomb function in the following partial wave representation

$$\psi(\mathbf{r}, \mathbf{k}) \quad = \quad (2\pi)^{-3/2}(kr)^{-1} \sum_{l=0}^{\infty} (2l + 1)i^l \, e^{i\sigma_l} \, F_l(k, r) \, P_l(\hat{\mathbf{k}} \cdot \hat{\mathbf{r}}). \tag{9.174}$$

The normalization has been determined by utilizing the orthogonality properties of the Legendre polynomials. As to be expected the expression (9.174) reduces to the well-known partial wave decomposition of the plane wave in the limit $\alpha_c \to 0$, i. e.

$$\psi(\mathbf{r}, \mathbf{k})\Big|_{\alpha_c \to 0} \longrightarrow (2\pi)^{-3/2} \sum_{l=0}^{\infty} (2l + 1)i^l j_l(kr) P_l(\hat{\mathbf{k}} \cdot \hat{\mathbf{r}}), \tag{9.175}$$

where $j_l(kr)$ is a spherical Bessel function.

9.9.2   Partial wave expansion of approximate three-body wave functions

Comparing the derivation of the two-body Coulomb wave function using the method of partial wave decomposition with the derivation in parabolic coordinates (section 2.3.2), it seems that the partial wave approach is more complicated and does not yield any further significant insight. In addition, as the Coulomb potential drops off very slowly the partial wave expansion of the scattering amplitude is expected to converge slowly (if at all). Nonetheless, the partial-wave decomposition method is useful when the potential contains a short-range part or when questions are addressed that concern the symmetry and the properties of excitation processes, such as the presence of selection rules. In fact in chapter 5 we have demonstrated how the selection rules for photoexcitation processes are derived using rotational symmetry properties. More importantly we have seen in section 5.4 (page 58) that this approach can be utilized to many-electron systems which is of great importance from a theoretical and experimental point of view (see also [239, 240] and references therein). The aim of this section is therefore to explore the possibility of expanding interacting three-body systems in partial waves.

The partial wave expansion in spherical harmonics is best suited for systems with a central isotropic potential force, as clearly seen from the derivation of the partial wave decomposition of the Coulomb wave function (9.174). Therefore, one may expect serious obstacles when carrying over the analysis of the preceding section to N correlated particle systems, because in this case, for a single particle, there will be more than one force center, even though the overall rotation symmetry is maintained.

In this section we inspect the possibility of expanding in partial waves the three-body wave functions we have derived previously, such as the 3C and the DS3C wave functions. These wave functions have the structure

$$\Psi = \psi^-_{\mathbf{k}_a}(\mathbf{r}_a)\psi^-_{\mathbf{k}_b}(\mathbf{r}_b)F^-_{\mathbf{k}_{ab}}(\mathbf{r}_a - \mathbf{r}_b), \tag{9.176}$$

where the function $\psi^-_{\mathbf{k}_j}(\mathbf{r}_j) = N_\psi e^{i\mathbf{k}_j \cdot \mathbf{r}_j} \, {}_1F_1(i\alpha_j, 1, -i[k_j r_j + \mathbf{k}_j \cdot \mathbf{r}_j]), \ (i = a, b)$ is a two-body Coulomb wave as it occurs for example in (9.45). As demonstrated by Eq. (9.174) the function $\psi^-_{\mathbf{k}_j}(\mathbf{r}_j), i = a, b$ can be expanded in terms of partial waves using the radial functions $f_l(k, r)$

$$
\begin{aligned}
\psi^-_{\mathbf{k}} &= \sum_l f_l(k, r) P_l(\hat{\mathbf{k}} \cdot \hat{\mathbf{r}}), \\
&= \sum_{l,m} f_l(k, r) C^*_{lm}(\hat{\mathbf{k}}) C_{lm}(\hat{\mathbf{r}}).
\end{aligned}
\tag{9.177}
$$

To simplify notation we introduced spherical harmonics normalized such that

$$C_{lm}(\hat{\mathbf{r}}) = \sqrt{\frac{4\pi}{2l+1}} Y_{lm}(\hat{\mathbf{r}}). \tag{9.178}$$

The (correlation) function $F^-_{\mathbf{k}_{ab}}(\mathbf{r}_a - \mathbf{r}_b)$ in Eq. (9.176) that depends on the interelectronic coordinate can be formally expanded as

$$F^-_{\mathbf{k}_{ab}}(\mathbf{r}_a - \mathbf{r}_b) = \sum_{l,m} g_l(k_{ab}, r_{ab}) C^*_{lm}(\hat{\mathbf{k}}_{ab}) C_{lm}(\hat{\mathbf{r}}_{ab}). \tag{9.179}$$

The key point is that the function $g_l(k_{ab}, r_{ab})$ depends only on the body-fixed coordinate $r_{ab} = |\mathbf{r}_a - \mathbf{r}_b|$ and carries therefore no total angular momentum. The factor $C_{lm}(\hat{\mathbf{r}}_{ab})$ can be expressed as a sum over bipolar harmonics [236]

$$C_{lm}(\hat{\mathbf{r}}_{ab}) = \sum_{\lambda=0}^{l} c_{l\lambda} \, r^{-l}_{ab} \, r^{\lambda}_a \, r^{l-\lambda}_b \, B^{\lambda \, (l-\lambda)}_{l \, m}(\hat{\mathbf{r}}_a, \hat{\mathbf{r}}_b), \tag{9.180}$$

where the coefficients $c_{l\,\lambda}$ are given by

$$c_{l\,\lambda} = (-)^{l\,\lambda} \sqrt{\frac{(2l)!}{(2l-2\lambda)!(2\lambda)!}}, \tag{9.181}$$

and the bipolar harmonics are obtained from the tensor product of two spherical harmonics, namely

$$B_{l\,m}^{\lambda\,(l-\lambda)}(\hat{\mathbf{r}}_a, \hat{\mathbf{r}}_b) = \sum_{\nu} \langle \lambda\nu\,(l-\lambda)(m-\nu)|lm\rangle\, C_{\lambda\nu}(\hat{\mathbf{r}}_a)C_{(l-\lambda)\,(m-\nu)}(\hat{\mathbf{r}}_b). \quad (9.182)$$

A formal expansion in partial waves of the three-body wave function Ψ is obtained upon inserting Eqs. (9.177, 9.179, 9.180) into Eq. (9.176) which leads to the series expansion

$$
\begin{aligned}
\Psi \;=\; &\sum_{l_1,l_2,l,\lambda} f_{l_1}(k_a, r_a)\, f_{l_2}(k_b, r_b)\, g_l(k_{ab}, r_{ab})\\
&\sum_{m_1,m_2,m} c_{l\,\lambda}\, r_{ab}^{-l}\, r_a^{\lambda}\, r_b^{l-\lambda}\, C_{l_1\,m_1}^{*}(\hat{\mathbf{k}}_a)\, C_{l_2\,m_2}^{*}(\hat{\mathbf{k}}_b)\\
&\qquad C_{l\,m}^{*}(\hat{\mathbf{k}}_{ab})\, C_{l_1\,m_1}(\hat{\mathbf{r}}_a)\, C_{l_2\,m_2}(\hat{\mathbf{r}}_b)\, B_{l\,m}^{\lambda\,(l-\lambda)}(\hat{\mathbf{r}}_a, \hat{\mathbf{r}}_b). \qquad (9.183)
\end{aligned}
$$

The next step is then to expand the expression

$$X := C_{l_1\,m_1}(\hat{\mathbf{r}}_a)\, C_{l_2\,m_2}(\hat{\mathbf{r}}_b)\, B_{l\,m}^{\lambda\,(l-\lambda)}(\hat{\mathbf{r}}_a, \hat{\mathbf{r}}_b) \qquad (9.184)$$

in terms of eigenfunctions of the total angular momentum. To this end we use the definition of the bipolar harmonics and deduce the coupling relation

$$
\begin{aligned}
C_{l_1\,m_1}(\hat{\mathbf{r}}_a)\, C_{\lambda\,\nu}(\hat{\mathbf{r}}_a) = \sum_{L_1\,M_1} &(-1)^{M_1}(2L_1+1)
\begin{pmatrix} l_1 & \lambda & L_1 \\ m_1 & \nu & -M_1 \end{pmatrix}\\
&\begin{pmatrix} l_1 & \lambda & L_1 \\ 0 & 0 & 0 \end{pmatrix} C_{L_1\,M_1}(\hat{\mathbf{r}}_a).
\end{aligned}
\qquad (9.185)
$$

Likewise, we write

$$
\begin{aligned}
C_{l_2\,m_2}(\hat{\mathbf{r}}_b)\, C_{(l-\lambda)\,(m-\nu)}(\hat{\mathbf{r}}_b) = \sum_{L_2\,M_2} &(-1)^{M_2}(2L_2+1)
\begin{pmatrix} l_2 & (l-\lambda) & L_2 \\ m_2 & (m-\nu) & -M_2 \end{pmatrix}\\
&\begin{pmatrix} l_2 & (l-\lambda) & L_2 \\ 0 & 0 & 0 \end{pmatrix} C_{L_2\,M_2}(\hat{\mathbf{r}}_b).
\end{aligned}
\qquad (9.186)
$$

With these relations we derive for the function X (given by Eq. (9.184)) the expression

$$X = \sum_{L_1 \, M_1 \, L_2 \, M_2 \, \nu} \langle \lambda \nu \, (l - \lambda)(m - \nu) | lm \rangle \, (-)^{M_1 + M_2} (2L_1 + 1)(2L_2 + 1)$$

$$\begin{pmatrix} l_1 & \lambda & L_1 \\ m_1 & \nu & -M_1 \end{pmatrix} \begin{pmatrix} l_1 & \lambda & L_1 \\ 0 & 0 & 0 \end{pmatrix}$$

$$\begin{pmatrix} l_2 & (l-\lambda) & L_2 \\ m_2 & (m-\nu) & -M_2 \end{pmatrix} \begin{pmatrix} l_2 & (l-\lambda) & L_2 \\ 0 & 0 & 0 \end{pmatrix}$$

$$C_{L_1 \, M_1}(\hat{\mathbf{r}}_a) C_{L_2 \, M_2}(\hat{\mathbf{r}}_b).$$

$$(9.187)$$

The spherical harmonics in Eq. (9.187) depending on the orbital angular momenta L_1 and L_2 are coupled to the total angular momentum L by virtue of the equation

$$C_{L_1 \, M_1}(\hat{\mathbf{r}}_a) C_{L_2 \, M_2}(\hat{\mathbf{r}}_b) = \sum_{L \, M} \langle L_1 M_1 \, L_2 M_2 | LM \rangle \, B_{LM}^{L_1 L_2}(\hat{\mathbf{r}}_a, \hat{\mathbf{r}}_b). \qquad (9.188)$$

Inserting this relation in Eq. (9.187) and performing the sums over ν, M_1 and M_2 (see [237] 3.21) we derive the relation

$$X = \sum_{L_1 \, L_2 \, L} (-)^{L+l+m} \sqrt{(2l+1)(2L+1)} \, (2L_1 + 1)(2L_2 + 1)$$

$$\begin{pmatrix} l_1 & \lambda & L_1 \\ 0 & 0 & 0 \end{pmatrix} \begin{pmatrix} l_2 & (l-\lambda) & L_2 \\ 0 & 0 & 0 \end{pmatrix} B_{LM}^{L_1 L_2}(\hat{\mathbf{r}}_a, \hat{\mathbf{r}}_b)$$

$$\sum_{l_{12}} (2l_{12} + 1) \begin{pmatrix} l_1 & l_2 & l_{12} \\ m_1 & m_2 & -(m_1 + m_2) \end{pmatrix} \begin{pmatrix} l_{12} & l & L \\ -(m_1 + m_2) & -m & M \end{pmatrix}$$

$$\begin{Bmatrix} l_1 & l_2 & l_{12} \\ \lambda & (l-\lambda) & l \\ L_1 & L_2 & L \end{Bmatrix}.$$

$$(9.189)$$

Substituting Eq. (9.189) into Eq. (9.183) we obtain the partial wave expansion of the wave function Ψ

$$\Psi = \sum_{\substack{l_1 \, m_1 \, l_2 \, m_2 \\ l \, m \, \lambda \\ L_1 \, L_2 \, L \, M \, l_{12}}} f_{l_1}(k_a, r_a) \, f_{l_2}(k_b, r_b) \, g_1(k_{ab}, r_{ab})$$

$$C^*_{l_1 \, m_1}(\hat{\mathbf{k}}_a) \, C^*_{l_2 \, m_2}(\hat{\mathbf{k}}_b) \, C^*_{l \, m}(\hat{\mathbf{k}}_{ab})$$

$$c_{l\lambda} \, r_{ab}^{-l} \, r_a^{\lambda} \, r_b^{(l-\lambda)}$$

$$(-)^{L+l+m} \sqrt{(2l+1)(2L+1)} \, (2L_1+1)(2L_2+1)(2l_{12}+1)$$

$$\begin{pmatrix} l_1 & \lambda & L_1 \\ 0 & 0 & 0 \end{pmatrix} \begin{pmatrix} l_2 & (l-\lambda) & L_2 \\ 0 & 0 & 0 \end{pmatrix}$$

$$\begin{pmatrix} l_1 & l_2 & l_{12} \\ m_1 & m_2 & -(m_1+m_2) \end{pmatrix} \begin{pmatrix} l_{12} & l & L \\ -(m_1+m_2) & -m & M \end{pmatrix}$$

$$\begin{Bmatrix} l_1 & l_2 & l_{12} \\ \lambda & (l-\lambda) & l \\ L_1 & L_2 & L \end{Bmatrix} B^{L_1 L_2}_{LM}(\hat{\mathbf{r}}_a, \hat{\mathbf{r}}_b).$$

$$(9.190)$$

To perform the sums over m_1, m_2 and m we consider the quantities

$$S = \sum_{m_1 m_2} \begin{pmatrix} l_1 & l_2 & l_{12} \\ m_1 & m_2 & -m_{12} \end{pmatrix} C_{l_1 m_1}(\hat{\mathbf{k}}_a) C_{l_2 m_2}(\hat{\mathbf{k}}_b),$$

$$U = \frac{(-)^{l_1 - l_2 + m_{12}}}{\sqrt{2l_{12}+1}} \begin{pmatrix} l_{12} & l & L \\ -m_{12} & -m & M \end{pmatrix} B^{l_1 l_2}_{l_{12} m_{12}}(\hat{\mathbf{k}}_a, \hat{\mathbf{k}}_b),$$

$$S' = \sum_{m_{12} m} \begin{pmatrix} l_{12} & l & L \\ -m_{12} & -m & M \end{pmatrix} C_{lm}(\hat{\mathbf{k}}_{ab}) B^{l_1 l_2}_{l_{12} m_{12}}(\hat{\mathbf{k}}_a, \hat{\mathbf{k}}_b). \quad (9.191)$$

Using Eq. (9.180) we deduce for the function S'

$$S' = \sum_k c_{lk} \, k_{ab}^{-l} \, k_a^k \, k_b^{(l-k)} \, S'',$$

$$S'' = \sum_{m_{12} m \nu} \langle k\nu \, (l-k)(m-\nu)|lm \rangle \begin{pmatrix} l_{12} & l & L \\ -m_{12} & -m & M \end{pmatrix}$$

$$C_{k\nu}(\hat{\mathbf{k}}_a) C_{(l-k) \, (m-\nu)}(\hat{\mathbf{k}}_b) B^{l_1 l_2}_{l_{12} m_{12}}(\hat{\mathbf{k}}_a, \hat{\mathbf{k}}_b),$$

$$S'' = \sum_{m_1 m_2 m_{12} m} \begin{pmatrix} l_{12} & l & L \\ -m_{12} & -m & M \end{pmatrix} \langle l_1 m_1 l_2 m_2 | l_{12} m_{12} \rangle$$

$$C_{l_1 m_1}(\hat{\mathbf{k}}_a) \, C_{l_2 m_2}(\hat{\mathbf{k}}_b) \, B^{k \, (l-k)}_{lm}(\hat{\mathbf{k}}_a, \hat{\mathbf{k}}_b).$$

$$(9.192)$$

For the factor

$$C_{l_1 m_1}(\hat{\mathbf{k}}_a) C_{l_2 m_2}(\hat{\mathbf{k}}_b) B^{k \, (l-k)}_{lm}(\hat{\mathbf{k}}_a, \hat{\mathbf{k}}_b)$$

that appears in Eq. (9.192) we employ the result (9.189) that we derived previously for X. This yields the expression

$$
S'' = \sum_{K_1 K_2} (-)^{L+M+l+l_1-l_2} \sqrt{\frac{(2l+1)(2l_{12}+1)}{2L+1}} (2K_1+1)(2K_2+1)
$$

$$
\begin{pmatrix} l_1 & k & K_1 \\ 0 & 0 & 0 \end{pmatrix} \begin{pmatrix} l_2 & (l-k) & K_2 \\ 0 & 0 & 0 \end{pmatrix}
$$

$$
\begin{Bmatrix} l_1 & l_2 & l_{12} \\ k & (l-k) & l \\ K_1 & K_2 & L \end{Bmatrix} B^{K_1 K_2}_{LM}(\hat{\mathbf{k}}_a, \hat{\mathbf{k}}_b). \tag{9.193}
$$

With this equation we conclude that three-body wave functions that can be written as a product of single particle wave functions and wave functions depending on the relative distance can be written in the most general form

$$
\Psi = \sum_{l_1 \, l_2 \, l_{12} \, \lambda \, l \, k \, L_1 \, L_2 \, K_1 \, K_2 \, L \, M} f_{l_1}(k_a, r_a) \, f_{l_2}(k_b, r_b) \, g_l(k_{ab}, r_{ab})
$$

$$
(k_{ab} r_{ab})^{-l} \, k_a^k \, r_a^\lambda \, k_b^{(l-k)} \, r_b^{(l-\lambda)}
$$

$$
(2l+1)(2l_{12}+1)(2L_1+1)(2L_2+1)(2K_1+1)(2K_2+1)
$$

$$
\begin{pmatrix} l_1 & \lambda & L_1 \\ 0 & 0 & 0 \end{pmatrix} \begin{pmatrix} l_1 & k & K_1 \\ 0 & 0 & 0 \end{pmatrix}
$$

$$
\begin{pmatrix} l_2 & (l-\lambda) & L_2 \\ 0 & 0 & 0 \end{pmatrix} \begin{pmatrix} l_2 & (l-k) & K_2 \\ 0 & 0 & 0 \end{pmatrix}
$$

$$
\begin{Bmatrix} l_1 & l_2 & l_{12} \\ \lambda & (l-\lambda) & l \\ L_1 & L_2 & L \end{Bmatrix} \begin{Bmatrix} l_1 & l_2 & l_{12} \\ l_2 & (l-k) & l \\ K_1 & K_2 & L \end{Bmatrix}
$$

$$
B^{K_1 K_2}_{LM}(\hat{\mathbf{k}}_a, \hat{\mathbf{k}}_b) \, B^{L_1 L_2}_{LM}(\hat{\mathbf{r}}_a, \hat{\mathbf{r}}_b).
$$

$$
\tag{9.194}
$$

9.9.3   Three-body S states

For S states, i.e. for $L = M = 0$ the $9 - j$ symbol reduces to

$$
\begin{Bmatrix} l_1 & l_2 & l_{12} \\ \lambda & (l-\lambda) & l \\ L_1 & L_2 & 0 \end{Bmatrix} = \frac{(-)^{l_2+\lambda+l+L_1} \delta_{L_1 L_2} \delta_{l_{12} l}}{\sqrt{(2l+1)(2L_1+1)}} \begin{Bmatrix} l_1 & l_2 & l \\ l-\lambda & \lambda & L_1 \end{Bmatrix}.
$$

$$
\tag{9.195}
$$

Furthermore, the bipolar harmonic simplifies to

$$
\begin{aligned}
B_{00}^{L_1 L_1}(\hat{\mathbf{r}}_a, \hat{\mathbf{r}}_b) &= \sum_m \langle L_1 m L_1 - m | 00 \rangle \; C_{L_1 m}(\hat{\mathbf{r}}_a) \, C_{L_1 - m}(\hat{\mathbf{r}}_b), \\
&= \sum_m \frac{(-)^{L_1 - m}}{\sqrt{2L_1 + 1}} \; C_{L_1 m}(\hat{\mathbf{r}}_a) \, C_{L_1 - m}(\hat{\mathbf{r}}_b), \\
&= \frac{(-)^{L_1}}{\sqrt{2L_1 + 1}} \; P_{L_1}(\hat{\mathbf{r}}_a, \hat{\mathbf{r}}_b).
\end{aligned}
\tag{9.196}
$$

With these simplifications we obtain for the wave function Ψ the expression

$$
\begin{aligned}
\Psi = \sum_{l_1 \, l_2 \, l \, \lambda \, k \, L_1 \, K_1} \quad & f_{l_1}(k_a, r_a) \, f_{l_2}(k_b, r_b) \, g_l(k_{ab}, r_{ab}) \\
& (k_{ab} r_{ab})^{-l} \, k_a^k \, r_a^\lambda \, k_b^{(l-k)} \, r_b^{(l-\lambda)} \\
& (-)^{\lambda + k} \, (2l+1)(2L_1+1)(2K_1+1) \\
& \begin{pmatrix} l_1 & \lambda & L_1 \\ 0 & 0 & 0 \end{pmatrix} \begin{pmatrix} l_1 & k & K_1 \\ 0 & 0 & 0 \end{pmatrix} \\
& \begin{pmatrix} l_2 & (l-\lambda) & L_1 \\ 0 & 0 & 0 \end{pmatrix} \begin{pmatrix} l_2 & (l-k) & K_1 \\ 0 & 0 & 0 \end{pmatrix} \\
& \begin{Bmatrix} l_1 & l_2 & l \\ (l-\lambda) & \lambda & L_1 \end{Bmatrix} \begin{Bmatrix} l_1 & l_2 & l \\ (l-k) & k & K_1 \end{Bmatrix} \\
& P_{K_1}(\hat{\mathbf{k}}_a \cdot \hat{\mathbf{k}}_b) \, P_{L_1}(\hat{\mathbf{r}}_a \cdot \hat{\mathbf{r}}_b).
\end{aligned}
\tag{9.197}
$$

9.9.4 Partial-wave expansion of two-center Coulomb functions

Having determined the formal structure of the partial wave decomposition for three-body wave functions we turn now to the actual determination of the radially dependent expansion coefficient functions. For the case of one charged particle moving in a central Coulomb field we already derived the explicit expression (9.174). Thus, it remains the task to determine a similar expansion for a Coulomb particle in a two-center potential. This situation is in fact the generic case for many-electron systems whereas Eq. (9.174) is valid only for two charged particle systems. In this section we consider only two-center Coulomb waves. To this end let us inspect the function

$$\frac{f_c(\mathbf{K}, \rho_\mathbf{r} - \mathbf{r})}{|\rho_\mathbf{r} - \mathbf{r}|} = {}_1F_1(-i\alpha, 1, i[K|\rho_\mathbf{r} - \mathbf{r}| + \mathbf{K} \cdot (\rho_\mathbf{r} - \mathbf{r})]), \tag{9.198}$$

$$= \frac{1}{\Gamma(-i\alpha)\Gamma(1 + i\alpha)|\rho_\mathbf{r} - \mathbf{r}|} \int_0^1 dt \, t^{-i\alpha-1}(1 - t)^{i\alpha}$$
$$\times \exp\{it[K|\rho_\mathbf{r} - \mathbf{r}| + \mathbf{K} \cdot (\rho_\mathbf{r} - \mathbf{r})]\}. \tag{9.199}$$

Eq. (9.199) follows from the integral representation of the hypergeometric function ${}_1F_1(a, b, z)$ [139, 99]. Now we convert $|\rho_\mathbf{r} - \mathbf{r}|$ in the exponent of Eq. (9.199) into a vector form by using the formula

$$\frac{e^{-\Lambda|\rho_\mathbf{r} - \mathbf{r}|}}{|\rho_\mathbf{r} - \mathbf{r}|} = \frac{1}{2\pi^2} \int d^3q \, \frac{e^{i\mathbf{q} \cdot (\rho_\mathbf{r} - \mathbf{r})}}{q^2 + \Lambda^2}. \tag{9.200}$$

In the following calculations we introduce the convergence factor λ_1 and use the abbreviation

$$\Lambda = \lambda_1 - itK. \tag{9.201}$$

The final result is achieved by performing the limit $\lambda_1 \to 0^+$.

Using the relation (9.200) on can write Eq. (9.199) as

$$\frac{f_c(\mathbf{K}, \rho_\mathbf{r} - \mathbf{r})}{|\rho_\mathbf{r} - \mathbf{r}|} = \frac{1}{2\pi^2\Gamma(-i\alpha)\Gamma(1 + i\alpha)} \int_0^1 dt \, t^{-i\alpha-1} (1 - t)^{i\alpha} I(t), \tag{9.202}$$

where the function I is expressed as the three-dimensional integral

$$I(t) = \int d^3q \, \frac{e^{i(\mathbf{q}+t\mathbf{K}) \cdot \rho_\mathbf{r} - i(\mathbf{q}+t\mathbf{K}) \cdot \mathbf{r}}}{q^2 + \Lambda^2}. \tag{9.203}$$

By introducing the substitution

$$\mathbf{Q} = \mathbf{q} + t\mathbf{K}$$

the function I is transformed into the form

$$I(t) = \int d^3Q \, \frac{e^{i\mathbf{Q} \cdot \rho_\mathbf{r}} \, e^{-i\mathbf{Q} \cdot \mathbf{r}}}{(\mathbf{Q} - t\mathbf{K})^2 + \Lambda^2}. \tag{9.204}$$

For the plane waves in this expression we employ the partial wave expansion (9.175) and conclude that

$$
\frac{f_c(\mathbf{K}, \rho_{\mathbf{r}} - \mathbf{r})}{|\rho_{\mathbf{r}} - \mathbf{r}|} = \frac{8}{\Gamma(-i\alpha)\Gamma(1+i\alpha)} \sum_{\lambda\mu\, lm} i^{\lambda - l}\, Y_{\lambda\mu}(\hat{\rho}_{\mathbf{r}})\, Y_{lm}(\hat{\mathbf{r}})
$$

$$
\int_0^1 dt\, t^{-i\alpha - 1}\, (1-t)^{i\alpha}
$$

$$
\int_0^\infty dQ\, Q^2\, j_\lambda(Q_{\rho_r})\, j_l(Q_r)
$$

$$
\int d^2\hat{\mathbf{Q}}\, \frac{Y_{\lambda\mu}^*(\hat{\mathbf{Q}})\, Y_{lm}^*(\hat{\mathbf{Q}})}{Q^2 + t^2 K^2 + \Lambda^2 - 2t\mathbf{Q}\cdot\mathbf{K}}.
$$

$$(9.205)$$

In this expression the product of spherical harmonics $Y_{\lambda\mu}^*(\hat{\mathbf{Q}})Y_{lm}^*(\hat{\mathbf{Q}})$ can be simplified using the relation

$$
Y_{\lambda\mu}^*(\hat{\mathbf{Q}})Y_{lm}^*(\hat{\mathbf{Q}}) = \sum_L \sum_M (-1)^{\lambda - l}\, \left[\frac{(2\lambda + 1)(2l + 1)}{4\pi}\right]^{1/2}
$$

$$
\times \begin{pmatrix} l & \lambda & L \\ 0 & 0 & 0 \end{pmatrix}
$$

$$
\times \langle l\lambda m\mu | LM \rangle\, Y_{LM}^*(\hat{\mathbf{Q}}).
$$

$$(9.206)$$

Furthermore, we couple $Y_{\lambda\mu}^*(\hat{\rho}_{\mathbf{r}})$ and $Y_{lm}^*(\hat{\mathbf{r}})$ to obtain the bipolar harmonics $\mathcal{Y}_{\lambda l}^{LM}(\hat{\rho}_{\mathbf{r}}, \hat{\mathbf{r}})$. This is done via the tensor product (see appendix A.1.2)

$$
\mathcal{Y}_{\lambda l}^{LM}(\hat{\rho}_{\mathbf{r}}, \hat{\mathbf{r}}) = \sum_{m\mu} \langle l\lambda m\mu | LM \rangle\, Y_{\lambda\mu}^*(\hat{\rho}_{\mathbf{r}})\, Y_{lm}^*(\hat{\mathbf{r}}).
$$

Using the above relations we can express $\frac{f_c(\mathbf{K},\rho_\mathbf{r}-\mathbf{r})}{|\rho_\mathbf{r}-\mathbf{r}|}$ in terms of the bipolar harmonics as

$$
\begin{aligned}
\frac{f_c(\mathbf{K},\rho_\mathbf{r}-\mathbf{r})}{|\rho_\mathbf{r}-\mathbf{r}|} = \frac{8}{\sqrt{4\pi}\Gamma(-i\alpha)\Gamma(1+i\alpha)} & \sum_{l\lambda LM} i^{l-\lambda}\,[(2\lambda+1)(2l+1)]^{1/2} \\
& \begin{pmatrix} l & \lambda & L \\ 0 & 0 & 0 \end{pmatrix} \mathcal{Y}^{LM}_{\lambda l}(\hat{\rho}_\mathbf{r},\hat{\mathbf{r}}) \\
& \int_0^1 dt\, t^{-i\alpha-1}\,(1-t)^{i\alpha} \\
& \int_0^\infty dQ\, Q^2\, j_\lambda(Q\rho_r)\, j_l(Qr) \\
& \int_{-1}^1 \frac{du}{Q^2+t^2 K^2+\Lambda^2-2tQ\,K\,u} \\
& \int_0^{2\pi} d\varphi\, Y^*_{LM}(\hat{\mathbf{Q}}).
\end{aligned}
$$

$$(9.207)$$

Here we choose the quantization axis such that $\hat{\mathbf{Q}}\parallel\hat{\mathbf{K}}$ and introduce the variable $u = \cos(\theta_Q)$. Noting that

$$
Y^*_{LM}(\hat{\mathbf{Q}}) = (-1)^M \left[\frac{(2L+1)}{4\pi}\frac{(L-M)!}{(L+M)!}\right]^{1/2} P_L^M(u)\, e^{-im\varphi} \tag{9.208}
$$

we evaluate one of the integral in Eq. (9.207) as

$$
\int_0^{2\pi} d\varphi\, Y^*_{LM}(\hat{\mathbf{Q}}) = 2\pi\, Y^*_{L0}(\hat{\mathbf{Q}})\,\delta_{M0} = \sqrt{\pi(2L+1)}\, P_L(u). \tag{9.209}
$$

Therefore, Eq. (9.207) can now be written in the form

$$
\begin{aligned}
\frac{f_c}{|\rho_\mathbf{r}-\mathbf{r}|} = & \sum_{L\lambda l} i^{l-\lambda}\,[(2\lambda+1)(2l+1)(2L+1)]^{1/2} \\
& \begin{pmatrix} l & \lambda & L \\ 0 & 0 & 0 \end{pmatrix} \mathcal{Y}^{L0}_{\lambda l}(\hat{\rho}_\mathbf{r},\hat{\mathbf{r}})F^L_{\lambda l}(\rho_r,r,K),
\end{aligned}
$$

$$(9.210)$$

where the functions $F^L_{\lambda l}(\rho_r,r,K)$ are defined as

$$
\begin{aligned}
F^L_{\lambda l}(\rho_r,r,K) = \frac{4}{\Gamma(-i\alpha)\Gamma(1+i\alpha)} & \int_0^1 dt\, t^{-i\alpha-1}\,(1-t)^{i\alpha} \\
& \int_0^\infty dQ\, Q^2\, j_\lambda(Q\rho_r)\, j_l(Qr) \\
& \int_{-1}^1 \frac{du\, P_L(u)}{Q^2+t^2 K^2+\Lambda^2-2tQ\,K\,u}.
\end{aligned}
$$

$$(9.211)$$

The integration over the t variable can be performed upon the reformulation

$$Q^2 + t^2 K^2 + \Lambda^2 - 2tQ\,K\,u = Q^2 + \lambda_1^2 - 2t(Q + i\lambda_1)\,K\,u,$$

which leads to the equation

$$F_{\lambda l}^L(\rho_r, r, K) = 4 \quad \int_0^\infty \frac{dQ\,Q^2\,j_\lambda(Q\rho_r)\,j_l(Qr)}{Q^2 + \lambda_1^2}$$

$$\int_{-1}^1 \frac{du\,P_L(u)}{\Gamma(-i\alpha)\,\Gamma(1 + i\alpha)}$$

$$\int_0^1 \frac{dt\,t^{-i\alpha-1}\,(1-t)^{i\alpha}}{1 - \frac{2t\,K\,u}{Q - i\lambda_1}}. \tag{9.212}$$

The t integration is readily performed by introducing the variables

$$b = -i\alpha, c = a = 1, \text{ and } z = \frac{2K\,u}{Q - i\lambda_1}. \tag{9.213}$$

From the integral representation of the hypergeometric function

$$\frac{1}{\Gamma(-i\alpha)\,\Gamma(1 + i\alpha)} \int_0^1 \frac{dt\,t^{-i\alpha-1}\,(1-t)^{i\alpha}}{1 - t\,z} = {}_2F_1(1, -i\alpha, 1, z) = (1 - z)^{i\alpha} \tag{9.214}$$

we then deduce the relation for $F_{\lambda l}^L$

$$F_{\lambda l}^L(\rho_r, r, K) = 4 \quad \int_0^\infty \frac{dQ\,Q^2\,j_\lambda(Q\rho_r)\,j_l(Qr)}{Q^2 + \lambda_1^2}$$

$$\int_{-1}^1 du\,P_L(u) \left(1 - \frac{2Ku}{Q - i\lambda_1}\right)^{i\alpha}. \tag{9.215}$$

Now let us examine the integral

$$I_L^{(\alpha)}(z) := \int_{-1}^1 du\,P_L(u)\,(1 - z\,u)^{i\alpha}.$$

For $P_L(u)$ the following relation

$$P_L(u) = \frac{1}{2^L\,L!}\,D^L\left[(u^2 - 1)^L\right]$$

applies. Therefore, the integral $I_L^{(\alpha)}(z)$ can be evaluated by integrating L times by parts. This leads us to the expression

$$I_L^{(\alpha)}(z) = \frac{1}{2^L\,L!} \left\{ \sum_{j=0}^{L-1} z^j\,(1 - zu)^{i\alpha-j}\,\frac{\Gamma(i\alpha + 1)}{\Gamma(i\alpha + 1 - j)}\,D^{L-j-1}\left[(u^2 - 1)^L\right] \Big|_{-1}^1 \right.$$

$$\left. + \frac{z^L\,\Gamma(i\alpha + 1)}{\Gamma(i\alpha + 1 - L)} \int_{-1}^1 du\,(1 - zu)^{i\alpha-L}\,(u^2 - 1)^L \right\}. \tag{9.216}$$

Making use of the equation

$$D^M \left[(u^2 - 1) \right]^L \Big|_{-1}^{1} = D^M (u-1)^L (u+1)^L \Big|_{-1}^{1}, \tag{9.217}$$

$$= \sum_{k=0}^{M} \binom{M}{k} D^k (u-1)^L D^{M-k} (u+1)^L \Bigg|_{-1}^{1} \tag{9.218}$$

we obtain the following relation

$$D^M [(u^2 - 1)]^L \Big|_{-1}^{1} = \sum_{k=0}^{M} \binom{M}{k} \frac{L!}{(L-k)!} (u-1)^{L-k}$$

$$\frac{L!}{(L+k-M)!} (u+1)^{L+k-M} \Bigg|_{-1}^{1} = 0, \text{ if } M < L. \tag{9.219}$$

Thus, the integral $I_L^{(\alpha)}(z)$ has the form

$$I_L^{(\alpha)}(z) = \frac{z^L \, \Gamma(i\alpha + 1)}{2^L \, L! \, \Gamma(i\alpha + 1 - L)} \int_{-1}^{1} du \, (u-1)^L \, (u+1)^L \, (1-zu)^{i\alpha - L}. \tag{9.220}$$

Shifting the variable according to

$$u = 1 - 2t \rightarrow u - 1 = -2t, \ u + 1 = 2(1-t)$$

we obtain

$$I_L^{(\alpha)}(z) = \frac{z^L \, \Gamma(i\alpha + 1) \, 2^{L+1}}{\Gamma(L+1) \, \Gamma(i\alpha + 1 - L)} \int_0^1 dt \, t^L \, (1-t)^L \, (1-z+2zt)^{i\alpha - L}. \tag{9.221}$$

The next step consists of introducing the abbreviations

$$b = L + 1, \ c = 2L + 2, \ a = L - i\alpha.$$

Recalling the integral representation (9.214) of the hypergeometric function $_2F_1(a, b, c, z)$ and comparing with (9.220) we conclude that

$$I_L^{(\alpha)}(z) = 2^{L+1} \left(\frac{z}{1-z} \right)^L \frac{(1-z)^{i\alpha} \, \Gamma(i\alpha + 1) \, \Gamma(L+1)}{\Gamma(i\alpha + 1 - L) \, \Gamma(2L + 2)}$$

$$\times \, _2F_1 \left(L - i\alpha, L + 1, 2L + 2, \frac{-2z}{1-z} \right). \tag{9.222}$$

If desired, this expression can be further simplified by making use of the fact that the hypergeometric function, occurring in (9.222), is related to the associated Legendre function of the second kind [238] $Q_\nu^\mu(z)$ where

$$\nu = L, \ \mu = -1 - i\alpha, \ \text{and}$$
$$\frac{2}{1-x} = \frac{-2z}{1-z} \Rightarrow 1 - x = \frac{1-z}{-z} = 1 - \frac{1}{z} \Rightarrow x = 1/z.$$

(9.223)

Therefore, the integral $I_L^{(\alpha)}(z)$ is expressible in terms of the associated Legendre functions of the second kind. The result is

$$I_L^{(\alpha)}(z) = \left(\frac{z}{1-z}\right)^L \frac{(1-z)^{i\alpha} \Gamma(i\alpha+1)2}{\Gamma(i\alpha+1-L)\Gamma(L-i\alpha)} \left(\frac{1+z}{z}\right)^{\frac{1+i\alpha}{2}}$$
$$\left(\frac{1-z}{z}\right)^{L+1-\frac{1+i\alpha}{2}} e^{(1+i\alpha)i\pi} Q_L^{-1-i\alpha}(1/z).$$

(9.224)

This expression is further simplified to yield

$$I_L^{(\alpha)}(z) = \frac{(-1)^{L+1} 2}{\Gamma(-i\alpha) z}(1-z^2)^{i\alpha/2+1/z} e^{-\alpha\pi} Q_L^{-1-i\alpha}(1/z).$$

(9.225)

Noting that the relation

$$(2L+1) Q_L^{-1-i\alpha}(1/z) \sqrt{\frac{1}{z^2}-1} = Q_{L+1}^{-i\alpha} - Q_{L-1}^{-i\alpha}$$

(9.226)

applies, we conclude that the value of the integral $I_L^{(\alpha)}(z)$ can be written as

$$I_L^{(\alpha)}(z) = \frac{2(-1)^{L+1} (1-z^2)^{i\alpha/2} e^{-\alpha\pi}}{\Gamma(-i\alpha)} \left[Q_{L+1}^{-i\alpha}(1/z) - Q_{L-1}^{-i\alpha}(1/z)\right].$$

(9.227)

With this expression for $I_L^{(\alpha)}(z)$ we end up with the final result for $F_{\lambda l}^L(\rho_r, r, K)$, namely

$$F_{\lambda l}^L(\rho_r, r, K) = 4 \int_0^\infty \frac{dQ \, Q^2}{Q^2 + \lambda_1^2} \, j_\lambda(Q\rho_r) \, j_l(Q_r) \, I_L^\alpha \left(\frac{2K}{Q - i\lambda_1}\right).$$

(9.228)

Therefore, the object of interest, i. e. Eq. (9.207), has the following final partial wave decomposition

$$\frac{f_c}{|\rho_{\mathbf{r}} - \mathbf{r}|} = 4 \sum_{L\,\lambda\,l} i^{l-\lambda} \int_0^\infty \frac{dQ\,Q^2}{Q^2 + \lambda_1^2} \left[(2\lambda + 1)(2l + 1)(2L + 1) \right]^{1/2}$$

$$\times \begin{pmatrix} l & \lambda & L \\ 0 & 0 & 0 \end{pmatrix}$$

$$\times \mathcal{Y}_{\lambda l}^{L0}(\hat{\rho}_{\mathbf{r}}, \hat{\mathbf{r}})\, j_\lambda(Q_{\rho_r})\, j_l(Q_r)\, I_L^\alpha \left(\frac{2K}{Q - i\lambda_1} \right).$$

$$(9.229)$$

The one-dimensional integral over Q has not been performed analytically yet, but it can be performed numerically.

10 Correlated continuum states of N-body systems

In chapter 8 we investigated in full details a number of aspects related to the three-body Coulomb problem. In particular, we have seen that the appropriate choice of the coordinate system renders possible the theoretical treatment in a transparent and a direct way. The question addressed in this chapter is, whether it is possible to extend the concepts developed for the three-body problem to deal with the excited states of N particle systems? A quick glance at the coordinate systems (9.13, 9.15, 9.17) (page 118) employed for the three-body problem leads to a negative answer to this question. Leaving aside the spin-degrees of freedom, in general, a system of interacting N particles can be treated using a center-of-mass coordinate system which has the dimension $n_{\mathrm{dim}} = 3N - 3$. The definitions of the coordinate systems (9.13, 9.15, 9.17) rely on the relative positions (momenta) $\mathbf{r}_{ij}$ ($\mathbf{k}_{ij}$) between the particles. The number of $\mathbf{r}_{ij}$ (or $\mathbf{k}_{ij}$) is given by the number of pairs which is $n_{\mathrm{pairs}} = N(N-1)/2$. Only for $N = 3$ we have $n_{\mathrm{pairs}} = n_{\mathrm{dim}}$ which makes clear the peculiarity of the three-body problem. Nevertheless, the structure of the three-body wave functions provides valuable hints on what to expect for N particle systems. For $N = 3$ we were able to split approximately the total Hamiltonian in a part [cf. Eq. (9.22)] having asymptotically oscillating eigenfunctions [cf. Eqs. (9.22, 9.37)] that incorporate the long-range inter-particle correlation. A second part [Eq. (9.31)] has eigenfunctions that decay exponentially [Eq. (9.35)] with distance and accounts for the short-range interactions. This situation bears a resemblance of the behaviour of a prototypical many-electron thermodynamical[1] system, the interacting homogeneous electron gas (EG) embedded in a neutralizing homogeneous background charge. In a series of fundamental papers Pines and co-workers [50, 84] pointed out that in an interacting electron gas the long-range part of the interelectronic Coulomb interactions is describable (quantum mechani-

---

[1]The term "thermodynamic" is used in this work for systems which have an infinitely large number of particles N and occupying an infinitely extended volume V, while the particle density $n_e = N/V$ remains finite, i. e. the limits apply $V \to \infty$, $N \to \infty$, $N/V \to n_e$.

cally) in terms of collective fields. These fields represent organized plasma oscillations of the electron gas. The total Hamiltonian is then written in terms of these collective modes and a set of individual electrons that interact with one another via short-range screened Coulomb potentials. The short-range part of the electron-electron interaction can be parameterized remarkably well by a Yukawa-type potential (exponentially screened Coulomb potential) with a screening length depending on the electron gas density. There is, in addition, a mixing term that couples the individual particles to the collective modes. This mixing term can be eliminated under certain conditions [50]. As we will see later on in the chapter this formal analogy between the EG and finite N electron systems ($N < \infty$) persists in the sense that the long-range part of the interaction can be disentangled from the short-range interactions, and a mixing term between these two types of interactions can be identified.

10.1   The interacting electron gas

A finite electronic system S with N_e electrons confined to the space volume V is expected to reveal features akin to the EG when N_e and V increases while the density $n_e = N_e/V$ remains finite. To clarify the analogies and the differences between the properties of S and EG it is instructive to recall the main aspects of the EG relevant to the present work. Only a brief introduction is given while a detailed treatment can be found elsewhere, e. g. [50, 84, 52, 85].

In the interacting homogeneous electron gas (jellium) model one considers N_e electrons confined to a volume V and interacting via the two-particle Coulomb potentials

$$h_{ee} = \sum_{j>i} V(\mathbf{r}_{ij}), \quad V(\mathbf{r}_{ij}) = \frac{1}{|\mathbf{r}_i - \mathbf{r}_j|}. \tag{10.1}$$

The complete system is neutral. The charge neutrality is guaranteed by a homogeneous background positive charge with a constant (ionic) density $n_{ion} = N_e/V$ that acts as a constant crystal potential. The total Hamiltonian is then

$$H = h_e + h_{ion} + h_{e,ion}, \tag{10.2}$$

where the electronic part has the form

$$h_e = -\frac{1}{2} \sum_j^{N_e} \Delta_j + h_{ee}. \tag{10.3}$$

The operator h_{ion} is the part related to the background charge and has the form

$$h_{ion} = \lim_{\lambda_i \to 0+} \frac{1}{2} \int d^3r d^3r' \frac{n_{ion}(\mathbf{r}) n_{ion}(\mathbf{r}')}{|\mathbf{r} - \mathbf{r}'|} \exp(-\lambda_i |\mathbf{r} - \mathbf{r}'|) = \frac{2\pi N_e^2}{\lambda_i^2 V}. \tag{10.4}$$

Here λ_i is a convergence factor. Furthermore, the interaction of the electrons with the ionic background is described by

$$h_{e,ion} = \lim_{\lambda_i \to 0+} -\sum_{j=1}^{N_e} \int d^3r \frac{n_{ion}(\mathbf{r})}{|\mathbf{r} - \mathbf{r}_j|} \exp(-\lambda_i |\mathbf{r} - \mathbf{r}'|) = -\frac{4\pi N_e^2}{\lambda_i^2 V}. \tag{10.5}$$

From the preceding it is clear that the most appropriate non-interacting single particle states labelled by the quantum numbers (wave vectors) $\mathbf{k}$ are the plane wave states

$$\xi_{\mathbf{k}}(\mathbf{r}) = \frac{1}{\sqrt{V}} e^{i\mathbf{k}\cdot\mathbf{r}}. \tag{10.6}$$

Assuming boundary conditions sets a quantization constraint on the wave vectors ($k = 2\pi(n_x, n_y, n_z)/L$, where $n_{x,y,z} \in \mathbb{Z}$ and $L^3 = V$). Using this basis, we can write the electron-electron interaction h_{ee} part in the second quantization (for clarity the spin indices have been suppressed)

$$h_{ee} = \frac{1}{2} \sum_{\mathbf{k}_1, \cdots, \mathbf{k}_4} v(\mathbf{k}_1, \cdots, \mathbf{k}_4) c_{\mathbf{k}_1}^+ c_{\mathbf{k}_2}^+ c_{\mathbf{k}_4} c_{\mathbf{k}_3}, \tag{10.7}$$

where $c_{\mathbf{k}}^+$ ($c_{\mathbf{k}}$) is the operator for the creation (annihilation) of the single particle state labelled by $\mathbf{k}$. The symmetry of the particles is taken into account by the (fermionic) commutation relations of $c_{\mathbf{k}}^+$ and $c_{\mathbf{k}}$. The matrix element $v(\mathbf{k}_1, \cdots, \mathbf{k}_4)$ of the (two-body) electron-electron interaction V is given by

$$
\begin{aligned}
v(\mathbf{k}_1, \cdots, \mathbf{k}_4) &= \lim_{\lambda_i \to 0+} \int d^3r_1 d^3r_2 \frac{e^{-\lambda_i |\mathbf{r}_1 - \mathbf{r}_2|}}{|\mathbf{r}_1 - \mathbf{r}_2|} \xi_{\mathbf{k}_1}^*(\mathbf{r}_1) \xi_{\mathbf{k}_2}^*(\mathbf{r}_2) \xi_{\mathbf{k}_3}(\mathbf{r}_1) \xi_{\mathbf{k}_4}(\mathbf{r}_2) \\
&= \frac{4\pi}{V(|\mathbf{k}_1 - \mathbf{k}_2|^2 + \lambda_i^2)} \delta([\mathbf{k}_1 - \mathbf{k}_3] - [\mathbf{k}_4 - \mathbf{k}_2]).
\end{aligned} \tag{10.8}
$$

With this relation Eq. (10.7) can be written as

$$
\begin{aligned}
h_{ee} &= \frac{2\pi}{V\lambda_i^2} \sum_{\mathbf{k}\mathbf{k}'} c_{\mathbf{k}}^+ c_{\mathbf{k}'}^+ c_{\mathbf{k}'} c_{\mathbf{k}} + \frac{2\pi}{V} \sum_{\mathbf{k}\mathbf{k}'\mathbf{k}''} \frac{1}{k''^2} c_{\mathbf{k}+\mathbf{k}''}^+ c_{\mathbf{k}'-\mathbf{k}''}^+ c_{\mathbf{k}'} c_{\mathbf{k}}, \\
&= \frac{2\pi}{\lambda_i^2 V}[-\mathbf{N}_e + \mathbf{N}_e \mathbf{N}_e] + \frac{2\pi}{V} \sum_{\mathbf{k},\mathbf{k}',\mathbf{k}''\neq 0} \frac{1}{k''^2} c_{\mathbf{k}+\mathbf{k}''}^+ c_{\mathbf{k}'-\mathbf{k}''}^+ c_{\mathbf{k}'} c_{\mathbf{k}},
\end{aligned} \tag{10.9}
$$

where $\mathbf{N}_e$ is the particle number operator. The term in (10.9) proportional to $\mathbf{N}_e \mathbf{N}_e$ cancels out with the two terms on the right-hand side of Eqs. (10.5, 10.4). Till this stage there were no limitations on the number of particles. To neglect the term (10.9) that is proportional to

$\mathbf{N}_e$ we need however to operate in the thermodynamic limit, in which case this term gives an energy per particle $E/N = -2\pi/(\lambda_i^2 V)$ that vanishes for large V. With this restriction the total Hamiltonian (10.2) is written in a second quantized form as a sum of single particle and two-particle terms

$$
\begin{aligned}
H &= \sum_{\mathbf{k}} \frac{k^2}{2}\, c_{\mathbf{k}}^+ c_{\mathbf{k}} + \frac{2\pi}{V} \sum_{\mathbf{k},\mathbf{k}',\mathbf{k}''\neq 0} \frac{1}{k''^2} c_{\mathbf{k}+\mathbf{k}''}^+ c_{\mathbf{k}'-\mathbf{k}''}^+ c_{\mathbf{k}'} c_{\mathbf{k}}, \\
&= H_0 + W_{ee}.
\end{aligned}
\tag{10.10}
$$

Here H_0 is the (kinetic energy) Hamiltonian of the non-interacting EG (the Sommerfeld model) which can be exactly diagonalized analytically. As a first attempt one can account for the electronic interaction term W_{ee} by means of perturbation theory. As clear from the structure of (10.10) this approach is reasonable and in fact yields good results [84] when the mean kinetic energy (or more generally the single particle energy contribution) is large as compared to the average electron-electron energy. This is the case for an electron gas or when the single particle confinement scatters strongly. This is for example the case for the ground state of helium where the mean kinetic energy is considerably (roughly three times) larger than the electron correlation energy contribution. For very slow electrons and shallow confining potentials, such as in quantum dots [241] the electron-electron interaction becomes the dominant factor and a preference to a single particle description is no more justified. Examples of the manifestations of the electron-electron interactions in this limit are for example the predicted Wigner crystallization [242] in EG or the striking effects of correlation on the structure of doubly excited resonance states in helium, that we discussed in section 5.6.

10.2   Excited states of N interacting electrons

From the preceding discussion it is clear that when considering N slow electrons moving in the field of an ion (which we will do in the next section), special attention should be given to the electron-electron interaction. Such states are realized experimentally by multiply exciting or multiply ionizing atomic and molecular systems by electrons [244, 245, 246, 247, 248, 249, 251, 252, 253, 254, 255, 256, 257, 258, 259, 260], photons [261] or other charged-particles [262, 263, 264, 265, 266, 267, 268]. Measurements of the cross sections for such reactions provide a tool to test the many-body excited state and may as well shed light on the physics of low-density electron gas. From a theoretical (numerical) point of view there are significant differences between the conventional treatment of the EG Hamiltonian (10.9) and that of N slow moving electrons in the field of an ion. Most importantly, in the latter case one has to deal with excited electronic states subject to a singular confining potential (the Coulomb potential of the ion). The singularity problem of the confinement is absent when considering the electrons in a harmonic trap, as usually done in the case of artificial atoms (quantum dots). However, even with this computationally convenient confining potential only the energetically low-lying excited states of a small number of electrons ($N < 20$) have been treated full numerically [269, 270] by means of the path integral Monte-Carlo method [243], that will be outlined briefly later on. The approach followed in this chapter is to find approximate solution of the N electron state by directly analyzing the properties of the many-body time-independent Schrödinger equation.

10.2.1   Formal development

Let us consider a system consisting of N charged particles of equal masses m and with charges Z_j, $j \in [1, N]$ subject to the Coulomb field of a residual massive charge Z with mass M and $M \gg m$. With the assumption $m/M \to 0$ the centre-of-mass system coincides with the laboratory frame of reference. We seek eigensolution $\Psi_N(\mathbf{r}_1, \cdots, \mathbf{r}_N)$ of the non-relativistic time-independent Schrödinger equation of the N-body system at the energy E, i.e.

$$\left[H_0 + \sum_{j=1}^{N} \frac{Z Z_j}{r_j} + \sum_{\substack{i,j \\ j>i=1}}^{N} \frac{Z_i Z_j}{r_{ij}} - E \right] \Psi_N(\mathbf{r}_1, \cdots, \mathbf{r}_N) = 0, \tag{10.11}$$

where $\mathbf{r}_j$ is the position vector of the particle j with respect to the residual charge Z and $\mathbf{r}_{ij} := \mathbf{r}_i - \mathbf{r}_j$. The kinetic energy operator H_0 is the sum of single particle operators

$$H_0 = -\sum_{\ell=1}^{N} \Delta_\ell / 2m,$$

where Δ_ℓ is the Laplacian with respect to the coordinate $\mathbf{r}_\ell$. The energy E is chosen to be above the complete fragmentation threshold. The N-particle state is characterized by N given asymptotic momenta $\hbar \mathbf{k}_\ell$, i.e. the wave function satisfies the boundary conditions

$$\left. \frac{-i\hbar \nabla_\ell \Psi_N(\mathbf{r}_1, \cdots, \mathbf{r}_N)}{\Psi_N(\mathbf{r}_1, \cdots, \mathbf{r}_N)} \right|_{r_\ell \to \infty} = \hbar \mathbf{k}_\ell. \tag{10.12}$$

The total energy of the system E is then given by

$$E = \sum_{l=1}^{N} E_l, \quad \text{where} \quad E_l = \frac{k_l^2}{2m}. \tag{10.13}$$

The structure of the three-body wave function (9.45) and the realization of the importance of electronic correlation for the desired states indicate that all interactions in the system should be treated equally. Therefore, we make for the wave function $\Psi_N(\mathbf{r}_1, \cdots, \mathbf{r}_N)$ the following ansatz

$$\Psi(\mathbf{r}_1, \cdots, \mathbf{r}_N) = \mathcal{N}\Phi_I(\mathbf{r}_1, \cdots, \mathbf{r}_N)\Phi_{II}(\mathbf{r}_1, \cdots, \mathbf{r}_N)\chi(\mathbf{r}_1, \cdots, \mathbf{r}_N). \tag{10.14}$$

The function Φ_I incorporates the interaction (to infinite order) of the particles with the ion in absence of the correlation, whereas Φ_{II} describes to infinite order all isolated two-body interactions. The function $\chi(\mathbf{r}_1, \cdots, \mathbf{r}_N)$ encompasses all higher order interactions and $\mathcal{N}$ is a normalization constant.

Mathematically this means that Φ_I satisfies the differential equation

$$\left(H_0 + \sum_{j=1}^{N} \frac{ZZ_j}{r_j} - E \right) \Phi_I(\mathbf{r}_1, \cdots, \mathbf{r}_N) = 0. \tag{10.15}$$

This equation is completely separable. The solution can be written in the product form

$$\Phi_I(\mathbf{r}_1, \cdots, \mathbf{r}_N) = \overline{\Phi}_I(\mathbf{r}_1, \cdots, \mathbf{r}_N) \prod_{j=1}^{N} \xi_j(\mathbf{r}_j), \tag{10.16}$$

where $\overline{\Phi}_I(\mathbf{r}_1, \cdots, \mathbf{r}_N)$ describes the effect of the ionic potential. The free-motion is characterized by the plane waves $\xi_j(\mathbf{r}_j) = \exp(i\mathbf{k}_j \cdot \mathbf{r}_j)$ (cf. Eq. (10.6)).

Substituting (10.16) into Eq. (10.15) one concludes that the regular solution Φ_I is given by the closed form

$$\Phi_I(\mathbf{r}_1, \cdots, \mathbf{r}_N) = \prod_{j=1}^{N} \xi_j(\mathbf{r}_j)\varphi_j(\mathbf{r}_j), \tag{10.17}$$

where the function $\varphi_j(\mathbf{r}_j)$ is a confluent hypergeometric function

$$\varphi_j(\mathbf{r}_j) = {}_1F_1[\alpha_j, 1, -i(k_j r_j + \mathbf{k}_j \cdot \mathbf{r}_j)]. \tag{10.18}$$

The independent particle solution Φ_I is a good approximation for the total wave function if $|ZZ_j| \gg |Z_j Z_i|$; $\forall i, j \in [1, N]$. If this is the case one can treat the correlation term perturbatively, but keeping in mind the convergence problems of the perturbation series when dealing with infinite range potentials (cf. chapter 11).

In the next step we capture the isolated two-body correlations to infinite order. This is done by making for Φ_{II} the ansatz

$$\Phi_{II}(\mathbf{r}_1,\cdots,\mathbf{r}_N) = \overline{\Phi}_{II}(\mathbf{r}_1,\cdots,\mathbf{r}_N) \prod_{j=1}^{N} \xi_j(\mathbf{r}_j). \tag{10.19}$$

The distortion factor due the inter-particle correlations is written as a product of pair correlation factors, i. e.

$$\overline{\Phi}_{II}(\mathbf{r}_1,\cdots,\mathbf{r}_N) = \prod_{j>i=1}^{N} \varphi_{ij}(\mathbf{r}_{ij}). \tag{10.20}$$

The correlation within a pair of charged particles is described by the function

$$\varphi_{ij}(\mathbf{r}_{ij}) = {}_1F_1[\alpha_{ij}, 1, -i(k_{ij}r_{ij} + \mathbf{k}_{ij} \cdot \mathbf{r}_{ij})].$$

Note that $\varphi_{ij}(\mathbf{r}_{ij})$ describes the scattering events between particles i and j to infinite order. In fact the function $\varphi_{ij}(\mathbf{r}_{ij}) \prod_{l=1}^{N} \xi_l(\mathbf{r}_l)$ solves for the Schrödinger Eq. (10.11) for extremely strong correlations between the particles i and particle j, i. e. $|ZZ_l| \ll |Z_i Z_j| \gg |Z_m Z_n|$, $\forall\, l, m, n \neq i, j$, i.e. the relation applies

$$\left(H_0 + \frac{Z_i Z_j}{r_{ij}} - E \right) \varphi_{ij}(\mathbf{r}_{ij}) \prod_{j=1}^{N} \xi_j(\mathbf{r}_j) = 0. \tag{10.21}$$

It is clear however that the function (10.19) does not solve for Eq. (10.11) when the correlations between the particles are of comparable strength and $Z = 0$, i.e. in a situation similar to the interacting EG. On the other hand from Eq. (10.21) it is readily concluded that the potential part is taken into account exactly by the function (10.19). Thus, higher order correlation terms, which are neglected by (10.19), originate from the kinetic energy operator. Therefore, we inspect the action of the Laplacian on the wave function $\overline{\Phi}_{II}$, as given by Eq. (10.19)

$$\Delta_m \overline{\Phi}_{II} = \sum_{l=1}^{m-1} \Delta_m \varphi_{lm} \prod_{\substack{j>i \\ i \neq l}}^{N} \varphi_{ij} + \sum_{n=m+1}^{N} \Delta_m \varphi_{mn} \prod_{\substack{j>i \\ j \neq n}}^{N} \varphi_{ij} + A_m, \quad m \in [1, N]. \tag{10.22}$$

The differential operator A_m is given by cross gradient terms in the following manner

$$
\begin{aligned}
A_m \;=\; & 2 \sum_{l=1}^{m-1} \left[(\nabla_m \varphi_{lm}) \cdot \Big(\sum_{n=m+1}^{N} \nabla_m \varphi_{mn} \Big) \right] \prod_{\substack{j>i \\ j\neq n, i\neq l}}^{N} \varphi_{ij} \\
& + \sum_{l=1}^{m-1} \left[(\nabla_m \varphi_{lm}) \cdot \Big(\sum_{l\neq s=1}^{m-1} \nabla_m \varphi_{sm} \Big) \right] \prod_{\substack{j>i \\ s\neq i\neq l}}^{N} \varphi_{ij} \\
& + \sum_{n=m+1}^{N} \left[(\nabla_m \varphi_{mn}) \cdot \Big(\sum_{\substack{t=m+1 \\ t\neq n}}^{N} \nabla_m \varphi_{mt} \Big) \right] \prod_{\substack{j>i \\ j\neq t\neq n}}^{N} \varphi_{ij}, \quad m \in [1, N].
\end{aligned}
$$

$$(10.23)$$

As pointed out above the wave function (10.19) encompasses all multiple scattering events within all (isolated) pairs of particles. What is left out are correlations between the pairs. To extract a defining equation for these higher order correlations we switch off the Coulomb field of the ion (i.e. we set $Z = 0$ in Eq. (10.11)) and insert the function (10.19) into the resulting Schrödinger equation (10.11). Making use of the relation (10.22) we identify the term which prevents complete separability to be

$$
A = \sum_{m=1}^{N} A_m,
\tag{10.24}
$$

where the functions A_m are given by Eq. (10.23). As clear from Eq. (10.23) the term A_m couples a pair of particles to all other pairs in the system. Thus, it is obvious that all the terms in the sum (10.23) are absent for a three-body system since in this case only one pair of interacting particle exist (recall we are considering the case $Z = 0$). Furthermore, Eq. (10.22) evidences that the coupling term A_m is a part of the kinetic energy operator and thus should fall off faster than the Coulomb potential. This hints on the existence of an asymptotic separability in the regime where $A_m \to 0$.

The above remarks apply to the case where $Z = 0$, i.e. in the limit of interacting electron gas (but without exchange). When switching on the Coulomb field of the residual ion the particles will not only interact with the ion but all pairs of particle will couple to other pairs formed by one particle and the ion. A mathematical description of this higher order coupling term is provided by the function $\chi(r_1, \cdots, r_N)$ that occurs in Eq. (10.14). A determining equation for this function is derived upon the substitution of the expressions (10.19, 10.17) in

(10.14). Inserting this in the Schrödinger equation (10.11) yields

$$\left\{ H_0 - \frac{A}{\Phi_{II}} - \sum_{\ell=1}^{N} \Big[(\nabla_\ell \ln \Phi_I + \nabla_\ell \ln \Phi_{II}) \cdot \nabla_\ell \right.$$
$$\left. + (\nabla_\ell \ln \Phi_I) \cdot (\nabla_\ell \ln \Phi_{II}) \Big] + E \right\} \chi(\mathbf{r}_1, \cdots, \mathbf{r}_N) = 0 . \quad (10.25)$$

Rewriting χ in the form

$$\chi(\mathbf{r}_1, \cdots, \mathbf{r}_N) = \prod_{j=1}^{N} \xi^*(\mathbf{r}_j)[1 - f(\mathbf{r}_1, \cdots, \mathbf{r}_N)], \qquad (10.26)$$

where $f(\mathbf{r}_1, \cdots, \mathbf{r}_N)$ is a function to be determined and inserting in Eq. (10.25) we obtain an inhomogeneous differential equation for the determination of f (and hence χ)

$$\left\{ H_0 - \sum_{\ell=1}^{N} [\nabla_\ell (\ln \Phi_I + \ln \Phi_{II}) + i\mathbf{k}_\ell] \cdot \nabla_\ell \right\} f + \mathcal{R}(1 - f) = 0, \qquad (10.27)$$

where the inhomogeneous term $\mathcal{R}$ is

$$\mathcal{R} := \sum_{m=1}^{N} \left\{ (\nabla_m \ln \overline{\Phi}_I) \cdot (\nabla_m \ln \overline{\Phi}_{II}) \right.$$
$$+ \sum_{l=1}^{m-1} \sum_{p=m+1}^{N} (\nabla_m \ln \varphi_{lm}) \cdot (\nabla_m \ln \varphi_{mp})$$
$$+ \frac{1}{2} \sum_{l=1}^{m-1} \sum_{s \neq l}^{m-1} (\nabla_m \ln \varphi_{lm}) \cdot (\nabla_m \ln \varphi_{sm})$$
$$\left. + \frac{1}{2} \sum_{n=m+1}^{N} \sum_{n \neq q=m+1}^{N} (\nabla_m \ln \varphi_{mn}) \cdot (\nabla_m \ln \varphi_{mq}) \right\}. \qquad (10.28)$$

Inspecting the structure of the function $\mathcal{R}$ it becomes clear that it contains the coupling terms between all the individual $N(N-1)/2$ two-particle subsystems. To explore the regime where these coupling terms are negligible we have to analyze the norm of the term $\mathcal{R}$. At first we note that

$$\nabla_\ell \ln \overline{\Phi}_I = \alpha_\ell k_\ell \, \mathbf{F}_\ell(\mathbf{r}_\ell), \qquad (10.29)$$

where the functions $\mathbf{F}_\ell(\mathbf{r}_\ell)$ is explicitly given by

$$\mathbf{F}_\ell(\mathbf{r}_\ell) \quad = \quad \frac{{}_1F_1 \left[1 + i\alpha_\ell, \, 2, \, -i(k_\ell \, r_\ell + \mathbf{k}_\ell \cdot \mathbf{r}_\ell) \right]}{{}_1F_1 \left[i\alpha_\ell, \, 1, \, -i(k_\ell \, r_\ell + \mathbf{k}_\ell \cdot \mathbf{r}_\ell) \right]} \, (\hat{\mathbf{k}}_\ell + \hat{\mathbf{r}}_\ell) . \qquad (10.30)$$

Furthermore we calculate for the gradient of logarithm of $\overline{\Phi}_{II}$ the relation

$$
\begin{aligned}
\nabla_m \ln \overline{\Phi}_{II} \;&=\; \sum_{n=m+1}^{N} \nabla_m \ln \varphi_{mn} + \sum_{l=1}^{m-1} \nabla_m \ln \varphi_{lm} \\[2mm]
&=\; \sum_{n=m+1}^{N} \alpha_{mn} \mathbf{k}_{mn} \mathbf{F}_{mn}(\mathbf{r}_{mn}) - \sum_{l=1}^{m-1} \alpha_{lm} \mathbf{k}_{lm} \mathbf{F}_{lm}(\mathbf{r}_{lm}).
\end{aligned}
$$

$$(10.31)$$

Here the functions $\mathbf{F}_{lm}(\mathbf{r}_{lm})$ are determined by

$$
\mathbf{F}_{ij}(\mathbf{r}_{ij}) \;:=\; \frac{{}_1F_1\left[1 + i\alpha_{ij},\, 2,\, -i(k_{ij}\, r_{ij} + \mathbf{k}_{ij}\cdot\mathbf{r}_{ij})\right]}{{}_1F_1\left[i\alpha_{ij},\, 1,\, -i(k_{ij}\, r_{ij} + \mathbf{k}_{ij}\cdot\mathbf{r}_{ij})\right]}\,(\hat{\mathbf{k}}_{ij} + \hat{\mathbf{r}}_{ij}).
$$

$$(10.32)$$

The function $\mathcal{R}$ (10.28) is written in terms of the generalized functions $\mathbf{F}_{ij}(\mathbf{r}_{ij})$, $\mathbf{F}_l(\mathbf{r}_l)$ as

$$
\begin{aligned}
\mathcal{R} \;=\; \sum_{m=1}^{N} &\left\{ \alpha_m k_m \mathbf{F}_m(\mathbf{r}_m) \cdot \left[\sum_{n=m+1}^{N} \alpha_{mn} k_{mn} \mathbf{F}_{mn}(\mathbf{r}_{mn}) \right.\right. \\[1mm]
&\left.\left. - \sum_{s=1}^{m-1} \alpha_{sm} k_{sm} \mathbf{F}_{sm}(\mathbf{r}_{sm}) \right] \right. \\[1mm]
&\left. - \sum_{l=1}^{m-1}\sum_{p=m+1}^{N} \alpha_{lm}\alpha_{mp} k_{lm} k_{mp} \mathbf{F}_{lm} \cdot \mathbf{F}_{mp} \right. \\[1mm]
&\left. + \frac{1}{2} \sum_{l=1}^{m-1}\sum_{s\neq l}^{m-1} \alpha_{lm}\alpha_{sm} k_{lm} k_{sm} \mathbf{F}_{lm} \cdot \mathbf{F}_{sm} \right. \\[1mm]
&\left. + \frac{1}{2} \sum_{n=m+1}^{N}\sum_{n\neq q=m+1}^{N} \alpha_{mn}\alpha_{mq} k_{mn} k_{mq} \mathbf{F}_{mn} \cdot \mathbf{F}_{mq} \right\}.
\end{aligned}
$$

$$(10.33)$$

The properties of $\mathcal{R}$ are governed by that of the generalized functions $\mathbf{F}_{ij}(\mathbf{r}_{ij})$ and $\mathbf{F}_l(\mathbf{r}_l)$. The asymptotic properties of these functions derive directly from the asymptotic expansion of the hypergeometric functions [11, 12] which are

$$
\lim_{r_{ij}\to\infty} |\mathbf{F}_{ij}(\mathbf{r}_{ij})| \;\to\; \left| \frac{\hat{\mathbf{k}}_{ij} + \hat{\mathbf{r}}_{ij}}{\mathbf{k}_{ij}\cdot(\hat{\mathbf{k}}_{ij} + \hat{\mathbf{r}}_{ij})\, r_{ij}} \right| + \mathcal{O}\left(|k_{ij}\, r_{ij} + \mathbf{k}_{ij}\cdot\mathbf{r}_{ij}|^{-2}\right),
$$

$$
\lim_{r_{j}\to\infty} |\mathbf{F}_{j}(\mathbf{r}_{j})| \;\to\; \left| \frac{\hat{\mathbf{k}}_{j} + \hat{\mathbf{r}}_{j}}{\mathbf{k}_{j}\cdot(\hat{\mathbf{k}}_{j} + \hat{\mathbf{r}}_{j})\, r_{j}} \right| + \mathcal{O}\left(|k_{j}\, r_{j} + \mathbf{k}_{j}\cdot\mathbf{r}_{j}|^{-2}\right).
$$

$$(10.34)$$

Since $\mathcal{R}$ is a sum of products of $\mathbf{F}_{ij}$ and $\mathbf{F}_l$ the expression $\mathcal{R}$ is of a finite range, and the leading order term in the asymptotic expansion of $\mathcal{R}$ behaves as

$$\lim_{\substack{r_{ij}\to\infty \\ r_l\to\infty}} \mathcal{R} \to \mathcal{O}\left(|k_{ij}\,r_{ij} + \mathbf{k}_{ij}\cdot\mathbf{r}_{ij}|^{-2}, |k_l\,r_l + \mathbf{k}_l\cdot\mathbf{r}_l|^{-2}\right), \quad \forall \; j > i, l \in [1, N].$$

(10.35)

From this equation we conclude that in the limit of large r_j and large r_{kl} the term $\mathcal{R}$ can be neglected asymptotically. Consequently, the leading order solution of the exact wave function of the Schrödinger equation (10.11) is

$$\Psi(\mathbf{r}_1, \cdots, \mathbf{r}_N) \approx \mathcal{N} \prod_{m>l,j=1}^{N} \xi_j(\mathbf{r}_j)\varphi_j(\mathbf{r}_j)\varphi_{lm}(\mathbf{r}_{lm}).$$

(10.36)

This wave function reduces to the wave function (9.45) which we have derived for the three-body case and possesses the asymptotic properties (10.12) and (9.37). In fact, the above procedure provides the mathematical foundation for the asymptotic form (9.37). From a physical point of view Eq. (10.36) states that the N particle system fragments asymptotically in $N(N-1)/2$ non-interacting pairs. The particles as such are not free because of multiple scattering within the pairs. Furthermore, it is worth noting that, how fast the limit (10.35) is approached is determined by the values of the momenta. For large inter-particle momenta the asymptotic region covers a large portion of the Hilbert space, as inferred from Eq. (10.34).

10.2.2 Normalizing the N-body wave functions

In the preceding section the normalization factor $\mathcal{N}$ remains undetermined. As a matter of definition $\mathcal{N}$ derives from a $3N$-dimensional integral over the norm of the function (10.36). For large N this procedure is inhibitable for correlated (continuum) systems. Therefore, we resort to a different method. We require that the flux generated by the wave function (10.36) through an asymptotic manifold defined by a constant large inter-particle separations have to be the same as that corresponding to normalized plane-waves, i.e.

$$\mathbf{J}_{PW} = \mathbf{J}_\Psi,$$

(10.37)

where the flux generated by the plane-wave is given by

$$\mathbf{J}_{PW} = -\frac{i}{2}(2\pi)^{-3N}\left[\prod_l^N \xi_l^*(\mathbf{r}_l)\nabla\prod_l^N \xi_l(\mathbf{r}_l) - \prod_l^N \xi_l(\mathbf{r}_l)\nabla\prod_l^N \xi_l^*(\mathbf{r}_l)\right],$$

$$= (2\pi)^{-3N}\sum_{l=1}^N \mathbf{k}_l,$$

(10.38)

where the total gradient $\boldsymbol{\nabla} := \sum_{l=1}^{N} \boldsymbol{\nabla}_l$ has been introduced. The flux associated with the wave function (10.36) is derived by utilizing Eqs. (10.29, 10.31) and writing the total gradient of the wave function (10.36) as

$$
\boldsymbol{\nabla}\Psi := \mathcal{N}\sum_{m=1}^{N}\left\{i\mathbf{k}_m\Psi + \alpha_m k_m \mathbf{F}_m \Psi + \left[\sum_{\substack{n=m+1}}^{N}\alpha_{mn}k_{mn}\overline{\mathbf{F}}_{mn}(\mathbf{r}_{mn})\prod_{\substack{j>i\\j\neq n}}^{N}\varphi_{ij}\right.\right.
$$
$$
\left.\left. -\sum_{l=1}^{m-1}\alpha_{lm}k_{lm}\overline{\mathbf{F}}_{lm}(\mathbf{r}_{lm})\prod_{\substack{j>i\\i\neq l}}^{N}\varphi_{ij}\right]\prod_{s=1}^{N}\xi_s(\mathbf{r}_s)\varphi_s(\mathbf{r}_s)\right\}.
$$

(10.39)

Here the shorthand notation

$$
\overline{\mathbf{F}}_{mn} = \mathbf{F}_{mn}\varphi_{mn}
$$

has been used. As far as the asymptotic flux is concerned only the first term of Eq. (10.39) contributes, as can be concluded from Eqs. (10.30, 10.32). Making use of the asymptotic expansion of the confluent hypergeometric function and accounting only for the leading-order terms in the interparticle distances, the flux $\mathbf{J}_\Psi$ is inferred to be

$$
\mathbf{J}_\Psi = \mathcal{N}^2\prod_{j=1}^{N}\frac{\exp(\pi\alpha_j)}{\Gamma(1-i\alpha_j)\Gamma^*(1-i\alpha_j)}\prod_{m>l=1}^{N}\frac{\exp(\pi\alpha_{lm})}{\Gamma(1-i\alpha_{lm})\Gamma^*(1-i\alpha_{lm})}\sum_{n=1}^{N}\mathbf{k}_n,
$$

(10.40)

where $\Gamma(x)$ is the Gamma function. From Eqs. (10.37, 10.38, 10.40) we derive the final result

$$
\mathcal{N} = (2\pi)^{-3N/2}\prod_{j=1,m>l=1}^{N}\exp[-\pi(\alpha_{lm}+\alpha_j)/2]\Gamma(1-i\alpha_j)\Gamma(1-i\alpha_{lm}).
$$

(10.41)

As to be expected, for a three-body system this normalization factor reduces to the normalization of the wave function (9.45) that has been derived exactly on page 124.

10.2.3   The two-body cusp conditions

Since all isolated two-body interactions are correctly accounted for by the wave function (10.36) it is to be expected that the Kato cusp conditions introduced in section 8.1.2 are satisfied by the wave function (10.36). To verify the validity of this statement, which ensures the regular behavior at the two-particle collision points, we inspect the relation

$$\left[\frac{\partial \, \tilde{\Psi}(\mathbf{r}_1, \cdots, \mathbf{r}_N)}{\partial \, r_i} \right]_{r_i=0} \;=\; k_i \alpha_i \Psi(\mathbf{r}_1, \cdots, r_i = 0, \cdots, \mathbf{r}_N),$$

$$\forall \; (r_i/r_j) \to 0, (r_i/r_{lm}) \to 0; \; m > l, \; i \neq j \in [1, N] \,.$$

$$(10.42)$$

The quantity $\tilde{\Psi}(\mathbf{r}_1, \cdots, \mathbf{r}_N)$ is the wave function $\Psi(\mathbf{r}_1, \cdots, \mathbf{r}_N)$ averaged over a sphere of small radius $r_\delta \ll 1$ around the singularity $r_i = 0$. An equivalent condition guarantees regularity at the coalescence points $r_{ij} \to 0$. To analyze whether or not the wave function (10.36) satisfies the conditions (10.42) we linearize $\Psi(\mathbf{r}_1, \cdots, \mathbf{r}_N)$ around $r_i = 0$ and average over a sphere of small radius $r_\delta \ll 1$. This leads to the equation

$$\tilde{\Psi}(\mathbf{r}_1, \cdots, \mathbf{r}_N) \;=\; N_\Psi \, D(\mathbf{r}_i) \prod_{\substack{i \neq j = 1 \\ l > m}}^{N} \xi_j \varphi_j(\mathbf{r}_j) \varphi_{lm}(\mathbf{r}_{lm}), \; \epsilon_{ilm} \neq 0,$$

$$(10.43)$$

where N_Ψ is a normalization factor and the form of the function $D(\mathbf{r}_i)$ can be calculated to be

$$
\begin{aligned}
D(\mathbf{r}_i) \;&=\; \frac{2\pi}{4\pi r_\delta^2} \int_{-1}^{1} r_\delta^2 \, d\cos\theta \; \Big[\, 1 + ik_i \cos\theta + \alpha_i k_i \, r_i (1 + \cos\theta) \Big], \\
&=\; 1 + \alpha_i \, k_i \, r_i \,,
\end{aligned}
$$

$$(10.44)$$

with $\cos\theta = \hat{\mathbf{k}}_i \cdot \hat{\mathbf{r}}_i$. From Eqs. (10.44, 10.43) one concludes that

$$
\begin{aligned}
\left[\frac{\partial \, \tilde{\Psi}(\mathbf{r}_1, \cdots, \mathbf{r}_N)}{\partial \, r_i} \right]_{r_i=0} \;&=\; \alpha_i k_i \, N_\Psi \prod_{\substack{i \neq j = 1 \\ l > m}}^{N} \xi_j \varphi_j(\mathbf{r}_j) \varphi_{lm}(\mathbf{r}_{lm}), \\
&=\; \alpha_i k_i \Psi(\mathbf{r}_1, \cdots, r_i = 0, \cdots, \mathbf{r}_N), \; \epsilon_{ilm} \neq 0.
\end{aligned}
$$

$$(10.45)$$

Similar considerations lead to the conclusion that the wave function (10.36) also satisfies the Kato cusp conditions at the collision points ($r_{ij} \to 0$). As for the case of the approximate three-body wave function (9.45), the N-body wave function (10.36) has a power series expansion around the triple collision point, a behaviour which is at variance with the Fock expansion (8.1).

Finally, it should be remarked that the wave function (10.36) has been successfully employed for the calculations of the scattering cross section of three and four-body collisions, e.g. in the case of electron-impact double ionization in which case a four-particle state is achieved in the final channel (three electrons and the residual ion) (cf. Ref. [214] and references therein).

11 Green's function approach at zero temperatures

In the preceding we described a stationary, non-relativistic quantum mechanical few-body system by diagonalizing its Hamolitonian H, i.e. by solving the eigenvalue problem

$$H \ket{\alpha_\nu} = \lambda_\nu \ket{\alpha_\nu},$$

$$\sum_\nu \ket{\alpha_\nu} \bra{\alpha_\nu} = \mathbb{1}. \tag{11.1}$$

Physical properties of the system are then deduced from the spectrum λ_ν. For example, thermodynamical properties can be obtained from the grand thermodynamical potential $\Omega_{therm} = \int_{-\infty}^\mu d\lambda (\lambda - \lambda_\nu)\, n(\lambda)$. Here the quantity

$$n(\lambda) = \sum_\nu \delta(\lambda - \lambda_\nu) \tag{11.2}$$

is the density of states and μ is the chemical potential. As demonstrated in the previous two chapters, with a growing number of particles, solving directly for the Schrödinger equation becomes increasingly difficult numerically, and only under some special conditions and limitations approximate wave functions can be found analytically. Therefore, alternative, more effective methods and techniques are desirable. One of these methods is the Green's function (GF) approach which is widely applied in virtually all branches of many-body theoretical physics. The GF strategy acknowledges, right from the outset the complexity of the many-body problem and hence develops a systematic and a powerful machinery for carrying out approximations. Consequently, the real power and usefulness of the GF becomes more prevalent when processes and properties in many-body compounds are addressed. The Green's function formalism has been developed for the treatment of systems at zero temperatures [52, 84, 85, 107, 279], at finite temperatures (Matsubara Green's function [277]) and at non-equilibrium conditions (Keldysh Green's function [278, 108]). There exist an extensive literature on the GF theory and its applications (e. g. [52, 84, 85, 107, 279]) here we provide a compact overview with special attention to aspects related to interacting few-body electronic systems and their excitations. Some selected properties and definitions of the finite-temperature and the non-equilibrium GF methods are included in the appendices A.4 and A.5. This chapter provides an overview of the single-particle, zero temperature GF and its use in excitation and collision processes.

11.1   Definitions and general remarks on the Green's functions

The GF approach is rooted in the theory of inhomogeneous differential equations. The GF G is introduced as a solution of a differential equation with a singular inhomogeneity (delta function), specifically

$$[z - H(\mathbf{r})]\, G(\mathbf{r}, \mathbf{r}', z) = \delta(\mathbf{r} - \mathbf{r}'), \tag{11.3}$$

which can be written in the representation-independent form

$$[z - H]\, G(z) = \mathbb{1}, \tag{11.4}$$

where $H(\mathbf{r})$ is a linear hermitian and time-independent differential operator with a spectrum given by (11.1) and $z \in \mathbb{C}$. If z is not an eigenvalue of H, which is the case for imaginary z (H is hermitian), then Eq. (11.3) can be inverted to yield

$$\begin{aligned} G(z) &= [z - H]^{-1}; \\ &= \sum_\nu \frac{|\alpha_\nu\rangle\, \langle\alpha_\nu|}{z - \lambda_\nu}. \end{aligned} \tag{11.5}$$

Since $\lambda_\nu \in \mathbb{R}$ we conclude that $G(z)$ is analytic in the entire complex z plane, except for the points on the real z axis that coincide with λ_ν. This means that the poles of $G(z)$ indicate the positions of the eigenvalues λ_ν. The Green's function and the eigenfunctions $\langle\, \mathbf{r}\, |\alpha_\nu\rangle$ satisfy the same boundary conditions on the surface of the domain where they are defined.

If λ_ν lies in the continuum the side limits

$$\lim_{\eta \to 0} G(\mathbf{r}, \mathbf{r}', \lambda_\nu \pm i|\eta|) =: G^\pm(\mathbf{r}, \mathbf{r}', \lambda_\nu);\ \eta \in \mathbb{R} \tag{11.6}$$

are defined but they are different depending on whether the real z axis is approached from the upper or from the lower half plane of the complex z plane, which signifies that GF has a branch cut along the real part of the z axis that coincides with the continuous spectrum. From

Eqs. (11.6, 11.5) it follows that

$$G^-(\mathbf{r},\mathbf{r}',z) = \left[G^+(\mathbf{r}',\mathbf{r},z)\right]^*; \Rightarrow \Re G^-(\mathbf{r},\mathbf{r},z) = \Re G^+(\mathbf{r},\mathbf{r},z),$$

$$\Im G^-(\mathbf{r},\mathbf{r},z) = -\Im G^+(\mathbf{r},\mathbf{r},z), \qquad (11.7)$$

$$G^*(\mathbf{r},\mathbf{r}',z) = G(\mathbf{r}',\mathbf{r},z^*); \Rightarrow G(\mathbf{r},\mathbf{r}',z) \in \mathbb{R}, \text{ for } \lambda_\nu \neq z \in \mathbb{R}. \qquad (11.8)$$

According to Eqs. (11.5, 11.6) we can write

$$G^\pm(\lambda) = \lim_{\eta \to 0+} \sum_\nu |\alpha_\nu\rangle \frac{1}{\lambda - \lambda_\nu \pm i\eta} \langle\alpha_\nu|.$$

$$= \sum_\nu |\alpha_\nu\rangle \left\{ \lim_{\eta \to 0+} \left[\frac{\lambda - \lambda_\nu}{(\lambda - \lambda_\nu)^2 + \eta^2} \mp i \frac{\eta}{(\lambda - \lambda_\nu)^2 + \eta^2} \right] \right\} \langle\alpha_\nu|. \qquad (11.9)$$

This equation can be simplified by means of the relations

$$\lim_{\eta \to 0+} x/(x^2 + \eta^2) = P(1/x), \quad \lim_{\eta \to 0+} \eta/(x^2 + \eta^2) = \pi\delta(x), \qquad (11.10)$$

where P stands for the principle value and we obtain

$$G^\pm(\lambda) = P \sum_\nu \frac{|\alpha_\nu\rangle \langle\alpha_\nu|}{\lambda - \lambda_\nu} \mp i\pi \sum_\nu |\alpha_\nu\rangle \delta(\lambda - \lambda_\nu) \langle\alpha_\nu|. \qquad (11.11)$$

Comparing with Eq. (11.2) we realize that the density of states $n(\lambda)$ has a simple expression in terms of the GF (or more precisely in terms of the discontinuity of GF at the branch cut), namely

$$n(\lambda) = \mp \frac{1}{\pi} \Im \mathrm{tr} G^\pm(\lambda),$$

$$= i \frac{1}{2\pi} \mathrm{tr} \left[G^+(\lambda) - G^-(\lambda) \right]. \qquad (11.12)$$

The diagonal elements of the operator

$$\rho(\lambda) = \sum_\nu |\alpha_\nu\rangle \delta(\lambda - \lambda_\nu) \langle\alpha_\nu| = \mp \frac{1}{\pi} \Im G^\pm(\lambda) = \frac{i}{2\pi} \left[G^+(\lambda) - G^-(\lambda) \right], \qquad (11.13)$$

which appear in Eq. (11.11) are a measure for the density of states per unit volume, as can be seen from $n(\lambda) = \int d^3 r \langle\mathbf{r}|\rho(\lambda)|\mathbf{r}\rangle$. Furthermore, the following relations are verified

$$\partial_z G(\mathbf{r},\mathbf{r}',z) = -\sum_\nu \frac{\langle\mathbf{r}|\alpha_\nu\rangle \langle\alpha_\nu|\mathbf{r}'\rangle}{(z - \lambda_\nu)^2},$$

$$= -\sum_\nu \frac{\langle\mathbf{r}|\alpha_\nu\rangle}{z - \lambda_\nu} \sum_{\nu'} \frac{\langle\alpha_{\nu'}|\mathbf{r}'\rangle}{z - \lambda_{\nu'}} \int d^3 r'' \langle\alpha_\nu|\mathbf{r}''\rangle\langle\mathbf{r}''|\alpha_{\nu'}\rangle,$$

$$= -\int d^3 r'' G(\mathbf{r},\mathbf{r}'',z) G(\mathbf{r}'',\mathbf{r}',z);$$

$$\partial_z G(\mathbf{r},\mathbf{r},z) = -\langle\mathbf{r}|(z - H)^{-2}|\mathbf{r}\rangle < 0.$$

$$(11.14)$$

Further straightforward considerations lead to the useful inequalities

$$\Im G(z) \lessgtr 0, \quad for \quad \Im(z) \gtrless 0.$$

Additional relations that are useful for practical applications, are obtained upon inspecting Eq. (11.5) and performing the following operations

$$G(z) = \int_{-\infty}^{\infty} d\lambda \frac{1}{z - \lambda} \sum_{\nu} |\alpha_{\nu}\rangle \, \delta(\lambda - \lambda_{\nu}) \, \langle \alpha_{\nu}|,$$

$$= \int_{-\infty}^{\infty} d\lambda \frac{\rho(\lambda)}{z - \lambda} = \mp \frac{1}{\pi} \int_{-\infty}^{\infty} d\lambda \frac{\Im G^{\pm}(\lambda)}{z - \lambda} \tag{11.15}$$

$$= \frac{i}{2\pi} \int_{-\infty}^{\infty} d\lambda \frac{[G^{+}(\lambda) - G^{-}(\lambda)]}{z - \lambda}.$$

The real and the imaginary parts of the Green's function are not independent. This is a consequence of the analyticity of $G(z)$ which imposes the Kramers-Kronig relations

$$\Re G^{+}(z) = \frac{1}{\pi} P \int_{-\infty}^{\infty} \frac{\Im G^{+}(\lambda)}{\lambda - z} d\lambda,$$

$$\Im G^{+}(z) = -\frac{1}{\pi} P \int_{-\infty}^{\infty} \frac{\Re G^{+}(\lambda)}{\lambda - z} d\lambda.$$

$$\tag{11.16}$$

As mentioned above the GF (11.5) possesses poles at the positions of the eigenvalues λ_{ν}. The residue of $G(\mathbf{r}, \mathbf{r}', z)$ at the pole position λ_{ν} is

$$\sum_{g}^{g_{\lambda_{\nu}}} \langle \mathbf{r} | \alpha_{\nu,g} \rangle \langle \alpha_{\nu,g} | \mathbf{r}' \rangle.$$

The summation runs over all degenerate eigenstates $|\alpha_{\nu,g}\rangle$ associated with the level λ_{ν}. The degeneracy $g_{\lambda_{\nu}}$ is obtained from the diagonal elements of $G(z)$, specifically

$$g_{\lambda_{\nu}} = \int d^3 \mathbf{r} \, \mathrm{Res} \left[G(\mathbf{r}, \mathbf{r}, \lambda_{\nu}) \right] = \mathrm{tr} \left\{ \mathrm{Res} \left[G(\lambda_{\nu}) \right] \right\}.$$

For non-degenerate states one readily obtains

$$\langle \mathbf{r} | \alpha_{\lambda_{\nu}} \rangle \langle \alpha_{\lambda_{\nu}} | \mathbf{r}' \rangle = \mathrm{Res} \left[G(\mathbf{r}, \mathbf{r}', \lambda_{\nu}) \right]^{1/2}. \tag{11.17}$$

The phase $\varphi_{\lambda_{\nu}}$ of the state $\langle \mathbf{r} | \alpha_{\lambda_{\nu}} \rangle$ is calculated from the relation

$$\varphi_{\lambda_{\nu}} = -i \ln \left\{ \frac{\mathrm{Res} \left[G(\mathbf{r}, 0, \lambda_{\nu}) \right]}{\left\{ \mathrm{Res} \left[G(\mathbf{r}, \mathbf{r}, \lambda_{\nu}) \right] \mathrm{Res} \left[G(0, 0, \lambda_{\nu}) \right] \right\}^{1/2}} \right\}. \tag{11.18}$$

Here the phase convention $\varphi_{\lambda_\nu} = 0$ at $r = 0$ has been made.

Having introduced the essential definitions and notations of the GF method we address now its power in treating perturbationally the coupling of the system to an external field.

11.1.1 The transition (T) operator

Suppose that the Hamiltonian H in Eq. (11.4) consists of a solvable part H_0 with a corresponding Green's operator $G_0(z) = (z - H_0)^{-1}$ and a residual interaction $V = H - H_0$. From

$$
\begin{aligned}
[z - H_0 - V]\,G(z) &= \mathbb{1}, \\
G_0(z)\,[z - H_0 - V]\,G(z) &= G_0(z),
\end{aligned}
\tag{11.19}
$$

one obtains the Lippmann-Schwinger equation [271]

$$
\begin{aligned}
G(z) &= G_0(z) + G_0(z)VG(z) = G_0(z) + G(z)VG_0(z), \\
&= G_0(z) + G_0(z)VG_0(z) + G_0(z)VG_0(z)VG_0(z) + \cdots.
\end{aligned}
\tag{11.20}
$$

This integral equation can be further reformulated as

$$
\begin{aligned}
G(z) &= [\mathbb{1} - G_0(z)V]^{-1}\,G_0(z), \\
&= V^{-1}\left\{ \underbrace{V\,[\mathbb{1} - G_0(z)V]^{-1}}_{T(z)} \right\} G_0(z), \\
&= V^{-1}\left\{ V + V\,[G_0(z) + G_0(z)VG_0(z) + \cdots]\,V \right\} G_0(z), \\
&= V^{-1}\,[V + VG(z)V]\,G_0(z).
\end{aligned}
\tag{11.21}
$$

Therefore, the relations apply

$$
\begin{aligned}
VG(z) &= T(z)G_0(z), \quad G(z)V = G_0(z)T(z), \\
\Rightarrow G(z) &= G_0(z) + G_0(z)T(z)G_0(z).
\end{aligned}
\tag{11.22}
$$

In the above equations we defined the transition operator

$$
T = V + VG(z)V = V + T(z)G_0(z)V = V + VG_0(z)T(z).
\tag{11.23}
$$

One of the reasons for introducing the T operator is that it disentangles the solvable part H_0 of the Hamiltonian from the residual interaction V (cf. (11.22)), i.e. $T = V\,[\mathbb{1} - G_0(z)V]^{-1}$. From this relation it is clear that the practical task of calculating the matrix elements of T

consists in inverting matrices which are limited in size if V is short ranged. This favorable feature hints on serious complications in the case of the infinite-range Coulomb potentials, a problem which will be discussed in the next sections.

11.1.2   The Møller and the scattering (S) operators

We examine now the continuum eigenstates $|\Psi_E\rangle$ of the Hamiltonian (11.19), i.e. we seek solution of the inhomogeneous differential equation

$$[E - H_0]|\Psi_E\rangle = V|\Psi_E\rangle. \tag{11.24}$$

The energy E belongs to the continuum spectrum. Furthermore, we assume the solution of the reference Hamiltonian H_0 to be known, i.e. the general solution of the homogeneous part of Eq. (11.24) is

$$[E - H_0]|\phi\rangle = 0. \tag{11.25}$$

We note in this context that usually H_0 and V are chosen such that the energy E lies in the continuum spectrum of both operators H_0 and $H = H_0 + V$. The general solution of (11.24) is then readily given in terms of the resolvent $G_0(E)^\pm$ of H_0, namely

$$|\Psi_E^\pm\rangle = |\phi\rangle + G_0^\pm(E)V|\Psi_E^\pm\rangle. \tag{11.26}$$

If E does not belong to the spectrum of H_0, this equation (11.26) turns homogeneous ($|\phi\rangle = 0$ in this case)

$$|\Psi_E\rangle = G_0(E)V|\Psi_E\rangle, \quad E < 0. \tag{11.27}$$

Eq. (11.26) can be utilized for finding the continuum eigenstates and eigenvalues of either H_0 and/or H. In contrast, Eq. (11.27) is appropriate for deriving the discrete eigenstates and discrete eigenvalues of H. To deduce the bound state wave functions and energies one determines at first the eigenvalues $\zeta(E) \in \mathbb{R}$ of the operator $G_0^\pm(E)V$, i.e. we solve for the eigenvalue problem

$$\zeta(E)|\Psi_E\rangle = G_0(E)V|\Psi_E\rangle, \quad E < 0. \tag{11.28}$$

Decomposing $|\Psi_E\rangle$ and V in partial waves we obtain from (11.28) a set of determining equations for the partial wave components $\Psi_\ell(p, E)$ of the wave function. In momentum space these equations read

$$\zeta(E)\Psi_\ell(p, E) = \int_0^\infty dp'\, p'^2 \frac{V_\ell(p, p')\Psi_\ell(p', E)}{E - p'^2/2}, \quad E < 0. \tag{11.29}$$

In practice the integral is replaced by a numerical sum and a matrix eigenvalue problem is obtained that can be solved numerically for a given E. In a second step Eq. (11.29) is solved for different $E < 0$ and one finds eventually the energy E_0 for which $\zeta(E_0) = 1$ applies. The above procedure can also be interpreted as finding the bound spectrum associated with the scaled potential $V/\zeta(E)$.

Now let us inspect more closely Eq. (11.26) which is related to the continuous spectrum and write it in the form

$$|\Psi_E^{\pm}\rangle = |\phi\rangle + G_0^{\pm}(E)V\,|\phi\rangle + G_0^{\pm}(E)VG_0^{\pm}(E)V\,|\phi\rangle + \cdots,$$

$$= \left\{\, \mathbb{1} + \left[\, G_0^{\pm}(E) + G_0^{\pm}(E)VG_0^{\pm}(E) + \cdots \right] V \,\right\}\, |\phi\rangle,$$

$$= \underbrace{\left[\mathbb{1} + G^{\pm}(E)V\right]}_{\Omega^{\pm}(E)} |\phi\rangle, \qquad (11.30)$$

$$= \left[\mathbb{1} + G_0^{\pm}(E)T\right]\,|\phi\rangle.$$

The operator $\Omega^{\pm}(E)$ is called the Møller operator [272]. It has the action of mapping the state of the system $|\phi\rangle$ (in the absence of the perturbation V) onto the state $|\Psi\rangle$ in the presence of V. The operator

$$S = \Omega^{-\dagger}\Omega^{+} \qquad (11.31)$$

is called the scattering operator or simply the S operator. To elucidate its physical meaning let us assume the system to be in a state ϕ_{E_i} before switching on the interaction V and to go over into the state ϕ_{E_f} upon the action of V. The matrix elements of S_{if} of the S operator can be expressed in terms of the discontinuity of the Green's function at the branch cut, i.e.

$$S_{if} = \langle \phi_{E_f}|S|\phi_{E_i}\rangle = \langle \phi_{E_f}|\Omega^{-\dagger}\Omega^{+}|\phi_{E_i}\rangle,$$

$$= \left\langle \Psi_{E_f}^{-}\middle|\Psi_{E_i}^{+}\right\rangle, \qquad (11.32)$$

$$= \delta_{f,i} + \left\langle \phi_{E_f}\middle|V\left[G^{+}(E_f) - G^{-}(E_f)\right]\middle|\Psi_{E_i}^{+}\right\rangle.$$

In deriving this relation we made use of the identities

$$\left[\langle\Psi_{E_i}^{+}|\Psi_{E_f}^{-}\rangle - \langle\Psi_{E_i}^{+}|\Psi_{E_f}^{+}\rangle\right]^{*} = \left[S_{if}^{*} - \delta_{i,f}\right]^{*}$$

$$= \left[\langle\Psi_{E_i}^{+}|\Omega^{-}(E_f) - \Omega^{+}(E_f)|\phi_{E_f}\rangle\right]^{*} \qquad (11.33)$$

$$= \left\{\left\langle\Psi_{E_i}^{+}\middle|\left[G^{-}(E_f) - G^{+}(E_f)\right]V\middle|\phi_{E_f}\right\rangle\right\}^{*}.$$

Using Eq. (11.11) the S matrix elements simplify to

$$
\begin{aligned}
S_{fi} &= \delta_{f,i} - i2\pi\delta(E_f - E_i) \, \langle \phi_{E_f} | V | \Psi_{E_i}^+ \rangle, \\
&= \delta_{f,i} - i2\pi\delta(E_f - E_i) \, \langle \phi_{E_f} | V\Omega^+(E_i) | \phi_{E_i} \rangle, \\
&= \delta_{f,i} - i2\pi\delta(E_f - E_i) \, \langle \phi_{E_f} | T^+(E_i) | \phi_{E_i} \rangle \\
&= \delta_{f,i} - i2\pi\delta(E_f - E_i) \, T_{if}^+(E_i), \\
&= \delta_{f,i} - i2\pi\delta(E_f - E_i) \, \langle \phi_{E_f} | V + VG_0^+(E_i)V \\
&\qquad\qquad\qquad\qquad + VG_0^+(E_i)VG_0^+(E_i)V + \cdots | \phi_{E_i} \rangle.
\end{aligned}
\tag{11.34}
$$

In deriving the last two equations we made use of Eqs. (11.30, 11.23) which also show that
$V\Omega^\pm(z) = T^\pm(z)$.

Performing similar steps we derive the equivalent expression

$$
S_{fi} = \delta_{f,i} - i2\pi\delta(E_f - E_i) \, \langle \Psi_{E_f}^- | V | \phi_{E_i} \rangle.
\tag{11.35}
$$

Eqs. (11.34, 11.35) illustrate the physical meaning of the S and T operators. The state $|\phi_{E_i}\rangle$ develops under the action of V into $|\phi_{E_f}\rangle$. The *on-shell* matrix elements of the T operator describes this process as an infinite, coherent, sequence of multiple scattering of $|\phi_{E_i}\rangle$ from V. The first term in the Eqs. (11.34, 11.35) stands for the unscattered flux. The first term in the multiple scattering expansion of the transition matrix elements in Eqs. (11.34) is identified as Fermi's golden rule.

11.1.3   Transition probabilities and cross sections

The S matrix elements S_{fi} or the on-shell part of T_{if} yield the transition probability amplitude. Thus, to obtain the probability P_{fi} we have to evaluate $|S_{fi}|^2$. For this purpose we switch to the time domain by utilizing the relation

$$
2\pi\delta(E_f - E_i) = \int_{-\infty}^{\infty} dt \; e^{i(E_f - E_i)t}
$$

and evaluating

$$
P_{fi} = S_{fi}S_{fi}^*.
$$

Upon elementary manipulations we derive for the transition rate, i. e. the transition probability per unit time, the relation

$$\frac{dP_{fi}}{dt} = 2\delta_{f,i}\Im T_{fi} + 2\pi\delta(E_f - E_i)|T_{fi}|^2, \tag{11.36}$$

$$= 2\pi\delta(E_f - E_i)|T_{fi}|^2. \tag{11.37}$$

This equation is valid for inelastic transitions, where $\delta_{f,i} = 0$.

In scattering processes one encounters often the situation, where an incoming projectile with initial velocity v_0 induces (via the projectile-target interaction V) a transition in the system from a well prepared state (described by H_0) to an infinitesimal group of final states centered around a certain level. For example, the final state of the projectile can be characterized by the density of states d^3k, centered around an asymptotically measured wave vector $\mathbf{k}$. In collision theory it is customary to normalize the transition rate for this reaction to the asymptotic probability flux density j_p of the incoming projectile (cf. Eq. (10.38) on page 179) and to call it the multiple differential "cross section"[1] $\sigma(\mathbf{k})$ which is determined by the equation

$$\sigma(\mathbf{k}) := \frac{dP_{fi}}{dt}\frac{1}{j_p}d^3k, \tag{11.38}$$

$$= (2\pi)^4\frac{1}{v_0}|T_{fi}|^2\delta(E_f - E_i)d^3k. \tag{11.39}$$

11.2 The Coulomb problem in momentum space

Having introduced the basic tools of the Green's function approach we turn in this section to the derivation of the Coulomb Green's function. For this purpose it is advantageous to consider at first the momentum space eigenfunction of a hydrogenic system consisting of an electron and a massive ion with charge Z.

In principle hydrogenic orbitals $\phi(\mathbf{p})$ in momentum space can be derived by Fourier transforming directly the real-space wave functions [273]. Another approach is to utilize the $O(4)$ symmetry of the Coulomb potential (discussed in chapter 2) for the derivation of $\phi(\mathbf{p})$, as done by Fock [5]. The connection between the $O(4)$ symmetry and the momentum-space wave function $\phi(\mathbf{p})$ is that $\phi(\mathbf{p})$ coincides (apart from a normalization factor) with the hyperspherical harmonics defined on the compact unit sphere embedded in the four dimensional space. These functions are also referred to as the momentum space Sturmians [280, 274, 275].

---

[1]This definition entails that the cross section does not always have area units.

We consider at first bound states with energy E where

$$E = -\frac{p_0^2}{2} < 0.$$

The result for the continuum spectrum can be inferred by analytic continuation as, e.g. done in section 2.3.3. In momentum space the Schrödinger equation for one electron in the field of a massive ion with charge Z has the form

$$\left[\mathbf{p}^2 + \mathbf{p}_0^2\right]\phi(\mathbf{p}) - \frac{Z}{\pi^2}\int d^3q \frac{\phi(\mathbf{q})}{|\mathbf{q} - \mathbf{p}|^2} = 0. \tag{11.40}$$

An expression for $\phi(\mathbf{q})$ is conveniently derived by scaling the momenta by p_0 and then embedding the three-dimensional momentum space $\mathbb{R}_3$ in a four dimensional space $\mathbb{R}_4$. The (scaled) space $\mathbb{R}_3$ is then projected stereographically onto the unit sphere $\mathcal{S}_3 \subset \mathbb{R}_4$ (see the illustration in Fig. 11.1) . In what follows we denote the elements of $\mathbb{R}_4$ by $\boldsymbol{\xi} = (\xi_0, \vec{\xi})$ and the vectors defining $\mathcal{S}_3$ by $\hat{u}$, i.e. the vectors $\hat{u}$ are those $\boldsymbol{\xi} = (\xi_0, \vec{\xi})$ which satisfy $\xi_0^2 + \vec{\xi}^2 = 1$, where $\vec{\xi} \in \mathbb{R}_3$ and $\xi_0 \in \mathbb{R}$. For $p > p_0$ we project onto the upper hemisphere (cf. Fig. 11.1), whereas for $p < p_0$ the space $\mathbb{R}_3$ is projected onto the lower hemisphere of $\mathcal{S}_3$.

From straightforward geometric considerations we express $\hat{u}$ by its components along the directions $\hat{n}$ and $\mathbf{p}/p_0$ (cf. Fig. 11.1), i. e. we write

$$\hat{u} = a\,\hat{n} + b\,\mathbf{p}/p_0$$
$$a^2 + b^2 p^2/p_0^2 = 1; \quad a + b = 1, \tag{11.41}$$

meaning that

$$a = \frac{p^2 - p_0^2}{p^2 + p_0^2}, \quad b = \frac{2p_0^2}{p^2 + p_0^2}. \tag{11.42}$$

In other words each point $\mathbf{p}/p_0$ in $\mathbb{R}_3$ is mapped onto $(\xi_0, \vec{\xi}) \in \mathbb{R}_4$, where

$$\xi_0 = \frac{p^2 - p_0^2}{p^2 + p_0^2}, \quad \vec{\xi} = \frac{2p_0\mathbf{p}}{p^2 + p_0^2}. \tag{11.43}$$

The element of area on the sphere is given by

$$d^3\Omega = \frac{d^3\vec{\xi}}{|\xi_0|} = \left(\frac{2p_0}{p^2 + p_0^2}\right)^3 d^3p, \quad \text{or} \quad \delta(\Omega - \Omega') = \left(\frac{p^2 + p_0^2}{2p_0}\right)^3 \delta(\mathbf{p} - \mathbf{p}'). \tag{11.44}$$

From Eq. (11.41) we conclude that the distance between two points

$$\hat{u} = \frac{p^2 - p_0^2}{p^2 + p_0^2}\,\hat{n} + \frac{2p_0}{p^2 + p_0^2}\,\mathbf{p}, \quad \text{and} \quad \hat{v} = \frac{q^2 - p_0^2}{q^2 + p_0^2}\,\hat{n} + \frac{2p_0}{q^2 + p_0^2}\,\mathbf{q} \tag{11.45}$$

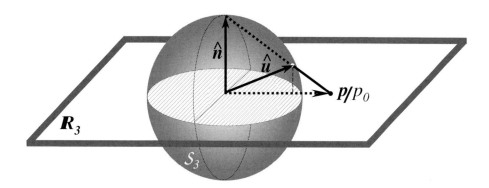

Figure 11.1: An illustration of the stereographic projection of the three-dimensional momentum space $\mathbb{R}_3$ onto the three-dimensional unit sphere $\mathcal{S}_3$ in $\mathbb{R}_4$.

on the unit sphere is

$$|\hat{\mathbf{u}} - \hat{\mathbf{v}}|^2 = \frac{4p_0^2}{(p^2 + p_0^2)(q^2 + p_0^2)}\,|\mathbf{p} - \mathbf{q}|^2 . \tag{11.46}$$

With this relation for the distance ($|\mathbf{p} - \mathbf{q}|^2$) and expressing the volume element in $\mathbb{R}_3$ in terms of the surface area on $\mathcal{S}_3$ [Eq. (11.44)] we can rewrite the Schrödinger equation (11.40) in the form

$$\left[\mathbf{p}^2 + \mathbf{p}_0^2\right]^2 \phi(\mathbf{p}) - \frac{Z}{2p_0\pi^2}\int d^3\Omega_{\hat{\mathbf{u}}}\,\frac{(q^2 + p_0^2)^2\phi(\mathbf{q})}{|\hat{\mathbf{u}} - \hat{\mathbf{v}}|^2} = 0. \tag{11.47}$$

Making the substitution

$$\frac{\left[\mathbf{p}^2 + \mathbf{p}_0^2\right]^2}{4p_0^{5/2}}\phi(\mathbf{p}) = \tilde{\phi}(\hat{\mathbf{u}}),$$

$$\frac{\left[\mathbf{q}^2 + \mathbf{q}_0^2\right]^2}{4p_0^{5/2}}\phi(\mathbf{q}) = \tilde{\phi}(\hat{\mathbf{v}}), \tag{11.48}$$

we obtain the integral equation

$$\tilde{\phi}(\hat{\mathbf{v}}) = \frac{Z}{2p_0\pi^2}\int d^3\Omega_{\hat{\mathbf{u}}}\,\frac{\phi(\hat{\mathbf{u}})}{|\hat{\mathbf{u}} - \hat{\mathbf{v}}|^2}. \tag{11.49}$$

Introducing the spherical harmonics $\mathcal{Y}_{\lambda l m_l}$ defined on the compact manifold $\mathcal{S}_3 \subset \mathbb{R}_4$ (and hence the quantum number $\lambda l m_l$ are discrete) and noting that

$$\frac{1}{4\pi^2 |\hat{u}-\hat{v}|^2} = \sum_{\lambda l m_l} \frac{1}{2\lambda+2}\mathcal{Y}_{\lambda l m_l}(\hat{v})\mathcal{Y}^*_{\lambda l m_l}(\hat{u}) =: D(\Omega_{\hat{u}},\Omega_{\hat{v}}), \qquad (11.50)$$

the eigenvalue problem Eq. (11.49) is solved by inserting (11.50) into (11.49) an exploiting the orthogonality properties of the hyperspherical harmonics to obtain

$$\phi(\hat{v}) = \mathcal{Y}_{\lambda l m_l}(\hat{v}). \qquad (11.51)$$

The eigenvalues are

$$p_0 = \frac{Z}{\lambda+1}, \quad E = -\frac{p_0^2}{2} = -\frac{Z^2}{2(\lambda+1)^2} \Rightarrow \lambda+1 = n. \qquad (11.52)$$

The structure of the eigenfunctions (11.51) evidences explicitly the $O(4)$ rotational symmetry of the Coulomb problem, as discussed in chapter 2^2.

Eq. (11.50) is a definition of the function $D(\Omega,\Omega')$ on the sphere. It is worth noting that Eq. (11.50) has a similar form as the Green's function of the four-dimensional Poisson equation, namely $(n = \lambda + 1)$

$$D(\boldsymbol{\xi},\boldsymbol{\xi}') = \frac{1}{4\pi^2}\frac{1}{(\boldsymbol{\xi}-\boldsymbol{\xi}')^2} = \sum_{\lambda l m_l} \frac{1}{2n}\mathcal{Y}_{\lambda l m_l}(\hat{v})\mathcal{Y}^*_{\lambda l m_l}(\hat{u}), \qquad (11.53)$$

$$-\Delta D(\boldsymbol{\xi},\boldsymbol{\xi}') = \delta(\boldsymbol{\xi}-\boldsymbol{\xi}'), \qquad (11.54)$$

where $\boldsymbol{\xi}$ and $\boldsymbol{\xi}'$ are being constrained to the sphere $\mathcal{S}_3$.

In four dimensions a vector $\boldsymbol{\xi} \in \mathcal{S}_3$ is uniquely characterized by three angles α, θ and φ (for vectors off the unit sphere $\mathcal{S}_3$ the radius ρ is also required), where

$$\xi_0 = \cos\alpha, \quad \vec{\xi} = \sin\alpha\,(\sin\theta\cos\varphi, \sin\theta\sin\varphi, \cos\theta).$$

The four dimensional spherical harmonics $\mathcal{Y}_{nlm_l}(\alpha,\theta,\varphi)$ are related to the standard three-dimensional spherical harmonics $Y_{lm_l}(\theta,\varphi)$ via the relation [276]

$$\mathcal{Y}_{nlm_l}(\alpha,\theta,\varphi) = R_{nl}(\alpha)Y_{lm_l}(\theta,\varphi)$$
$$= N_{nl}(\sin\alpha)^l C^{l+1}_{n-l-1}(\cos\alpha)\,Y_{lm_l}(\theta,\varphi), \qquad (11.55)$$

---

^{2}Note that the four-dimensional spherical harmonics $\mathcal{Y}_{nlm}$ are eigenfunctions of the Laplace operator acting in the domain $\mathcal{S}_3$. Thus, these wave functions describe a free particle moving on $\mathcal{S}_3$. The principal quantum number n is in fact a (hyper) angular-momentum quantum number. This demonstrates explicitly that the origin of the l (accidental) degeneracy in $\mathbb{R}_3$ has its origin in a four-dimensional rotational symmetry.

where C_{n-l-1}^{l+1} are the Gegenbauer polynomials [99]. The normalization constant is chosen such that

$$\int_0^\pi d\alpha\, R_{nl}(\alpha) R_{n'l}(\alpha) \sin^2 \alpha = \delta_{nn'}.$$

11.3 The Coulomb two-body Green's functions

Similar to Eq. (11.40) the Green's function $G(\mathbf{p}, \mathbf{p}')$ of the Coulomb potential in momentum space satisfies the equation

$$-\frac{1}{2}\left[\mathbf{p}^2 + p_0^2\right] G(\mathbf{p}, \mathbf{p}') + \frac{Z}{2\pi^2} \int d^3 \mathbf{p}'' \frac{1}{|\mathbf{p} - \mathbf{p}''|^2} G(\mathbf{p}'', \mathbf{p}') = \delta(\mathbf{p} - \mathbf{p}'). \quad (11.56)$$

Performing for the Green's function the steps that lead to Eqs. (11.46, 11.47, 11.48) one transforms the Green's function $G(\mathbf{p}, \mathbf{p}')$ to $\Gamma(\Omega, \Omega')$ where the function $\Gamma(\Omega, \Omega')$ is given by

$$\Gamma(\Omega, \Omega') = -\frac{1}{2(2p_0)^3}(p_0^2 + p^2)^2\, G(\mathbf{p}, \mathbf{p}')\, (p_0^2 + p'^2)^2. \quad (11.57)$$

As in the case of the wave function (cf. Eq. (11.49)) one obtains then the following integral equation for $\Gamma(\Omega, \Omega')$ [7]

$$\Gamma(\Omega, \Omega') - 2\nu \int d\Omega''\, D(\boldsymbol{\xi}, \boldsymbol{\xi}'')\, \Gamma(\Omega'', \Omega') = \delta(\Omega, \Omega'), \quad (11.58)$$

where

$$\nu = Z/p_0$$

The poles of the Green's function can be obtained by using Eq. (11.50), i.e. we constrain $\boldsymbol{\xi}, \boldsymbol{\xi}''$ to $\mathcal{S}_3$. In this case we readily derive the solution of Eq. (11.58) to be

$$\Gamma(\Omega, \Omega') = \sum_{\lambda l m_l} \frac{\mathcal{Y}_{\lambda l m_l}(\Omega) \mathcal{Y}_{\lambda l m_l}^*(\Omega')}{1 - \nu/n}. \quad (11.59)$$

From this equation we recover the (bound) spectrum as the poles of $\Gamma(\Omega, \Omega')$ that occur at $Z/p_0 = \nu = n$. The energy levels are thus enumerated by n and occur at

$$E_n = -p_0^2/2 = -Z^2/(2n^2). \quad (11.60)$$

The wave functions are recovered as the residua related to these poles in the energy plane as

$$\phi(\mathbf{p}) = \frac{4p_0^{5/2}}{[\mathbf{p}^2 + \mathbf{p}_0^2]^2} \mathcal{Y}_{nlm_l}(\Omega). \quad (11.61)$$

To obtain a general expression for the Green's function we have to employ the solution (11.53) of the Poisson equation instead of Eq. (11.50), i.e. we have to relax the condition of constraining the variables in Eq. (11.50) to $\mathcal{S}_3$. This yields, as in the case of the solution of the Poisson equation in three dimensions, a function of the form

$$D(\boldsymbol{\xi} - \boldsymbol{\xi}') = \sum_{nlm_l} \frac{\rho_<^{n-1}}{\rho_>^{n+1}} \frac{1}{2n} \mathcal{Y}_{nlm_l}(\Omega) \mathcal{Y}^*_{nlm_l}(\Omega'), \tag{11.62}$$

where ρ is the length of $\boldsymbol{\xi}$. If $|\boldsymbol{\xi}|$ is larger (smaller) than $|\boldsymbol{\xi}'|$ then $\rho_>$ ($\rho_<$) is the length of $|\boldsymbol{\xi}|$ ($|\boldsymbol{\xi}'|$). With Eq. (11.62) we deduce for Eq. (11.58) a representation in terms of a one-dimensional integral

$$\Gamma(\Omega, \Omega') = \delta(\Omega - \Omega') + \frac{\nu}{2\pi^2} \frac{1}{(\boldsymbol{\xi} - \boldsymbol{\xi}')^2} + \frac{\nu^2}{2\pi^2} \int_0^1 \frac{d\rho \, \rho^{-\nu}}{(1-\rho)^2 + \rho(\boldsymbol{\xi} - \boldsymbol{\xi}')^2}. \tag{11.63}$$

To obtain this relations the expansion has been utilized

$$\frac{1}{(1-\rho)^2 + \rho(\boldsymbol{\xi} - \boldsymbol{\xi}')^2} = 2\pi^2 \sum_{n=1, lm_l}^{\infty} \frac{\rho^{n-1}}{n} \mathcal{Y}_{nlm_l}(\Omega) \mathcal{Y}^*_{nlm_l}(\Omega'),$$

$$\frac{1}{1 - \nu/n} = 1 + \frac{\nu}{n} + \frac{\nu^2}{n} \int_0^1 d\rho \, \rho^{-\nu} \rho^{n-1}; \quad \forall \, \nu < 1. \tag{11.64}$$

From Eq. (11.63) we obtain the Schwinger expression for the Coulomb Green's function [7]

$$G(\mathbf{p}, \mathbf{p}', E) = \frac{\delta(\mathbf{p} - \mathbf{p}')}{E - T} + \frac{1}{E - T} \left\{ -\frac{Z}{2\pi^2 |\mathbf{p} - \mathbf{p}'|^2} \left[1 - \right. \right.$$

$$\left. \left. -4i\nu \int_0^1 d\rho \frac{\rho^{-i\nu}}{f(1-\rho)^2 - 4\rho} \right] \right\} \frac{1}{E - T'}. \tag{11.65}$$

In this equation the kinetic energies T and T' are given by

$$T = p^2/2, \quad T' = p'^2/2. \tag{11.66}$$

The function f has the form

$$f = \frac{(p_0^2 - p^2)(p_0^2 - p'^2)}{p_0^2 |\mathbf{p} - \mathbf{p}'|^2}. \tag{11.67}$$

The expression (11.65) is valid for bound states only ($E < 0$) due to the restriction $i\nu < 0$. As Schwinger pointed out [7], this restriction is circumvented by analytic continuation. This is achieved by replacing the real integral over ρ in Eq. (11.65) by a contour integral. The integration is performed along the path $\mathcal{C}$ that begins at $\rho = 1 + i0^+$ with a zero phase of ρ and then moves to the origin encircling it once and then terminates at $\rho = 1 - i0^+$. The Green's function has then the general form

$$G(\mathbf{p}, \mathbf{p}', E) = \frac{\delta(\mathbf{p} - \mathbf{p}')}{E - T} + \frac{1}{E - T} \left\{ \frac{-Z}{2\pi^2 |\mathbf{p} - \mathbf{p}'|^2} \left[1 - \right. \right.$$

$$\left. \left. - \frac{4i\nu}{e^{2\pi\nu} - 1} \int_{\mathcal{C}} d\rho \frac{\rho^{-i\nu}}{f(1 - \rho)^2 - 4\rho} \right] \right\} \frac{1}{E - T'}.$$

$$(11.68)$$

Alternative representations are obtained from this expressions upon performing partial integration yielding

$$G(\mathbf{p}, \mathbf{p}', E) = \frac{\delta(\mathbf{p} - \mathbf{p}')}{E - T} + \frac{1}{E - T} \left\{ \frac{-Z}{2\pi^2 |\mathbf{p} - \mathbf{p}'|^2} \right.$$

$$\left. \times \frac{4}{1 - e^{2\pi\nu}} \int_{\mathcal{C}} d\rho \, \rho^{-i\nu} \partial_\rho \left[\frac{\rho}{f(1 - \rho)^2 - 4\rho} \right] \right\} \frac{1}{E - T'}.$$

$$(11.69)$$

and

$$G(\mathbf{p}, \mathbf{p}', E) = \frac{1}{E - T} \left\{ \frac{-Z}{2\pi^2 |\mathbf{p} - \mathbf{p}'|^2} \right.$$

$$\left. \times \frac{4if}{\nu(e^{2\pi\nu} - 1)} \int_{\mathcal{C}} d\rho \, \rho^{-i\nu} \partial_\rho \left[\frac{\rho(1 - \rho^2)}{[f(1 - \rho)^2 - 4\rho]^2} \right] \right\} \frac{1}{E - T'}.$$

$$(11.70)$$

As discussed above the Green's function as function of the energy has poles at bound-state energies and branch cut along the continuum spectrum of the system.

Properties of the Coulomb Green's function

The Coulomb GF has some peculiar features. In addition to the complicated asymptotic behaviour due to the long-range tail of the Coulomb potentia, the Coulomb Green's function does not have a well-defined on-shell limit, i.e. the limits $p \to \pm\sqrt{2E}$, $p' \to \pm\sqrt{2E}$, or $p \to p' \to \pm\sqrt{2E}$ are not well-defined [7, 281]. As deduced from Eq. (11.23), this fact is reflected in a similar anomalous behaviour of the T matrix elements which are directly related to the scattering amplitude. As demonstrated by Schwinger [7] the amplitude for the Rutherford scattering can nevertheless be extracted from Eq. (11.70) by first removing singular factors and then taking the on-shell limit. Later on a similar procedure has been applied when evaluating the (physical) on-shell T matrix elements for scattering processes involving Coulomb interaction (see e.g.[282] and references therein).

The derivation of the above expressions for the Coulomb GF, which is due to Schwinger [7] (see also [283]), utilizes the $O(4)$ symmetry of the bound Coulomb system and then the continuum spectrum is incorporated via analytical continuation. Equivalently, one can employ the $O(1,3)$ (rotation plus translation) of the continuum spectrum and obtains directly the GF [284, 285, 286, 287]. It should be also noted that Hostler [288] has provided an alternative derivation of the GF starting from the spectral representation (11.5) and showing that the continuum spectrum may be integrated over, such that the sum of the discrete states is cancelled. The expression derived by Hostler [288] can be retrieved from Eq. (11.68) upon the substitution $x = (1 + \rho)/(1 - \rho)$.

A further approach to obtain the GF has been put forward by Okubo and Feldman [281] who solved the integral equation (cf. 11.20)

$$G(\mathbf{p}, \mathbf{p}', E) = \frac{\delta(\mathbf{p} - \mathbf{p}')}{E - T'} + \frac{1}{E - T} \int d^3\mathbf{p}'' \, V(\mathbf{p}, \mathbf{p}'') G(\mathbf{p}'', \mathbf{p}', E) \qquad (11.71)$$

using the integral transform method. The resulting GF is expressed in terms of a one-dimensional integral

$$G(\mathbf{p}, \mathbf{p}', E) = \frac{\delta(\mathbf{p} - \mathbf{p}')}{E - T'} + \frac{2}{E - T'} \int_0^\infty d\xi \frac{Q(\xi, p', E)}{[\xi(p^2 - p_0^2) + |\mathbf{p} - \mathbf{p}'|^2]^2}, \qquad (11.72)$$

where the function Q is a solution of the differential equation

$$\partial_\xi Q(\xi, p', E) = \frac{iZ}{\sqrt{2}} \frac{Q(\xi)}{\sqrt{\xi(\xi + 1)E - \xi T'}}. \qquad (11.73)$$

The solution of this equation is

$$Q(\xi, p', E) = \frac{Z}{2\pi^2} \left\{ \frac{\sqrt{\xi(\xi + 1)E - \xi T'} + \xi\sqrt{E}}{\sqrt{\xi(\xi + 1)E - \xi T'} - \xi\sqrt{E}} \right\}^{i\nu}. \qquad (11.74)$$

Inserting (11.74) in (11.72) results in an expression for the Green's function that is shown to be identical [289] to the Green's functions (11.70).

A closed form expression of the Coulomb Green's function can be obtained [290] by decomposing the integrand in (11.68) in partial fraction and expressing the contour integral in terms of hypergeometric functions. This is done as follows: in (11.68) the contour integral is rewritten as

$$\frac{4i\nu}{e^{2\pi\nu}-1}\int_C d\rho \frac{\rho^{-i\nu}}{f(1-\rho)^2-4\rho}$$
$$= \frac{4i\nu}{f(f_+-f_-)}\int_0^1 dt \left\{\frac{t^{-1-i\nu}}{1-t\,f_+}-\frac{t^{-1-i\nu}}{1-t\,f_-}\right\}, \quad (11.75)$$

where the functions $f_\pm$ are defined as

$$f_\pm = \left(1+\frac{2}{f}\right) \mp \frac{2\sqrt{1+f}}{f}. \tag{11.76}$$

Recalling the integral representation of the hypergeometric function ${}_2F_1(a,b,c,z)$ [99]

$$_2F_1(a,b,c,z) = \frac{\Gamma(c)}{\Gamma(b)\Gamma(c-b)}\int_0^1 dt \frac{t^{b-1}(1-t)^{c-b-1}}{(1-t\,z)^a}, \quad \Re(c) > \Re(b) > 0, \tag{11.77}$$

we can evaluate the integrals (11.75) in terms of the hypergeometric functions ${}_2F_1(a,b,c,z)$ and insert them (provided $\Re(-i\nu) > 0$) in Eq. (11.68) to achieve a closed form expression for G. It should be noted that this same result has been also obtained [285] by expressing the GF in terms of Gegenbauer polynomials and utilizing the relation between the Gegenbauer Polynomials and the hypergeometric functions, see [289, 291, 292, 293, 294] for further details and references.

11.4   The off-shell T-matrix

From Eq. (11.22) it is clear that once the Green's function is obtained the off-shell T matrix can be derived according to the equation

$$G(\mathbf{p},\mathbf{p}',E) = \frac{\delta(\mathbf{p}-\mathbf{p}')}{E-T'} + \frac{T(\mathbf{p},\mathbf{p}',E)}{[E-T][E-T']}. \tag{11.78}$$

The expressions for the Green's function given in the previous sections can now be utilized for the derivation of T. E.g. using Eqs. (11.75, 11.77, 11.68) we derive for the off-shell transition matrix elements under the condition that $\Re(i\nu) < 0$

$$T(\mathbf{p}, \mathbf{p}', E) = \frac{-Z}{2\pi^2 |\mathbf{p} - \mathbf{p}'|^2} \left\{ 1 + \frac{1}{\sqrt{1+f}} \left[{}_2F_1(1, -i\nu, 1 - i\nu, f_-) \right. \right.$$

$$\left. \left. - {}_2F_1(1, -i\nu, 1 - i\nu, f_+) \right] \right\}. \tag{11.79}$$

The convergence properties of this expression are readily obtained from those for the hypergeometric function ${}_2F_1(a, b, c, z)$ which converges on the entire unit circle $|z| = 1$.

From Eq. (11.22) we conclude that, similarly to GF, the matrix elements $T(\mathbf{p}, \mathbf{p}', E + i\eta)$, $1 \gg \eta \in \mathbb{R}^+$ possess an infinite set of simple poles for $i\nu = n \in \mathbb{N}^+$ associated with the bound states of the spectrum. This follows directly from the properties of hypergeometric functions ${}_2F_1$ occurring (11.79) which have poles at [99]

$$\lim_{\eta \to 0^+} \left\{ \frac{Z}{\sqrt{-2(E + i\eta)}} \right\} = n \in \mathbb{N}^+.$$

11.4.1   The on-shell limit of the Coulomb T matrix elements

As illuded to above the Coulomb Green's function does not approach a well-defined limit on the energy shell. Due to the inter-relation between G and T it is expected that the T matrix elements will show similar anomalous behaviour. This tendency can be explored [295, 293] by transforming the hypergeometric functions in (11.79) according to [99]

$$\begin{aligned} {}_2F_1(a, b, c, z) = {} & \frac{\Gamma(c)\Gamma(b-a)}{\Gamma(b)\Gamma(c-a)} (-z)^{-a} \, {}_2F_1(a, 1-c+a, 1-b+a, z^{-1}) \\ & + \frac{\Gamma(c)\Gamma(a-b)}{\Gamma(a)\Gamma(c-b)} (-z)^{-b} \, {}_2F_1(b, 1-c+b, 1-a+b, z^{-1}). \end{aligned} \tag{11.80}$$

With this relation Eq. (11.79) can be rewritten in the form

$$T(\mathbf{p}, \mathbf{p}', E) = \frac{-Z}{2\pi^2 |\mathbf{p} - \mathbf{p}'|^2} \left\{ \tau_a(\mathbf{p}, \mathbf{p}', E) + \tau_b(\mathbf{p}, \mathbf{p}', E) \right\}, \tag{11.81}$$

where

$$\tau_a(\mathbf{p}, \mathbf{p}', E) = 1 - \frac{1}{\sqrt{1+f}} \left\{ \frac{i\nu f_+}{1+i\nu} {}_2F_1(1, 1+i\nu, 2+i\nu, f_+) \right.$$

$$\left. + {}_2F_1(1, -i\nu, 1-i\nu, f_+) \right\},$$

$$= 1 - \frac{1}{\sqrt{1+f}} \left\{ 1 + \sum_{j=1}^{\infty} \frac{2\nu^2}{j^2 + \nu^2} f_+^j \right\}.$$

(11.82)

The function τ_b is given by

$$\tau_b(\mathbf{p}, \mathbf{p}', E) = \frac{(-f_+)^{-i\nu}}{\sqrt{1+f}} \Gamma(1 - i\nu)\Gamma(1 + i\nu),$$

$$= \frac{f_+^{-i\nu}}{\sqrt{1+f}} \frac{\pi\nu}{\sinh \pi\nu} (-)^{-i\nu},$$

$$= \frac{f_+^{-i\nu}}{\sqrt{1+f}} \frac{2\pi\nu}{e^{2\pi\nu} - 1}.$$

(11.83)

To explore the on-shell limit we calculate, that near the energy shell the functions f and $f_\pm$, given by Eq. (11.67) and Eq. (11.76), are related as

$$f_+ \approx f/4, \quad f_+^* \approx f^*/4.$$

(11.84)

In addition, the function τ_a (11.82) tends to zero in the half-energy-shell limit, i. e. for $p \to p_0$ or $p' \to p_0$. This is inferred from the definition of f (11.67), which also tends to zero half on-shell, so that the function τ_a (11.82) can be expanded in powers of f as [296]

$$\tau_a = \frac{f}{2(1+\nu^2)} - \frac{3f^2}{2(1+\nu^2)(4+\nu^2)} + O(f^3).$$

(11.85)

With the help of Eqs. (11.82, 11.83, 11.84, 11.85) we infer the behaviour of $T(\mathbf{p}, \mathbf{p}', E)$ in the half-energy shell as well as in the on-shell limits. At first let us inspect the half-energy-shell limit, i. e.

$$\lim_{\mathbf{p}' \to \mathbf{p}_0} T(\mathbf{p}, \mathbf{p}', p_0^2/2 + i0^+) = \frac{-Z}{2\pi^2 |\mathbf{p} - \mathbf{p}_0|^2} \frac{2\pi\nu}{e^{2\pi\nu} - 1} \left[\frac{f}{4}\right]^{-i\nu}.$$

(11.86)

From a physical point of view this (half) on-shell expression should coincide with the Coulomb scattering amplitude[3] $-\frac{1}{4\pi^2} f(\hat{\mathbf{p}} \cdot \hat{\mathbf{p}}_0)$, which obviously does not do. To isolate the source of this

---

[3]In the (first-order) Born approximation Wentzel [297] and Oppenheimer [298] derived the Coulomb scattering amplitude as

$$f_B(\hat{\mathbf{p}} \cdot \hat{\mathbf{p}}_0) = \frac{2Z}{|\mathbf{p} - \mathbf{p}_0|^2},$$

(11.87)

problem we rewrite Eq. (11.86) taking advantage of the definitions of f and τ_b [Eqs. (11.67, 11.82)]

$$\lim_{\mathbf{p}'\to\mathbf{p}_0} T(\mathbf{p},\mathbf{p}',p_0^2/2+i\eta)\Big|_{\eta\to 0^+} = \lim_{\mathbf{p}'\to\mathbf{p}_0} g(p)\left[-\frac{1}{4\pi^2}f(\hat{\mathbf{p}}\cdot\hat{\mathbf{p}}_0)\right]g(p'), \qquad (11.89)$$

with the off-shell factors $g(p_j)$ being defined as

$$g(p) = \left[\frac{p_0^2-p^2+i\eta}{4p_0^2}\right]^{-i\nu}\Gamma(1-i\nu)\,e^{-\pi\nu/2};$$

$$g(p') = \left[\frac{p_0^2-p'^2+i\eta}{4p_0^2}\right]^{-i\nu}\Gamma(1-i\nu)\,e^{-\pi\nu/2}. \qquad (11.90)$$

In the on-shell limit $\mathbf{p}\to\mathbf{p}_0$ these functions are singular and contain divergent phase factors. This is readily inferred from the relation

$$\lim_{\mathbf{p}\to\mathbf{p}_0,\,\mathbf{p}'\to\mathbf{p}_0,} g(p_0) = (4p_0^2)^{i\nu}\,\Gamma(1-i\nu)\,e^{-i\nu\ln\eta}. \qquad (11.91)$$

Obviously these factors diverge for $\eta\to 0$. This means on the other hand that the physical on-shell Coulomb scattering amplitude and the off-shell T matrix elements are related to each others via singular factors that account for the absence of free asymptotic states in the initial and in the final channels in two-body Coulomb scattering. This fact has already been noted by Schwinger [7] who suggested to incorporate these singular off-shell factors into the free Green's function G_0.

Interestingly, in the high energy limit ($p_0 \gg 1$), where conventionally the Born series (11.20) is expected to perform well, Eq. (11.86) reduces to

$$\lim_{\mathbf{p}'\to\mathbf{p}_0} T(\mathbf{p},\mathbf{p}',p_0^2/2+i\eta)\Big|_{\eta\to 0^+,\,p_0\to\infty} \longrightarrow \frac{-Z}{2\pi^2|\mathbf{p}-\mathbf{p}_0|^2}\,e^{-i\nu\ln(f/4)}. $$

$$(11.92)$$

which yielded the correct Rutherford formula for the differential cross section

$$\frac{d\sigma_B}{d\Omega} = \frac{d\sigma_{exact}}{d\Omega} = \frac{Z^2}{4p_0^4\sin^4(\theta/2)}.$$

The polar angle θ is measured with respect to the incoming beam. On the other hand a number of studies [299, 300, 301, 302, 303] have shown that this coincidence of the Born and the exact result is due to the fact that the exact Coulomb scattering amplitude contains only additional phase factors

$$f_{exact}(\hat{\mathbf{p}}\cdot\hat{\mathbf{p}}_0) = \frac{2Z}{|\mathbf{p}-\mathbf{p}_0|^2}\left[\frac{4p_0^2}{|\mathbf{p}-\mathbf{p}_0|^2}\right]^{-i\nu}e^{2i\sigma_0};$$

$$\sigma_0 = \arg\Gamma(1+i\nu). \qquad (11.88)$$

This relation follows from

$$\lim_{p_0 \gg 1} \frac{2\pi\nu}{e^{2\pi\nu} - 1} \to 1. \tag{11.93}$$

This means summing up the complete Born series $T = V + VG_0V + VG_0VG_0V + \cdots$ produces merely a modification of the Coulomb scattering amplitude by a divergent phase factor which is irrelevant for the evaluation of the cross section (provided the on-shell limit is taken after performing all calculations). This highlights at the same time the danger of dealing with the perturbation expansions for Coulomb systems, e.g. such as evaluating only certain terms in (11.20): when Coulomb potentials are involved one has to deal with divergent factors that sum up to a single (irrelevant) phase in the final result. Thus, despite the fact that the first order term produces the correct cross section (cf. Eq. (11.87), a truncation of the expansion after certain terms may well lead to a situation where the divergent terms do not sum up correctly and hence divergent cross sections are obtained [301].

12 Operator approach to finite many-body systems

In the preceding section we discussed the subtle features and difficulties related to the infinite-range tail of the Coulomb potential. For "well-behaved" potentials, however, the Lippmann-Schwinger integral equations (11.20) for T or G are powerful tools for the treatment of two-particle systems, because the kernel $K = G_0 V$ can be made square integrable[1] and hence standard results of the Fredholm theory of integral equations [308] can be utilized. For three-body problem (involving "well-behaved" potentials) the situation changes. Even though we can formally define the resolvent of the three-body Schrödinger equation as $G = G_0 + G_0 U G$ (where U is the total potential consisting of pair interactions) two difficulties arise when attempting to derive state vectors or evaluate the matrix elements: 1.) In contrast to the two-particle case, for a three-body system the Lippmann-Schwinger equations for the state vectors do not have a unique solution [306, 307]. 2.) As proved by Faddeev [304, 305] the kernel $K = G_0 U$ of the Lippmann-Schwinger integral equations is not a square integrable operator for $N \geq 3$, i.e. the norm $\|K(E)\| = \left\{ \mathrm{tr}\left[K(E) K^\dagger(E) \right] \right\}^{1/2}$ is not square integrable, even after iterations. The source of this problem is the occurrence of disconnected diagrams where one of the particles is a spectator and does not interact with the other particles. Faddeev and Weinberg [304, 305, 309] also addressed the question of the compactness of K and found that the kernel K is not compact. In view of this situation Faddeev and others [304, 305, 309, 310, 311, 312, 313, 314, 318] proposed a new formulation of the three-body problem in terms of convergent integral equations. In particular, Faddeev has shown in details that the kernel of his equations is connected after one iteration. In fact, with certain (weak) requirements on the potential it can be shown that the kernel is compact for all but physical values of E, which implies that the solution of the Faddeev equations is unique at energies below the three-body breakup threshold.

---

[1] In fact the kernel $K(E) = G_0(E) V$ is not square integrable, but it can be made so by iterating the integral equation once or by multiplying by $V^{1/2}$.

In the next section we will outline briefly without going into the mathematical details the main elements of the Faddeev theory.

12.1   Faddeev approach to the three-body problem

Let us consider a non-relativistic three-body system. The total potential V is a sum of pairwise interactions V_{ij}, i.e.

$$V = \sum_{j>i} V_{ij} = \sum_{k} V_k. \tag{12.1}$$

Here we introduced the following notation for the pair potentials

$$V_k = V_{ij}, \quad \epsilon_{ijk} \neq 0; \quad V_0 \equiv 0.$$

Accordingly, we introduce the total transition operator T, the auxiliary operators T_j and the transition operator t_j of the pair j by the following equations

$$
\begin{aligned}
T &= V + V G_0 T, \\
T_j &= V_j + V_j G_0 T, \\
t_j &= V_j + V_j G_0 t_j, \quad j = 1, 2, 3.
\end{aligned}
\tag{12.2}
$$

$$\tag{12.3}$$

From these definitions it is readily verified that

$$T = \sum_{j=1}^{3} T_j. \tag{12.4}$$

Since $V_j = t_j - t_j G_0 V_j$ we can deduce for the operator T_j the following relations

$$
\begin{aligned}
T_j &= t_j - t_j G_0 V_j + t_j G_0 T - t_j G_0 V_j G_0 T, \\
&= t_j + t_j G_0 (T - T_j), \\
T_j &= t_j + t_j G_0 (T_k + T_l), \quad \epsilon_{jkl} \neq 0.
\end{aligned}
\tag{12.5}
$$

This relation constitutes the Faddeev equations that can be written in the matrix form

$$
\begin{pmatrix} T_1 \\ T_2 \\ T_3 \end{pmatrix} = \begin{pmatrix} t_1 \\ t_2 \\ t_3 \end{pmatrix} + [\mathbf{K}] \begin{pmatrix} T_1 \\ T_2 \\ T_3 \end{pmatrix}.
\tag{12.6}
$$

The kernel $[\mathbf{K}]$ is a matrix operator that depends on the (off-shell) two-body transition matrix elements t_j, i.e.

$$[\mathbf{K}] = \begin{pmatrix} 0 & t_1 & t_1 \\ t_2 & 0 & t_2 \\ t_3 & t_3 & 0 \end{pmatrix} G_0. \tag{12.7}$$

While this kernel seems to contain disconnected terms, any further iteration of (12.6) is free of expressions with disconnected diagrams. A careful inspection of the kernel matrix showed [304, 305] that it is square integrable, i.e. $\operatorname{tr} \sum_{ij=1}^{3} \left\{ [\mathbf{K}]_{ij} [\mathbf{K}]_{ij}^{\dagger} \right\} < \infty$. As clear from (12.7), the building blocks of the Faddeev equations are the off-shell two-body t_j matrix elements. The unitarity of t_j ensures the unitarity of the solution T.

Since the T operators and the Green's operator $G(E)$ are related to each other via relation $G = G_0 + G_0 T G_0$, we can write $G(E)$ in the form

$$G = G_0 + \sum_{j=1}^{3} G_j. \tag{12.8}$$

Similar to T_j the auxiliary operators G_j are expressed in terms of the pair Green's operators

$$g_j = G_0 + G_0 V_j g_j \tag{12.9}$$

via the relation that follows directly from (12.6)

$$\begin{pmatrix} G_1 \\ G_2 \\ G_3 \end{pmatrix} = \begin{pmatrix} g_1 - G_0 \\ g_2 - G_0 \\ g_3 - G_0 \end{pmatrix} + [\tilde{\mathbf{K}}] \begin{pmatrix} G_1 \\ G_2 \\ G_3 \end{pmatrix}. \tag{12.10}$$

The kernel of this matrix integral equation derives from the kernel $[\tilde{\mathbf{K}}]$ of (12.6) as

$$[\tilde{\mathbf{K}}] = G_0 [\mathbf{K}] G_0^{-1}.$$

Having established the relations for G and T we can write equivalent relations for the state vectors [317], in a similar way as demonstrated in the preceding chapter.

12.1.1   The Lovelace operator, Lovelace equations and AGS equations

In a three-body system a variety of interaction channels exist. For example, let us consider the situation that one of the particles, say particle 1, to be initially decoupled[2] from the other two particles, which are bound to each other. This asymptotic three-body state we denote by

---

[2] We assume all interactions in the system to be of a finite range.

$|\phi_\alpha\rangle$, where α labels all the quantum numbers needed for the characterization of the state of the system. $|\phi_\alpha\rangle$ is an eigenstate of the "channel" Hamiltonian H_α, i.e.

$$H_\alpha |\phi_\alpha\rangle \;=\; [H_0 + V_\alpha] = E_\alpha |\phi_\alpha\rangle. \tag{12.11}$$

A collision of particle 1 from the bound system (23) may leave (23)*, which labels an excited state of the system (23). It may also lead to three unbound particles (break-up channel). Particle 1 may as well be elastically scattered from (23) or may substitute for one of the bound particles (rearrangement channel). We designate the three-body state achieved upon the collision by $|\phi_\beta\rangle$ and the corresponding channel Hamiltonian by H_β where

$$H_\beta |\phi_\beta\rangle \;=\; [H_0 + V_\beta] |\phi_\beta\rangle = E_\beta |\phi_\beta\rangle. \tag{12.12}$$

As in Eq. (11.30), we define channel Møller operators that select the (interacting) three-body state that develops from (to) the asymptotic state $|\phi_\alpha\rangle$ ($|\phi_\beta\rangle$) under the action of the perturbation $\bar{V}_\alpha$ ($\bar{V}_\alpha$)

$$|\Psi_\alpha^+\rangle \;=\; \Omega_\alpha^+ |\phi_\alpha\rangle = (1 + G^+ \bar{V}_\alpha)|\phi_\alpha\rangle, \tag{12.13}$$

$$|\Psi_\beta^-\rangle \;=\; \Omega_\beta^- |\phi_\beta\rangle = (1 + G^- \bar{V}_\beta)|\phi_\beta\rangle. \tag{12.14}$$

The interaction potentials $\bar{V}_\alpha$ and $\bar{V}_\alpha$ are those parts of the total potential that are not contained in H_α and H_β, i.e.

$$\bar{V}_\alpha = H - H_\alpha, \quad \bar{V}_\beta = H - H_\beta, \tag{12.15}$$

where H is the total three-body Hamiltonian.

As in the two-body case (cf. 11.31) we define the S operator as

$$S = \Omega_\beta^{-\dagger} \Omega_\alpha^+.$$

The matrix elements $S_{\beta,\alpha}$ of the S operator provide a measure for the transition probability from the channel α to β. $S_{\beta,\alpha}$ are determined from

$$\langle \Psi_\beta^-|\Psi_\alpha^+\rangle = \langle \phi_\beta|\Omega_\beta^{-\dagger}\Omega_\alpha^+|\phi_\alpha\rangle = \langle \phi_\beta|S|\phi_\alpha\rangle. \tag{12.16}$$

Reformulating $S_{\beta,\alpha}$ according to the steps (11.32)-(11.34) we obtain the post form for the matrix elements $S_{\beta,\alpha}$ [319, 315]

$$\langle \phi_\beta|S|\phi_\alpha\rangle \;=\; \delta_{\beta,\alpha} - 2\,i\,\pi\,\delta(E_\beta - E_\alpha)\,\langle \phi_\beta|\bar{V}_\beta|\Psi_\alpha^+\rangle, \tag{12.17}$$

as well as the prior form

$$\langle \phi_\beta | S | \phi_\alpha \rangle = \delta_{\beta,\alpha} - 2\,i\,\pi\,\delta(E_\beta - E_\alpha)\,\langle \Psi_\beta^- | \bar{V}_\alpha | \phi_\alpha \rangle. \tag{12.18}$$

Comparing these two relations for $S_{\beta,\alpha}$ with Eqs. (11.34, 11.35) makes clear that the matrix elements $T_{\beta\alpha} := \langle \phi_\beta | \bar{V}_\beta | \Psi_\alpha^+ \rangle$ and $T_{\beta\alpha} \dot{=} \langle \Psi_\beta^- | \bar{V}_\alpha | \phi_\alpha \rangle$ play the role of the transition matrix elements, where

$$T_{\beta\alpha} = \langle \phi_\beta | \bar{V}_\beta\,\Omega_\alpha^+ | \phi_\alpha \rangle = \langle \phi_\beta | \bar{V}_\beta\,(1 + G^+ \bar{V}_\alpha) | \phi_\alpha \rangle = \langle \phi_\beta | U_{\beta\alpha}^+ | \phi_\alpha \rangle. \tag{12.19}$$

Similarly we deduce the relations

$$T_{\beta\alpha} = \langle \phi_\beta | \Omega_\beta^{-\dagger}\,\bar{V}_\alpha | \phi_\alpha \rangle = \langle \phi_\beta | (1 + \bar{V}_\beta\,G^+)\,\bar{V}_\alpha | \phi_\alpha \rangle = \langle \phi_\beta | U_{\beta\alpha}^- | \phi_\alpha \rangle. \tag{12.20}$$

The two-channel operators

$$U_{\beta\alpha}^+ = \bar{V}_\beta\,(1 + G^+ \bar{V}_\alpha) = \bar{V}_\beta + \bar{V}_\beta\,G_0^+\,\bar{V}_\alpha + \bar{V}_\beta\,G_0^+\,V\,G_0^+\,\bar{V}_\alpha + \cdots, \tag{12.21}$$

$$U_{\beta\alpha}^- = (1 + \bar{V}_\beta\,G^+)\,\bar{V}_\alpha = \bar{V}_\alpha + \bar{V}_\beta\,G_0^+\,\bar{V}_\alpha + \bar{V}_\beta\,G_0^+\,V\,G_0^+\,\bar{V}_\alpha + \cdots, \tag{12.22}$$

are called the Lovelace-operators [311]. From the above derivation it is evident that $U_{\beta\alpha}^+$ and $U_{\beta\alpha}^-$ play the role of a two-channel transition operator.

The channel transition operators T_γ, $(\gamma = \alpha, \beta)$, defined as

$$T_\gamma^\pm = V_\gamma + T_\gamma^\pm G_0^\pm (E) V_\gamma, \tag{12.23}$$

yield in combination with Eqs. (12.21, 12.22) the Faddeev-Lovelace equations

$$U_{\beta\alpha}^+ = \bar{V}_\beta + \sum_{\gamma \neq \alpha} U_{\beta\gamma}^+ G_0^+ T_\gamma^+,$$

$$U_{\beta\alpha}^- = \bar{V}_\alpha + \sum_{\gamma \neq \beta} T_\gamma^+ G_0^+ U_{\beta\gamma}^-. \tag{12.24}$$

Using Eq. (12.23) the potential V_γ can be expressed in terms of T_γ, and if inserted in the Faddeev-Lovelace equations, one obtains the so-called Faddeev-Watson multiple scattering expansion [315]. A further variant of the Faddeev equations for the Lovelace operators (12.24) are the so-called Alt-Grassberger-Sandhas (AGS) equations

$$U_{\beta\alpha}^+ = (E - H_0)(1 - \delta_{\beta\alpha}) + \sum_{\gamma \neq \beta} T_\gamma^+ G_0^+ U_{\gamma\alpha}^+. \tag{12.25}$$

From this equation we also obtain a multiple scattering expansion by iterating the T operator (12.23) that occurs in (12.25) to obtain

$$U_{\beta\alpha}^+ = (E - H_0)(1 - \delta_{\beta\alpha}) + \sum_{\gamma \neq \beta, \gamma \neq \alpha} T_\gamma^+ + \sum_{\gamma \neq \beta, \delta \neq \gamma} T_\gamma^+ G_0^+ T_\delta^+ G_0^+ U_{\delta\alpha}^+, \tag{12.26}$$

i.e. successive T operators in the last term of this equation should be different. The AGS equations (12.25) link the operators for the elastic $U^+_{\beta\beta}$ and the rearrangement reactions $U^+_{\beta\alpha}, (\beta \neq \alpha)$. The break-up processes are also incorporated (the operator $T^+_\gamma G^+_0$ requires the knowledge of $U^+_{\gamma\alpha}$ in the full Hilbert space). More details on the AGS scheme as well as on the numerical realization of the Faddeev equations and of the various problems that arise in this context can be found in Ref. [317].

The Faddeev approach to the three-body problem can be generalized to finite many-particle systems. This has been done by Yakubovsky [316]. In the next section we outline an alternative route to the four and many-body finite systems [101].

12.2   Reduction scheme of the Green's operator of N-particle systems

The Faddeev theory of the three-particle problem delivered the formula (12.10) for the determination of the three-body Green's operator in terms of the reference Green's function G_0 and two-particle quantities, the off-shell two-body Green's and transition operators g_j and t_j, which are the known input in the theory. In this section we address the question of how to generalize this scheme such that we obtain the transition and the Green's operators of an interacting N-body system from the solutions of the $N - M$ body problem, where $M = 1, 2 \cdots, N - 2$.

For this purpose let us consider a nonrelativistic system consisting of N particles interacting via pair-wise interactions v_{ij}, i. e. the total potential $U^{(N)}$ can be written as

$$U^{(N)} = \sum_{j>i=1}^{N} v_{ij}. \tag{12.27}$$

At this stage there is no need to specify constraints on the individual potentials v_{ij}, because we are going to perform only algebraic manipulations.

The essential point of what follows is that the potential $U^{(N)}$ satisfies the recurrence relations

$$U^{(N)} = \frac{1}{N-2} \sum_{j=1}^{N} u_j^{(N-1)}, \tag{12.28}$$

$$u_j^{(N-1)} = \frac{1}{N-3} \sum_{k=1}^{N-1} u_{jk}^{(N-2)}, \quad j \neq k. \tag{12.29}$$

The interaction term $u_j^{(N-1)}$ is the total potential of the N particle system in which only $N-1$ particles are interacting, while the particle labelled j is free. This means all interaction lines to particle j are switched off. In terms of the pair potentials v_{mn} the interaction $u_j^{(N-1)}$ casts

$$u_j^{(N-1)} = \sum_{m>n=1}^{N} v_{mn}, \quad m \neq j \neq n. \tag{12.30}$$

To illustrate the meaning of Eqs. (12.28, 12.29) we consider in Fig. 12.1 the interaction potential of a system consisting of six particles. According to Eqs. (12.28, 12.29) the total potential of the six interacting particles can be linearly expanded in terms of the total potentials of five correlated particles. In the latter potentials the number of interaction lines can be further reduced by mapping onto the total potentials of four interacting particles (cf. Eq. (12.29)).

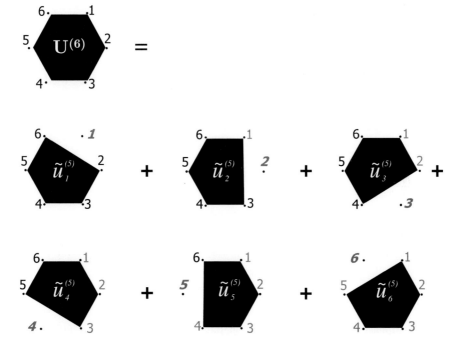

Figure 12.1: A pictorial representation of the total potential expansion (12.28) for six interacting particles enumerated and marked by the full dots at the corners of the hexagon. The hexagon stands for the full potential $U^{(6)}$ of the six correlated particles and is reduced according to (12.28) to five pentagons. Each pentagon symbolizes the full five body potential $\tilde{u}_j^{(5)} = \left(u_j^{(5)}\right)/4$ of those five particles that are at the corners of the pentagon. Particles that reside not at a corner of a pentagon are free (disconnected).

This recursive procedure can be repeated to reach the case of the pair-interactions. From

Fig. 12.1 it is clear that all interactions are treated on equal footing. Furthermore, the N-body potential is reduced systematically to sums of $N - M$ potentials with $M = 1, 2, \cdots, N - 2$.

The total Hamiltonian can be written as

$$H^{(N)} = H_0 + U^{(N)} = H_0 + \frac{1}{N-2} \sum_{j=1}^{N} u_j^{(N-1)} \, .$$

The key question is now whether the N particle transition operator

$$T^{(N)} = U^{(N)} + U^{(N)} G_0 T^{(N)} \tag{12.31}$$

and the interacting N-body Green's operators

$$G^{(N)} = G_0 + G_0 U^{(N)} G^{(N)} \tag{12.32}$$

satisfy recursion relations similar to Eq. (12.28) [in Eqs. (12.31, 12.32) we used $G_0(z) = (z-H_0)^{-1}$]. To answer this question we proceed as follows: Making use of the decomposition (12.28), we write the integral equation for the transition operator as

$$
\begin{aligned}
T^{(N)} &= \sum_{j=1}^{N} T_j^{(N-1)} \, , &\tag{12.33} \\
T_j^{(N-1)} &= \tilde{u}_j^{(N-1)} + T^{(N)} G_0 \tilde{u}_j^{(N-1)}, \quad j = 1, \cdots, N. &\tag{12.34}
\end{aligned}
$$

The scaled potentials $\tilde{u}_j^{(N-1)}$ are defined as

$$\tilde{u}_j^{(N-1)} = \left(u_j^{(N-1)} \right) / (N-2).$$

It should be emphasized that the (auxiliary) operators $T_j^{(N-1)}$, given by Eq. (12.34), are not the T operators of a system in which only $N - 1$ particles are interacting. This is evident from the fact that on the right-hand side of Eq. (12.34) the N-particle T operator reappears.

The transition operator of the N particle system in which all interaction lines to particle j are switched off, while the remaining $N - 1$ particles are interacting via the scaled potential $\tilde{u}_j^{(N-1)}$, has the form

$$t_j^{(N-1)} = \tilde{u}_j^{(N-1)} + \tilde{u}_j^{(N-1)} G_0 t_j^{(N-1)}. \tag{12.35}$$

Using this relation we reformulate Eq. (12.34) in the following manner

$$
\begin{aligned}
T_j^{(N-1)} &= t_j^{(N-1)} + t_j^{(N-1)}G_0 T^{(N)} - t_j^{(N-1)}G_0\left(\tilde{u}_j^{(N-1)} + \tilde{u}_j^{(N-1)}G_0 T^{(N)}\right), \\
&= t_j^{(N-1)} + t_j^{(N-1)}G_0\left(T^{(N)} - T_j^{(N-1)}\right), \\
&= t_j^{(N-1)} + t_j^{(N-1)}G_0 \sum_{k\neq j}^{N} T_k^{(N-1)}.
\end{aligned}
$$

$$(12.36)$$

In matrix form this integral equation can be expressed as

$$
\begin{pmatrix} T_1^{(N-1)} \\ T_2^{(N-1)} \\ \vdots \\ T_{N-1}^{(N-1)} \\ T_N^{(N-1)} \end{pmatrix}
=
\begin{pmatrix} t_1^{(N-1)} \\ t_2^{(N-1)} \\ \vdots \\ t_{N-1}^{(N-1)} \\ t_N^{(N-1)} \end{pmatrix}
+ [\mathbf{K}^{(N-1)}]
\begin{pmatrix} T_1^{(N-1)} \\ T_2^{(N-1)} \\ \vdots \\ T_{N-1}^{(N-1)} \\ T_N^{(N-1)} \end{pmatrix}.
\tag{12.37}
$$

The kernel $[\mathbf{K}^{(N-1)}]$ is a matrix operator consisting of the (off shell) transition operators with $N-1$ interacting particles. It has the explicit form

$$
[\mathbf{K}^{(N-1)}] =
\begin{pmatrix}
0 & t_1^{(N-1)} & t_1^{(N-1)} & \cdots\cdots & t_1^{(N-1)} \\
t_2^{(N-1)} & 0 & t_2^{(N-1)} & \cdots\cdots & t_2^{(N-1)} \\
\cdots\cdots\cdots\cdots\cdots\cdots\cdots\cdots\cdots\cdots \\
t_{N-1}^{(N-1)} & \cdots\cdots & t_{N-1}^{(N-1)} & 0 & t_{N-1}^{(N-1)} \\
t_N^{(N-1)} & \cdots\cdots & t_N^{(N-1)} & t_N^{(N-1)} & 0
\end{pmatrix}
G_0.
\tag{12.38}
$$

According to the scheme (12.37) the lowest order approximation to the $T^{(N)}$ operator of the N interacting particles is given by

$$
T^{(N)} = \sum_{j=1}^{N} t_j^{(N-1)}.
\tag{12.39}
$$

For the potential depicted in Fig. 12.1 the first iteration of the integral equation (12.37) is shown diagrammatically in Fig. 12.2. From this illustration it is obvious that the lowest order approximation is a sum of N elements each representing separate systems with different interacting particles. Thus, the matrix elements of the operators $t_j^{(N-1)}$ can be evaluated individually and independent of each other, which brings about a substantial simplification in the numerical realization.

From the defining equation of $t_j^{(N-1)}$ (12.35) and from the structure of the related potential Eq. (12.29) it is evident that the above procedure can be repeated and the operator $t_j^{(N-1)}$ can

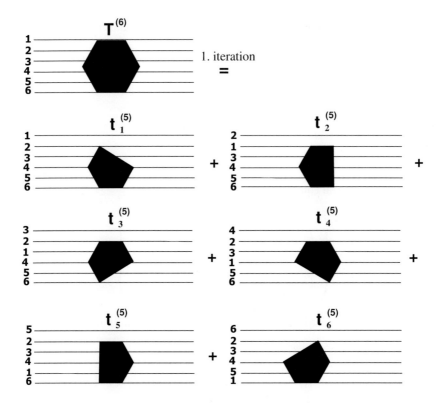

Figure 12.2: A diagrammatic representation of Eq. (12.39) for a system consisting of six corre-lated particles (cf. Fig. 12.1). The hexagons and the pentagons (with a specific orientation) label the same potentials as explained in Fig. 12.1. Each of the pictures stands for a transition opera-tor of the six-body system (the particles are labelled by straight lines). The respective transition operator is indicated on each of the diagrams.

also be expressed in terms of the transition operators of systems in which only $N-2$ particles are interacting, i.e.

$$t_j^{(N-1)} = \sum_{k \neq j}^{N-1} T_k^{(N-2)}$$

and the operators $T_k^{(N-2)}$ are deduced from a relation similar to Eq. (12.37) with N being replaced by $N-1$.

Now we focus on the corresponding relations for the N particle Green's operator. Since the T and the G operators are linked to each others via

$$G^{(N)} = G_0 + G_0 T^{(N)} G_0$$

we can deduce from the recursive scheme for the $T^{(N)}$ operators that the Green's operator $G^{(N)}$ can be cast in the form

$$G^{(N)} = G_0 + \sum_{j=1}^{N} G_j^{(N-1)}. \tag{12.40}$$

As in the case for the transition operators $T^{(N)}$, the (auxiliary) operators $G_j^{(N-1)}$ are related (but not identical) to the Green's operators $g_j^{(N-1)}$ of systems in which only $N-1$ particles are correlated by virtue of $\tilde{u}_j^{(N-1)}$, more precisely

$$\begin{pmatrix} G_1^{(N-1)} \\ G_2^{(N-1)} \\ \vdots \\ G_{N-1}^{(N-1)} \\ G_N^{(N-1)} \end{pmatrix} = \begin{pmatrix} g_1^{(N-1)} - G_0 \\ g_2^{(N-1)} - G_0 \\ \vdots \\ g_{N-1}^{(N-1)} - G_0 \\ g_N^{(N-1)} - G_0 \end{pmatrix} + [\tilde{\mathbf{K}}^{(N-1)}] \begin{pmatrix} G_1^{(N-1)} \\ G_2^{(N-1)} \\ \vdots \\ G_{N-1}^{(N-1)} \\ G_N^{(N-1)} \end{pmatrix}, \tag{12.41}$$

where

$$[\tilde{\mathbf{K}}^{(N-1)}] = G_0 [\mathbf{K}^{(N-1)}] G_0^{-1}.$$

From Eqs. (12.37, 12.41) the following picture emerges. Assuming that the Green's operator of the interacting $N-1$ body system is known, the Green's operator of the N particles can be evaluated by solving a set of N linear, coupled integral equations (namely Eqs. (12.37, 12.41)). This case we encountered in Fig. 12.2 for the first-order iteration of the operator $T^{(N)}$. If, on the other hand, only the solution of the $N - M$ problem is known, where $M = 2, \cdots, N-2$, we have to perform a hierarchy of calculations starting by obtaining the solution for the $N - M + 1$ problem and repeating the procedure to reach the solution of the N-body problem. Thus, the lowest order approximation within this strategy [first iteration of (12.41)] is schematically shown in Fig. 12.3 for N particles, where only the solution of the two-body problem is exactly known (numerically or analytically). The first step (lower part of Fig. 12.3) consists in constructing the three-particle Green's function and from that the four-particle Green's function, going up in the hierarchy till the Green's function of the interacting N particle system is derived.

Figure 12.3 as well as equations (12.37, 12.41) suggest the existence of disconnected terms. It can be shown however that any iteration of the kernels of Eqs. (12.37, 12.41) is free of disconnected terms since the disconnected terms occur only in the off-diagonal elements of

$[\mathbf{K}^{N-M}]$ and $[\tilde{\mathbf{K}}^{N-M}]$. Furthermore, it is clear from Eqs. (12.37, 12.41) and from Fig. 12.3 that for $N = 3$ the present method reduces to the Faddeev approach.

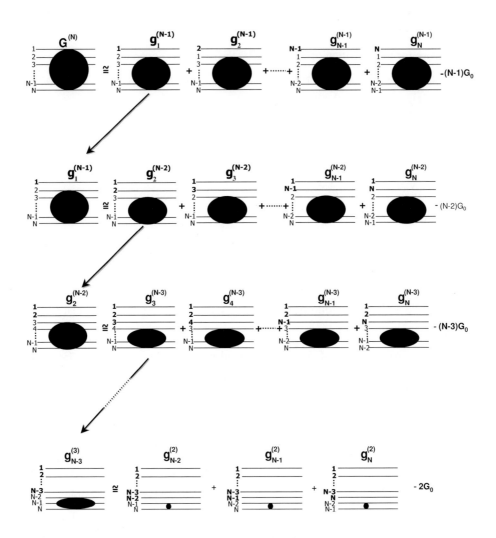

Figure 12.3: A diagrammatic representation of the lowest order approximation to Eqs. (12.37, 12.41) that yields the Green's operator $G^{(N)}$ of N interacting particles. The total potential is labelled by the black circle. The interactions between $N - 1$, $N - 2$, and $N - 3$ particles are indicated by ellipses with different eccentricities. The particles are symbolized by the solid lines. Only those particles that cross an ellipse are interacting. Each diagram stands for the Green's operator (specified on top of the diagram). The corresponding potential is indicated by the circles and/or the ellipses.

12.3   Thermodynamics of interacting N-particle systems

Strictly speaking, finite systems do not expose phase transitions [102]. For example, in a finite system thermodynamical quantities have naturally an upper bound and so do their fluctuations (which grow infinite in the thermodynamic limit). Nevertheless, we expect to observe the onset of a critical behaviour when the thermodynamic limit is approached. The traditional theory that addresses these questions is the finite-size scaling theory (e.g. [104] and references therein).

The incremental scheme (12.37, 12.41) outlined in the previous section for the derivation of the Green's function $G^{(N)}$ of an interacting N-particle system can be utilized to deal with thermodynamic problems in finite correlated systems. In this context it is important to note that our Green's function expansion (12.41) is derived for a fixed number of particles N. Its essence is to dilute the interaction strength by successively mapping the interacting system onto a non-interacting one. Therefore, it is appropriate to operate within the canonical ensemble.

The canonical partition function[3] can be expressed in terms of the electronic density of states (DOS) [defined by Eq. (11.12)] as [84]

$$Z(\beta) = \int dE \, \Omega(E) \, e^{-\beta E}. \tag{12.42}$$

The DOS, denoted by $\Omega(E)$, can be deduced from the imaginary part of the trace of $G^{(N)}$ via [cf. Eq. (11.12)], i.e.

$$\Omega(E) = -\frac{1}{\pi} \, \text{tr} \left[\Im G^{(N)}(E) \right].$$

An important feature of the recurrence scheme (12.41), and in particular of its first iteration shown in Fig. 12.3, is that the Green's function $G^{(N)}(E)$ is given as a sum of less correlated GF's. This is decisive for the calculation of the trace (and hence of $\Omega(E)$), since in this case the trace of $G^{(N)}(E)$ is directly linked to the sum of traces of the GF's for systems with a lower number of interacting particles [cf. Fig. 12.3]. This immediately leads (in a first order approximation) to the recursion relation for the canonical partition function $Z^{(N)}$

$$Z^{(N)} = \sum_{j=1}^{N} Z_j^{(N-1)} - (N-1)Z_0, \tag{12.43}$$

---

[3]For the rest of this section β denotes the inverse temperature; in the units used, the Boltzmann constant is unity.

where Z_0 is the canonical partition function of a reference system consisting of independent particles. This can be taken as the non-interacting homogeneous electron gas described by the Hamiltonian H_0, but H_0 and the associated Z_0 can as well be chosen as any known reference system. In (12.43) $Z_j^{(N-1)}$ is the canonical partition function of a system containing N particles. The interaction strength in this system is however diluted by cutting all interaction lines that connect to particle j.

Equation (12.43) that relies on the GF expansion (12.41) renders possible the study of thermodynamic properties of finite systems on a microscopic level. In particular, relation (12.43) offers a tool for the investigation of the inter-relation between the thermodynamics and the strength of correlations in a finite system. For the study of the onset of critical behaviour we can utilize the ideas put forward by Yang and Lee [102, 103]. As an example let us consider the onset of condensation in a quantum Bose gas, the ground-state occupation number $\eta_0(N, \beta)$ is given by

$$\eta_0(N, \beta) = -\frac{1}{\beta}\frac{\partial_{\epsilon_0} Z^{(N)}(\beta)}{Z^{(N)}(\beta)} = -\frac{1}{\beta}\frac{\sum_{j=1}^{N} \partial_{\epsilon_0} Z_j^{(N-1)} - (N-1)\partial_{\epsilon_0} Z_0}{Z^{(N)}},$$

$$(12.44)$$

where ϵ_0 is the ground-state energy. Eq. (12.44) offers the possibility to study systematically the influence of the inter-particle interaction strength on the onset of the critical regime. As customary in the theory of Yang and Lee [102, 103], one may as well opt to investigate in the complex β plane the roots of of the partition function, i. e. of Eq. (12.43). Zero points of $Z^{(N)}(\beta)$ that approach systematically the real β axis signify a possible occurrence of critical behaviour in the thermodynamic limit.

12.4   Incremental product expansion of the many-body Green's function

The Green's operator expansion (12.41) has been derived using formal relations between the Hamilton operator and its resolvent as well as by utilizing an exact recursive relation of the total potential. Thus, the expansion (12.41) is formally exact. From a calculational point of view however, the first order approximation, depicted in Fig. 12.3 is of prime interest in particular as it allows for a direct study of the thermodynamics of finite correlated systems.

A close look at this first order term reveals however several deficiencies, e.g. 1.) The first-order term [shown in Fig. 12.3] does not yield the exact solution in the case where the total potential (12.28) is separable (for example, if the potential shown in Fig. 12.1) has the form $U^{(N)} = v_{12} + v_{34} + v_{56}$). This is clearly seen from the structure of the GF. For separable Hamiltonians the total GF (and the state vectors) has to be a product of the GF's (and state vectors) of the individual separable parts. In contrast, we obtain from Fig. 12.1 an expansion as a sum for the Green's function (12.41). 2.) The applicability of our scheme to Coulomb potentials will be discussed below. Here we note that the first order term of (12.41) does not have the correct asymptotic behaviour (8.10) that results in the asymptotic separability. This is evident from the three-particle case where, in a first-order approximation, the three-body state vector is expressed as a sum of three two-body Coulomb states (and a free particle state). This is at variance with the asymptotic form (8.10). To remedy this deficiency we develop in this section a recursive product expansion scheme for the total Green's function.

12.4.1   Expansion of the Green's operator of uncorrelated clusters

To illustrate the idea of the product expansion let us inspect the properties of the GF of a system consisting of N particles that are distributed between L clusters. The clusters which are labelled by the index l contain each m_l interacting particles, i.e.

$$\sum_l^L m_l = N.$$

We consider the case where the particles within different clusters do not interact with each other. However within each subdivision l, the m_l particles are correlated via the m_l interaction potential $v_l^{(m_l)}$. The total interaction potential $U^{(N)}$ is then

$$U^{(N)} = \sum_l^L v_l^{(m_l)}. \tag{12.45}$$

The total Hamiltonian $H^{(N)}$ of the system has the form

$$H^{(N)} = H_0 + U^{(N)}, \tag{12.46}$$

where H_0 is a non-interacting operator. Now we introduce the Green's operator $\mathsf{G_m}$ of a system with the total potential $\sum_{j=1}^m v_j^{(m_j)}$, $m \in [1, L]$, i.e.

$$\left[z - H_0 - \sum_{j=1}^m v_j^{(m_j)} \right] \mathsf{G_m}(z) = \mathbb{1}, \quad m \in [1, L]. \tag{12.47}$$

Thus, as clear from Eq. (12.45) the following relations apply.

$$\mathsf{G_L} \equiv G^{(N)}, \quad \text{where} \quad G^{(N)}(z) = \left[z - H^{(N)} \right]^{-1}$$

is the Green's operator of the total system. On the other hand Eq. (12.46) can be reformulated as

$$\underbrace{\left(H_0 + \sum_{j=1}^{m-1} v_j^{(m_j)} \right) + \sum_{j=m}^L v_j^{(m_j)}}_{H^{(m-1)}} = H^{(N)},$$

$$\left[z - H^{(m-1)} \right] \mathsf{G_{m-1}}(z) = \mathbb{1}. \tag{12.48}$$

Furthermore, we conclude that

$$\left[z - H^{(L-1)} - v_L^{(m_L)} \right] \mathsf{G_L}(z) = \mathbb{1},$$

$$\Rightarrow \left[\mathbb{1} - \mathsf{G_{L-1}}(z) \, v_L^{(m_L)} \right] \mathsf{G_L}(z) = \mathsf{G_{L-1}}(z). \tag{12.49}$$

This equation leads to the recurrence relations

$$G^{(N)} \equiv \mathsf{G_L} = \mathsf{G_{L-1}} \left[\mathbb{1} + v_L^{(m_L)} \mathsf{G_L} \right],$$

$$\mathsf{G_{L-1}} = \mathsf{G_{L-2}} \left[\mathbb{1} + v_{L-1}^{(m_{L-1})} \mathsf{G_{L-1}} \right]. \tag{12.50}$$

The Green's operator $G^{(N)}(z)$ can thus be written in the exact product form

$$G^{(N)} = \mathsf{G_0} \left[\mathbb{1} + v_1^{(m_1)} \mathsf{G_1} \right] \cdots \left[\mathbb{1} + v_{L-1}^{(m_{L-1})} \mathsf{G_{L-1}} \right] \left[\mathbb{1} + v_L^{(m_L)} \mathsf{G_L} \right]. \tag{12.51}$$

12.4.2 Green's operator expansion of correlated finite systems

The decomposition (12.51) is valid for a system consisting of decoupled clusters each of which contains a certain number of interacting particles. The question to be addressed now is, whether the same or a similar product expansion for $G^{(N)}$ holds true when the interaction between the clusters is switched on. To shed light on this problem we study again a system with N particles that interact via pair potentials v_{ij} and formulate the total potential $U^{(N)} = \sum_{j>i=1}^{N} v_{ij}$ according to the recursion relation (12.28, 12.29). Furthermore we introduce the auxiliary Green's operator $G_m^{(N-1)}$ of the system that involves the potential $\sum_{j=1}^{m} \tilde{u}_j^{(N-1)}$, $m \in [1, N]$, i.e.

$$\left[z - H_0 - \sum_{j=1}^{m} \tilde{u}_j^{(N-1)} \right] G_m^{(N-1)}(z) = \mathbb{1}, \quad m \in [1, N].$$

(12.52)

In this notation we have

$$G_N^{(N-1)} \equiv G^{(N)}.$$

Since the total hamiltonian can be written as

$$H^{(N)} = \underbrace{\left[K + \sum_{j=1}^{N-1} \tilde{u}_j^{(N-1)} \right]}_{H_{N-1}^{(N-1)}} + \tilde{u}_N^{(N-1)},$$

(12.53)

we obtain for the resolvent of $H^{(N)}$

$$\left[z - H_{N-1}^{(N-1)} - u_N^{(N-1)} \right] G^{(N)}(z) = \mathbb{1}.$$

(12.54)

Employing Eq. (12.52) we obtain

$$\left[\mathbb{1} - G_{N-1}^{(N-1)} \tilde{u}_N^{(N-1)} \right] G^{(N)}(z) = G_{N-1}^{(N-1)}(z).$$

(12.55)

These reformulations lead us to the recurrence relations

$$G^{(N)} \equiv G_N^{(N-1)} = G_{N-1}^{(N-1)} \left[\mathbb{1} + \tilde{u}_N^{(N-1)} G_N^{(N-1)} \right],$$

(12.56)

$$G_{N-1}^{(N-1)} = G_{N-2}^{(N-1)} \left[\mathbb{1} + \tilde{u}_{N-1}^{(N-1)} G_{N-1}^{(N-1)} \right].$$

(12.57)

Thus, the total Green's operator is expressible exactly in the product expansion

$$G^{(N)} = G_0 \left[\mathbb{1} + \tilde{u}_1^{(N-1)} G_1^{(N-1)} \right] \cdots \left[\mathbb{1} + \tilde{u}_{N-1}^{(N-1)} G_{N-1}^{(N-1)} \right] \left[\mathbb{1} + \tilde{u}_N^{(N-1)} G_N^{(N-1)} \right].$$

(12.58)

While this relation is formally exact it is of little practical use since on the right hand side the total Green's operator appears again. Thus, a systematic approximation scheme is needed for numerical realization. For this purpose we recall that according to Eq. (12.52) the operator $G_l^{(k)}$, $k, l \in [2, N]$ is the Green's operator of a system involving l potentials ($\sum_{j=1}^{l} \tilde{u}_j^{(k)}$). Each of these potentials ($\tilde{u}_j^{(k)}$) describes the interaction between k particles. Therefore, according to Eq. (12.40) the Green's operator $G_l^{(k)}$ has the series expansion

$$G_l^{(k)} = G_0 + \sum_{j=1}^{l} \Gamma_j^{(k)}.$$

In an analogous way to relation (12.41) the operators $\Gamma_j^{(k)}$ are related to the Green's operators $g_j^{(k)}$ associated with the potentials ($\tilde{u}_j^{(k)}$) via linear coupled integral equations

$$\begin{pmatrix} \Gamma_1^{(k)} \\ \Gamma_2^{(k)} \\ \vdots \\ \Gamma_{l-1}^{(k)} \\ \Gamma_l^{(k)} \end{pmatrix} = \begin{pmatrix} g_1^{(k)} - G_0 \\ g_2^{(k)} - G_0 \\ \vdots \\ g_{l-1}^{(k)} - G_0 \\ g_l^{(k)} - G_0 \end{pmatrix} + [\tilde{\mathbf{K}}^{(k)}] \begin{pmatrix} \Gamma_1^{(k)} \\ \Gamma_2^{(k)} \\ \vdots \\ \Gamma_{l-1}^{(k)} \\ \Gamma_l^{(k)} \end{pmatrix}.$$

$$(12.59)$$

As for the case of Eq. (12.41) the kernel $[\tilde{\mathbf{K}}^{(k)}]$ of the integral equation (12.59) contains only Green's operators with a reduced number of interactions. For simplicity and clarity let us consider the first order term (first iteration) of Eq. (12.59) from which follows

$$G_l^{(k)} = G_0 + \left[\sum_{j=1}^{l} g_j^{(k)} \right] - lG_0. \qquad (12.60)$$

The central quantity in the product expansions (12.56, 12.57) of the total Green's operator has the form $\mathbb{1} + u_l^{(k)} G_l^{(k)} = G_0^{-1}(G_0 + G_0 u_l^{(k)} G_l^{(k)})$. The structure of this quantity is unravelled by Eq. (12.60) which indicates that

$$\begin{aligned} G_0 + G_0 u_l^{(k)} G_l^{(k)} &= G_0 + G_0 u_l^{(k)} g_l^{(k)} + G_0 u_l^{(k)} g_{l-1}^{(k)} + \cdots \\ &\quad \cdots + G_0 u_l^{(k)} g_1^{(k)} - (l-1) G_0 u_l^{(k)} G_0 , \\ &= g_l^{(k)} + G_0 u_l^{(k)} \left[G_0 + G_0 u_{l-1}^{(k)} g_{l-1}^{(k)} \right] + \cdots \\ &\quad \cdots + G_0 u_l^{(k)} \left[G_0 + G_0 u_1^{(k)} g_1^{(k)} \right] - (l-1) G_0 u_l^{(k)} G_0 , \\ G_0 + G_0 u_l^{(k)} G_l^{(k)} &= g_l^{(k)} + G_0 u_l^{(k)} \left[G_0 u_{l-1}^{(k)} g_{l-1}^{(k)} + \cdots + G_0 u_1^{(k)} g_1^{(k)} \right]. \end{aligned}$$

$$(12.61)$$

From this relation it follows

$$
\begin{aligned}
\mathbb{1} + u_l^{(k)} G_l^{(k)} &= \mathbb{1} + u_l^{(k)} g_l^{(k)} + u_l^{(k)} \left[G_0 u_{l-1}^{(k)} g_{l-1}^{(k)} + \cdots + G_0 u_1^{(k)} g_1^{(k)} \right], \\
G_l^{(k)} &= g_l^{(k)} + G_0 u_{l-1}^{(k)} g_{l-1}^{(k)} + \cdots + G_0 u_1^{(k)} g_1^{(k)}.
\end{aligned}
$$

$$(12.62)$$

The leading term of equation (12.62) is identified as the Green's operator $g_l^{(k)}$. All other terms are higher order multiple scattering between different subdivisions of the total system. Hence, the first-order terms in the exact expansions (12.56, 12.57) attain the forms

$$
G^{(N)} = G_N^{(N-1)} = g_{N-1}^{(N-1)} \left[\mathbb{1} + \tilde{u}_N^{(N-1)} g_N^{(N-1)} \right],
\tag{12.63}
$$

$$
g_{N-1}^{(N-1)} = g_{N-2}^{(N-1)} \left[\mathbb{1} + \tilde{u}_{N-1}^{(N-1)} g_{N-1}^{(N-1)} \right].
\tag{12.64}
$$

The Green's operator $G^{(N)}$ can then be expressed in the explicit form

$$
G^{(N)} = G_N^{(N-1)} = g_1^{(N-1)} \left[G_0^{-1} g_2^{(N-1)} \right] \cdots \left[G_0^{-1} g_{N-1}^{(N-1)} \right] \left[G_0^{-1} g_N^{(N-1)} \right].
$$

$$(12.65)$$

The whole set of (flow) equations reads

$$
\begin{aligned}
G_n^{(N-1)} &= G_0 \left[\mathbb{1} + \tilde{u}_1^{(N-1)} g_1^{(N-1)} \right] \cdots \\
&\cdots \left[\mathbb{1} + \tilde{u}_{N-2}^{(N-1)} g_{N-2}^{(N-1)} \right] \left[\mathbb{1} + \tilde{u}_{N-1}^{(N-1)} g_{N-1}^{(N-1)} \right] \left[\mathbb{1} + \tilde{u}_N^{(N-1)} g_N^{(N-1)} \right],
\end{aligned}
$$

$$(12.66)$$

$$
g_{N-1}^{(N-2)} = G_0 \left[\mathbb{1} + \tilde{u}_1^{(N-2)} g_1^{(N-2)} \right] \cdots \left[\mathbb{1} + \tilde{u}_{N-2}^{(N-2)} g_{N-2}^{(N-2)} \right] \left[\mathbb{1} + \tilde{u}_{N-1}^{(N-2)} g_{N-1}^{(N-2)} \right],
$$

$$
\vdots
$$

$$(12.67)$$

$$
\vdots
$$

$$
g_3^{(2)} = G_0 \left[\mathbb{1} + \tilde{u}_1^{(2)} g_1^{(2)} \right] \left[\mathbb{1} + \tilde{u}_2^{(2)} g_2^{(2)} \right] \left[\mathbb{1} + \tilde{u}_3^{(2)} g_3^{(2)} \right].
$$

$$(12.68)$$

The recursive scheme (12.68) is the main result of this subsection. Eqs. (12.68) can be visualized by means of Fig. 12.1 and 12.2. In a first step the total potential, say for six particles as in the case of Fig. 12.1, is reduced to a set of six potentials $\tilde{u}_j^{(5)}$, $(j = 1 \cdots 6)$ involving five interacting particles. For each of these potentials $\tilde{u}_j^{(5)}$ the (off-shell) Green's operator $g_j^{(5)}$

is calculated. If this last step is possible the six particle Green's operator is readily given by Eq. (12.66). If the five-particle Green's operator should be simplified we express each of the five-particle potentials $\tilde{u}_j^{(5)}$ in terms of four-particle potentials $\tilde{u}_{jk}^{(4)}$, $k \neq j$, evaluate the four-particle Greens function $g_{jk}^{(4)}$, $k \neq j$ and obtain the five particle Green's operator from (12.67). This procedure can be repeated until we reach Eq. (12.68) where the three-body Green's operator is written in terms of two-body Green's operator which are generally amenable to numerical (or analytical) calculations. In fact, for the three-body problem the present scheme contains only one term (12.67) which is identical to the Møller operator expansion suggested in Ref. [73]. Furthermore, it is straightforward to show that on the two-body energy shell the approximation (12.68) for the three-particle problem produces a state vector which is a product of three (isolated) two-body states. For Coulomb potentials this state is identical to the eigenfunctions of the operator (9.22) (on page 119) and possesses thus the correct asymptotic behaviour (8.10) as detailed in chapter (8).

The expansion (12.68) has favorable features from the point of view of perturbation theory: for clarity let us consider the three-body case, i.e. Eq. (12.68). Same arguments hold true for the general case. For a three body system, in which the interaction $\tilde{u}_2^{(2)}$ strength between two particles, say particle 1 and particle 3, is small compared to the other interactions in the system. Eqs. (12.68) can then by expanded perturbationally as

$$
\begin{aligned}
g_3^{(2)} &= G_0 \left[\mathbb{1} + \tilde{u}_1^{(2)} g_1^{(2)} \right] \left[\mathbb{1} + \tilde{u}_2^{(2)} G_0 + \tilde{u}_2^{(2)} G_0 \tilde{u}_2^{(2)} G_0 + \cdots \right] \left[\mathbb{1} + \tilde{u}_2^{(2)} g_3^{(2)} \right] , \\
&\approx g_1^{(2)} G_0^{-1} g_3^{(2)} + g_1^{(2)} \tilde{u}_2^{(2)} g_3^{(2)} + \cdots .
\end{aligned}
$$

$$(12.69)$$

This means in the extreme case of $\tilde{u}_2^{(2)} = 0$ we obtain a product of two two-body Green's operators, which is the correct result, for in this case the three-body Hamiltonian is separable in two two-body Hamiltonians. While this results seems plausible it is not reproduced by the first iteration of the Faddeev equations, where for $\tilde{u}_2^{(2)} = 0$ we obtain a solution as a sum of two two-body Green's operators. Further terms of the perturbation expansion (12.69) regard the two interacting two-body subsystems as quasi single particles which are perturbationally coupled by the small parameter $\tilde{u}_2^{(2)}$.

12.4.3   Remarks on the applicability to Coulomb potentials

Originally the Faddeev approach has been developed for short-range potentials in which case the kernel of the Faddeev equations is connected (after one iteration). Hence, according to the Fredholm alternative [308], either the homogeneous or the inhomogeneous equations possess unique solutions [317, 305]. We recall that the solution of the homogeneous Faddeev equation yields the bound or the resonant states depending on whether the energy eigenvalues are real of complex. With this solid mathematical background the Faddeev approach has been successfully applied to a variety of problems involving short-range interactions, in particular in nuclear-physics the Faddeev equations have found numerous applications (see for example, Refs. [321, 322] and references therein).

Recalling that the kernel (12.7) of the Faddeev equations is given in terms of off-shell two-body T-operators we expect serious difficulties when the infinite-range Coulomb potentials are involved, for in this case the two-body T operators contains divergent terms, as explicitly shown in section 11.4. The same arguments apply to the integral matrix equations with the kernel (12.38). In view of the importance of the Coulomb interaction for physical systems several techniques have been put forward to circumvent this situation (see e.g. [323, 324, 325, 326, 327, 328] and references therein). Here we mention some of these attempts.

For a three-body system governed by short-range and Coulomb forces Noble [323] suggested to include in the reference Green's function G_0 in Eq. (12.10) all the Coulomb interactions, i.e. in this case G_0 is a (three-body) Coulomb Green's operator whose exact form is generally not known. Having encapsulated the Coulomb potential peculiarities in G_0 one obtains the "Faddeev-Noble" equations which are mathematically well behaved. To show briefly the basics of this idea let us assume that the total potential (12.1) is a sum of two-body interactions, each consisting of a repulsive Coulomb part V_k and a short range part $V_k^{(s)}$, i. e.

$$V = \sum_{j>i} V_{ij} + V_{ij}^{(s)} = \sum_k V_k + V_k^{(s)}. \tag{12.70}$$

In the spirit of the Faddeev approach, the three-body state vector is split into a sum of three vectors

$$|\Psi\rangle = \sum_{j=1}^{3} |\psi_k\rangle, \tag{12.71}$$

where the state vector components are derived according to

$$|\psi_k\rangle = G_{Coul}(z)V_k^{(s)}|\Psi\rangle . \tag{12.72}$$

The Coulomb Green's operator is deduced from

$$G_{Coul}(z) = \left[z - H_0 - \sum_{k=1}^{3} V_k \right]^{-1} .$$

Each of the Faddeev components $|\psi_k\rangle$ (12.72) of the wave function carries only one type of two-body asymptotics. That is the case if $G_{Coul}(z)$ does not possess bound state poles[4]. The integral equations deduced by Noble for the components $|\psi_k\rangle$ (12.72) are

$$|\psi_k\rangle = G_{Coul,k}(z)V_k^{(s)}\sum_{j\neq k}|\psi_j\rangle , \tag{12.73}$$

where the channel Green's operator $G_{Coul,k}(z)$ is given by

$$G_{Coul,k}(z) = \left[z - H_0 - \sum_{j=1}^{3} V_j - V_k^{(s)} \right]^{-1} .$$

Obviously, this procedure reduces to the standard Faddeev approach in the absence of Coulomb interactions. However, it has the disadvantage that the (unknown) Coulomb Green's function is used as an input in the theory. Therefore, Bencze [329] has suggested to use instead of the (unknown) three-body Coulomb Greens operator the channel-distorted Coulomb Greens operator. This approximation brings indeed significant simplification in the practical implementation [331].

Using similar ideas employed in Noble's and Bencze work, the recursive scheme (12.37) can be applied to Coulomb problems. Recently, this has been realized numerically for a four-particle system [101].

As briefly outlined above, Noble's approach relies on a splitting (in the two-body configuration space) of the total potential in Coulomb-type short-range interactions. One can however perform the separation in the three-body configuration space, as done by Merkuriev and co-workers [324, 325]. This renders possible the treatment of attractive as well as repulsive Coulomb interactions. The integral equations (with connected kernels) derived in this

---

[4]This is valid for repulsive Coulomb interactions. This presents a limitation of the range of applicability of the original work of Noble.

case can be transformed [324, 325] into differential equations with specified (asymptotic) boundary conditions. From the numerical point of view the resulting "Faddeev-Merkuriev" integral equations are less favorable since their kernel contains the Green's operators with a complicated structure.

A further method to deal with the long-range tail of the Coulomb interaction is to screen or to cut it off at some large distance R_0. A re-normalization procedure is then used to obtain the results for the unscreened case ($R_0 \to \infty$) [330].

12.5   Path-Integral Monte-Carlo method

In previous chapters we described the properties of a quantum finite system by certain approximate solutions of the Schrödinger equation. The Green's function can be represented using these solutions. Equivalently, one can transform the differential Schrödinger equation satisfying certain boundary conditions, into an integral equation and seek suitable expressions for the Green's function. An alternative route to obtain numerically the propagator and hence the Green's function has been developed by Feynman using the path integral method [335, 332, 333]. Here we outline the main feature of this method with emphasis on numerical realization by means of the Monte-Carlo technique.

Using Eq. (12.42) we were able to describe the (equilibrium) thermodynamic of quantum systems, after having determined the corresponding density of states using (approximate) expressions for Green's function. On the other hand the expectation value of any observable O derives from the density operator ρ as

$$\langle O \rangle = \frac{\mathrm{tr}(O\rho)}{Z}, \quad \rho = e^{-\beta H},$$

where Z is the partition function given by $Z = \mathrm{tr}(\rho)$, and $H = H_0 + V$ is the Hamiltonian with a non-interacting part H_0 and V is the potential energy. As mentioned in section (12.3), the desired physical quantities are readily inferred from the partition function, e.g. the internal energy E derives as $E = -\partial_\beta(\ln Z)$. Since in general V and H_0 do not commute the relation for the density operator

$$e^{-\beta(H_0+V)} = e^{-\beta H_0}\, e^{-\beta V}\, e^{-\beta^2[H_0,V]/2} \tag{12.74}$$

applies.

For a system of distinguishable particles the partition function has the explicit form

$$
\begin{aligned}
Z(\beta) &= \int d^3 \mathbf{r}_1 \langle \mathbf{r}_1 | e^{-\beta H} | \mathbf{r}_1 \rangle, \\
&= \int d^3 \mathbf{r}_1 \cdots \int d^3 \mathbf{r}_M \prod_{j=1}^{M} \langle \mathbf{r}_j | e^{-\beta H/M} | \mathbf{r}_{j+1} \rangle.
\end{aligned}
$$

(12.75)

For N fermions the partition function $Z_F(\beta)$ reads

$$
Z_F(\beta) = \int d^3 \mathbf{r}_1 \cdots \int d^3 \mathbf{r}_M \frac{1}{N!} \sum_P (-)^{\delta P} \langle \mathbf{r}_1 | e^{-\beta H/M} | \mathbf{r}_2 \rangle \cdots
$$

$$
\cdots \langle \mathbf{r}_{M-1} | e^{-\beta H/M} | \mathcal{P}(\mathbf{r}_M) \rangle, \quad (12.76)
$$

where $\mathbf{r}_j$, $j = 2 \cdots M$ are intermediate coordinates and M is the number of "time slices". Furthermore, $\mathcal{P}$ is the N-particle permutation (exchange) operator and the sum in (12.76) runs over all permutations. The advantage of writing the partition function in this way is that the contribution of the commutator in (12.74) becomes of a less importance with growing M (or/and) increasing temperature [334], namely

$$
\begin{aligned}
e^{-\beta(H_0+V)} &= \left\{ e^{-\frac{\beta}{M} H_0} e^{-\frac{\beta}{M} V} e^{-\frac{\beta^2}{2M^2}[H_0,V]} \right\}^M, \\
&\overset{M \to \infty}{\longrightarrow} \left\{ e^{-\frac{\beta}{M} H_0} e^{-\frac{\beta}{M} V} \right\}^M + \mathcal{O}(\beta^2/M^2).
\end{aligned}
$$

(12.77)

Thus, for the evaluation of the partition function of a system consisting of N polarized fermions in D space dimensions one has to perform $D \times N \times M$ dimensional integrals. In absence of any spin dependent interaction in the Hamiltonian Takahashi and Imada [337] have shown that for N interacting electrons with position vectors $\mathbf{r}_j$ the partition function can be written as

$$
Z = \left(\frac{1}{N!} \right)^M \int \left[\prod_{i=1}^{M} \prod_{j=1}^{N} d^3 \mathbf{r}_{j,i} \right] \prod_{k=1}^{M} \det(\mathbf{A}(k, k+1))
$$

$$
\times \quad \exp\left(-\frac{\beta}{M} \sum_{i=1}^{M} V(\mathbf{r}_{1,i}, \ldots, \mathbf{r}_{N,i}) \right) + \mathcal{O}\left(\frac{\beta^3}{M^2} \right).
$$

(12.78)

The matrix elements of **A** are given by

$$[A(k, k+1)]_{i,j} = \left(\frac{M}{2\pi\beta}\right)^{D/2} \exp\left[-\frac{M}{2\beta}(\mathbf{r}_{i,k} - \mathbf{r}_{j,k+1})^2\right]. \tag{12.79}$$

Here $\mathbf{r}_{i,j}$, $j = 1 \dots M$ are the intermediate coordinates of the position vectors $\mathbf{r}_i$ satisfying the boundary condition $\mathbf{r}_{i,M+1} = \mathbf{r}_{i,1}$. The numerical task is then to evaluate the $(D \times N \times M)$-dimensional integrals (12.78). This is usually done using standard Metropolis Monte Carlo techniques [338, 336]. Accurate results are achieved with increasing the number of time slices M and/or for higher temperatures.

With decreasing temperatures an enormous amount of time steps is required. An additional difficulty is the so-called the "fermion sign problem" [339, 340]: the integrand (12.78) is not always positive because of its dependence on the determinant $\det(\mathbf{A})$. Therefore, the expectation value of a position dependent observable O is to be evaluated as

$$\langle O \rangle = \frac{\sum_{j=1}^{G} O_j \mathrm{sign}(I_j)}{\sum_{j=1}^{G} \mathrm{sign}(I_j)}. \tag{12.80}$$

In the jth Monte Carlo step O_g refers to the value of the observable O and I_j is the integrand in (12.78) evaluated at the step j. The ratio between the integrands in (12.78) having positive signs (I^+) and negative sign (I^-) is given approximately by [341, 342]

$$\frac{I^+ - I^-}{I^+ + I^-} \sim e^{-\beta(E_F - E_B)}, \tag{12.81}$$

where the energies E_F and E_B are the ground state of the fermionic (antisymmetric ground state) and bosonic (symmetric ground state) system, respectively. Obviously, for higher temperatures the sign problem is of less importance but with decreasing temperatures the statistical error in (12.80) grows rapidly. Despite these difficulties (which can be partly circumvented, see e.g. [343]) the path integral Monte-Carlo method has been applied successfully to a number of electronic systems with a limited number of electrons ($N < 20$), e.g. [269, 270, 343, 344].

13 Finite correlated systems in a multi-center potential

In the previous parts of this book we studied the bound and the continuum spectrum of a finite electronic system in a single external potential. Here we focus on the description of the scattering of a finite correlated system from a multi-center potential, such as the crystal potential of solids or surfaces. We begin with a brief summary of the relevant tools for the treatment of the single-particle scattering in periodic potential.

13.1 Single-particle scattering from a multi-center potential

The treatment of the spectrum and in particular of the ground state of a particle in an ordered and disordered multi-center potential is a wide field with well-established techniques [51, 347, 348]. One of the important tools for the description of the scattering of a particle from a crystal potential is the so-called scattering-path formalism [350, 351], which is outlined below. The scattering path operator offers a convenient way for the evaluation of the T matrix elements. Of particular interest in the context of the present work is the generalization of this single-particle (multi-scattering-center) scheme as to deal with the scattering of correlated compounds (atoms, molecules, ...) from multi-center potentials.

13.1.1 Scattering-path formalism

Let us consider a non-relativistic particle scattered from a multi-center potential w_{ext}. We assume that the potential w_{ext} is a superposition of M non-overlapping potentials w_i each acting only in the domain Ω_i centered around the position $\mathbf{R}_i$ i. e.

$$w_{\text{ext}}(\mathbf{r}) = \sum_{i}^{M} w_i(\mathbf{r}_i), \quad \Omega_i \cap \Omega_j = 0, \ \forall \, j \neq i, \quad \mathbf{r}_i = \mathbf{r} - \mathbf{R}_i \qquad (13.1)$$

As discussed in details in the previous chapter, the complete information on the scattering dynamics is carried by the T operator

$$t_{\text{ext}}^{\pm}(z) = w_{\text{ext}} + w_{\text{ext}}G_0^{\pm}(z)t_{\text{ext}}^{\pm}, \tag{13.2}$$

where $G_0^{\pm}$ is the Green's operator in absence of w_{ext} and the $\pm$ signs stands for incoming or outgoing wave boundary conditions. For brevity we will suppress the $\pm$ signs and the energy argument z. As done in the context of the recursive scheme (12.36) we employ the expansion (13.1) to deduce for (13.2)

$$t_{\text{ext}} = \sum_{k=1}^{M} q^{(k)}, \tag{13.3}$$

$$q^{(k)} = w_k + w_k G_0 t_{\text{ext}} \tag{13.4}$$

$$q^{(k)} = w_k + w_k G_0 q^{(k)} + \sum_{l \neq k}^{M} w_k G_0 q^{(l)}. \tag{13.5}$$

The single site transition operator is given by

$$t_k = w_k + w_k G_{\text{int}}^{(N)} t_k.$$

Therefore, Eq. (13.5) is expressible as

$$q^{(k)} = t_k + \sum_{l \neq k}^{M} t_k G_0 q^{(l)}. \tag{13.6}$$

As done in (12.38) we can express this relation in a matrix integral equation as

$$\begin{pmatrix} q^{(1)} \\ q^{(2)} \\ \vdots \\ q^{(M-1)} \\ q^{(M)} \end{pmatrix} = \begin{pmatrix} t_1 \\ t_2 \\ \vdots \\ t_{M-1} \\ t_M \end{pmatrix} + [\mathbf{K}] \begin{pmatrix} q^{(1)} \\ q^{(2)} \\ \vdots \\ q^{(M-1)} \\ q^{(M)} \end{pmatrix}. \tag{13.7}$$

The kernel $[\mathbf{K}]$ is given in terms of single-site operators t_k

$$[\mathbf{K}] = \begin{pmatrix} 0 & t_1 & t_1 & .. & t_1 \\ t_2 & 0 & t_2 & .. & t_2 \\ \multicolumn{5}{c}{\dotfill} \\ t_{M-1} & . & t_{M-1} & 0 & t_{M-1} \\ t_M & . & t_M & t_M & 0 \end{pmatrix} G_0. \tag{13.8}$$

Alternatively, we can write

$$t_{\text{ext}} = \sum_{k}^{M} \left\{ t_k + \sum_{l \neq k}^{M} t_k G_0(t_l + w_l G_0 t_{\text{ext}}) \right\}. \tag{13.9}$$

As pointed out by Gyorffy [350, 351] a particularly useful way to represent t_{ext} is achieved by introducing the *scattering path operators* τ^{ij} as

$$\tau^{ij} = t_i \delta_{ij} + \sum_{k \neq i}^{M} t_i G_0 \tau^{ik} = t_i \delta_{ij} + \sum_{k \neq j}^{M} \tau^{ik} G_0 t_j. \qquad (13.10)$$

A comparison with Eq. (13.6) yields

$$q^{(i)} = \sum_{j}^{M} \tau^{ij}, \quad t_{\text{ext}} = \sum_{i}^{M} q^{(i)} = \sum_{ij}^{M} \tau^{ij}. \qquad (13.11)$$

The operator $q^{(k)}$ describes the collision process of the particle from the site k in the presence of all other scatterers. The operators τ^{kl} contain the details on the scattering and the propagation process of the particle under the influence of the potential w_k following an encounter of the particle with the potential centered around the site l. The matrix elements of the operators τ^{ij} can be written in a suitable form for numerical realization, taking into account that the matrix elements of the single site operators t_k vanish outside the domain Ω_k (note that $t_k = w_k + w_k G w_k$ and $w_k(\mathbf{r}) = 0$ for $\mathbf{r} \notin \Omega_k$)

$$\langle \mathbf{r} | \tau^{ij} | \mathbf{r}' \rangle = \langle \mathbf{r}_i | \tau^{ij} | \mathbf{r}'_j \rangle = \langle \mathbf{r}_i | t_i | \mathbf{r}'_i \rangle +$$

$$+ \sum_{k \neq i} \int d^3 \mathbf{r}''_i d^3 \mathbf{r}'''_k \langle \mathbf{r}_i | t_i | \mathbf{r}''_i \rangle \langle \mathbf{r}''_i + \mathbf{R}_i | G_0 | \mathbf{r}'''_k + \mathbf{R}_k \rangle \langle \mathbf{r}'''_k | \tau^{kj} | \mathbf{r}'_j \rangle,$$

$$\mathbf{r}'_j = \mathbf{r}' - \mathbf{R}_j, \quad \mathbf{r}_i = \mathbf{r} - \mathbf{R}_i. \qquad (13.12)$$

The matrix elements of the free Green's operator $G_0^{ik} = \langle \mathbf{r}''_i + \mathbf{R}_i | G_0 | \mathbf{r}'''_k + \mathbf{R}_k \rangle$ are obviously dependent on the structure of the crystal formed by the scattering potentials. They can be evaluated, e.g. in an angular momentum basis which is particularly convenient if the potentials w_k are spherically symmetric [352]. An important feature of the scattering path operators (that follows from the special form of the scattering potential (13.1)) is that the on-shell matrix elements of τ^{ij} are directly related to the matrix elements of the T operator (and hence to experimental observables). The above relations deduced for the T operator can be employed for the derivation of corresponding equations of the Green's function of a particle in a multi-center potential.

13.1.2 Scattering of correlated compounds from multi-center potentials

The expression of the single-particle T operator in terms of scattering path operators plays a central role in the calculation of electron scattering in ordered and disordered materials. Such calculations are required for the treatment of a variety of important processes in condensed matter physics, such as the low, high and medium energy electron diffraction and scattering from crystals [346, 345], and single photoelectron emission from solids and surfaces [111].

For the description of the scattering of correlated compounds from multi-center potentials the derivation for the single-particle case, as outline in the previous section, has to be reconsidered. This situation arises for example in the treatment of colliding atoms, molecules and correlated electrons from surfaces (see [353, 349] and references therein). In this section we will generalize the scattering path operator concept to the scattering of particles with internal structure and expose how the internal motion of the compound is influenced by the scattering from the external multi-center potential and vice versa. To this end we consider a compound consisting of N particles that interact via the total internal potential $U_{int}^{(N)}$. This compound is scattered from the external potential W_{ext} with M centers, i.e.

$$W_{ext} = \sum_k^M \sum_l^N w_{kl}. \tag{13.13}$$

The interaction of particle l with the scattering site k is described by the potential w_{kl}.

The total Hamiltonian of the system can thus be written as

$$\mathcal{H} = H_{int}^{(N)} + W_{ext}, \tag{13.14}$$

$$H_{int}^{(N)} = H_0 + U_{int}^{(N)}. \tag{13.15}$$

H_0 is the Hamilton operator of the compound in absence of $U_{int}^{(N)}$ and W_{ext}. In the previous chapter we discussed a number of methods for the description of the Green's operator

$$G_{int}^{(N)}(z) = \left[z - H_{int}^{(N)}\right]^{-1} \tag{13.16}$$

of the correlated compound in absence of W_{ext}. We therefore isolate the problem of treating $G_{int}^{(N)}(z)$ from that of describing the scattering of N particles from a M center scattering potential. This is achieved by writing for the transition operator T_{ext}

$$T_{ext} = W_{ext} + W_{ext}G_{int}^{(N)}T_{ext}. \tag{13.17}$$

Now we define the interaction of the compound with the scattering site k as

$$\bar{w}_k = \sum_{l=1}^N w_{kl}. \tag{13.18}$$

With this definition we write for T_{ext}

$$T_{\text{ext}} = \sum_{k=1}^{M} Q_k, \tag{13.19}$$

$$Q_k = \bar{w}_k + \bar{w}_k \, G_{\text{int}}^{(N)} \, T_{\text{ext}}. \tag{13.20}$$

Defining the T operator $\bar{q}_k$ for the scattering of the compound from the site k as

$$\bar{q}_k = \bar{w}_k + \bar{q}_k \, G_{\text{int}}^{(N)} \, \bar{w}_k \tag{13.21}$$

we can write for the auxiliary operators Q_k

$$Q_k = \bar{q}_k + \bar{q}_k \, G_{\text{int}}^{(N)} \left[T_{\text{ext}} - \left(\bar{q}_k + \bar{q}_k G_{\text{int}}^{(N)} T_{\text{ext}} \right) \right]$$

$$= \bar{q}_k + \bar{q}_k G_{\text{int}}^{(N)} \sum_{j \neq k}^{M} Q_j. \tag{13.22}$$

From Eqs. (13.18, 13.21) and introducing the T operator of the particle l from the scattering site k as

$$t_{kl} = w_{kl} + w_{kl} G_{\text{int}}^{(N)} t_{kl} \tag{13.23}$$

we can write furthermore

$$\bar{q}_k = \sum_{l=1}^{N} \left[w_{kl} + w_{kl} G_{\text{int}}^{(N)} \bar{q}_k \right]$$

$$= \sum_{l=1}^{N} q_{kl} \tag{13.24}$$

$$q_{kl} := w_{kl} + w_{kl} G_{\text{int}}^{(N)} \bar{q}_k \tag{13.25}$$

$$q_{kl} = t_{kl} + t_{kl} G_{\text{int}}^{(N)} \left[\bar{q}_k - \left(w_{kl} + w_{kl} G_{\text{int}}^{(N)} \bar{q}_k \right) \right] \tag{13.26}$$

$$q_{kl} = t_{kl} + t_{kl} G_{\text{int}}^{(N)} \left[\bar{q}_k - q_{kl} \right] \tag{13.27}$$

$$q_{kl} = t_{kl} + t_{kl} G_{\text{int}}^{(N)} \sum_{j \neq l}^{N} q_{kj} \tag{13.28}$$

The above equation can be written in the compact form

$$T_{\text{ext}} = \sum_{k=1}^{M} Q_k,$$

$$\begin{pmatrix} Q_1 \\ Q_2 \\ \vdots \\ Q_{M-1} \\ Q_M \end{pmatrix} = \begin{pmatrix} \bar{q}_1 \\ \bar{q}_2 \\ \vdots \\ \bar{q}_{M-1} \\ \bar{q}_M \end{pmatrix} + [\mathbf{K}_{\bar{q}}] \begin{pmatrix} Q_1 \\ Q_2 \\ \vdots \\ Q_{M-1} \\ Q_M \end{pmatrix},$$

$$\bar{q}_k = \sum_{l=1}^{N} q_{kl},$$

$$\begin{pmatrix} q_{k1} \\ q_{k2} \\ \vdots \\ q_{kN-1} \\ q_{kN} \end{pmatrix} = \begin{pmatrix} t_{k1} \\ t_{k2} \\ \vdots \\ t_{kN-1} \\ t_{kN} \end{pmatrix} + [\mathbf{K}_k] \begin{pmatrix} q_{k1} \\ q_{k2} \\ \vdots \\ q_{kN-1} \\ q_{kN} \end{pmatrix},$$

$$t_{kl} = w_{kl} + w_{kl} G_{\text{int}}^{(N)} t_{kl}. \tag{13.29}$$

Thus, an essential part of the calculations is the evaluation of the single site transition operator t_{kl}. This operator describes the scattering of the particle l from the scattering site k under the influence of the internal correlations in the compound, which are contained in $G_{\text{int}}^{(N)}$. Depending on the nature of $U_{\text{int}}^{(N)}$ one chooses the appropriate method for the treatment of $G_{\text{int}}^{(N)}$. The kernels of the integral equations (13.29) depend on (off-shell) T matrices and on $G_{\text{int}}^{(N)}$

$$[\mathbf{K}_{\bar{q}}] = \begin{pmatrix} 0 & \bar{q}_1 & \bar{q}_1 & \cdots & \bar{q}_1 \\ \bar{q}_2 & 0 & \bar{q}_2 & \cdots & \bar{q}_2 \\ \cdots\cdots\cdots\cdots\cdots\cdots\cdots\cdots \\ \bar{q}_{M-1} & \cdots & \bar{q}_{M-1} & 0 & \bar{q}_{M-1} \\ \bar{q}_M & \cdots & \bar{q}_M & \bar{q}_M & 0 \end{pmatrix} G_{\text{int}}^{(N)}, \tag{13.30}$$

$$[\mathbf{K}_k] = \begin{pmatrix} 0 & t_{k1} & t_{k1} & \cdots & t_{k1} \\ t_{k2} & 0 & t_{k2} & \cdots & t_{k2} \\ \cdots\cdots\cdots\cdots\cdots\cdots\cdots\cdots \\ t_{kN-1} & \cdots & t_{kN-1} & 0 & t_{kN-1} \\ t_{kN} & \cdots & t_{kN} & t_{kN} & 0 \end{pmatrix} G_{\text{int}}^{(N)}. \tag{13.31}$$

As done in the preceding section one can formulate the above relation as well in terms of many-particle scattering path operators [354].

The above formulations can be extended to the treatment of the single particle [347, 51] and the multi-particle [355] scattering from disordered (alloyed) potentials.

14 Excitations in extended electronic systems

14.1 Time-dependent single-particle Green's functions

In chapter 11 we introduced the Green's function associated with the time-independent Schrö-dinger equation for a single-particle. Analogously, we define the single particle Green's function $g(\mathbf{r}, \mathbf{r}', t, t')$ corresponding to the time-dependent Schrödinger equation with a time-dependent source term $f(r, t)$

$$[i\partial_t - H(\mathbf{r})]\, \phi(\mathbf{r}, t) = f(r, t) \tag{14.1}$$

as the solution of an equivalent equation with a point source, i. e.

$$[i\partial_t - H(\mathbf{r})]\, g(\mathbf{r}, \mathbf{r}', t, t') = \delta(\mathbf{r} - \mathbf{r}')\delta(t - t'). \tag{14.2}$$

The GF $g(\mathbf{r}, \mathbf{r}', t, t')$ and $\phi(\mathbf{r}, t)$ satisfy the same boundary conditions. From the homogeneity of the time space we conclude that $g(\mathbf{r}, \mathbf{r}', t, t')$ depends only on the time difference $\tau = t - t'$. The relation between $g(\mathbf{r}, \mathbf{r}', t, t')$ and the time-independent GF (11.4) is unravelled by expressing $g(\mathbf{r}, \mathbf{r}', t, t')$ in terms of its frequency (ω) components

$$g(\mathbf{r}, \mathbf{r}', t - t' = \tau) = \frac{1}{2\pi} \int_{-\infty}^{\infty} d\omega' e^{-i\omega'\tau} g(\mathbf{r}, \mathbf{r}', \omega') \tag{14.3}$$

and inserting into Eq. (14.2). This yields for $g(\mathbf{r}, \mathbf{r}', \omega)$ the determining equation

$$[\omega - H(\mathbf{r})]\, g(\mathbf{r}, \mathbf{r}', \omega) = \delta(\mathbf{r} - \mathbf{r}').$$

Comparing this equation with Eq. (11.4) we see that $g(\mathbf{r}, \mathbf{r}', \omega)$ is equal to $G(\mathbf{r}, \mathbf{r}', z = \omega)$. On the other hand, as discussed in chapter 11, $G(\mathbf{r}, \mathbf{r}', z = \omega)$ is an analytical function of ω ($\omega \in \mathbb{C}$) with isolated singularities or a continuum branch cut along the real ω axis. Therefore, the integral (14.3) along the real axis is not well-defined and one has to resort to a (side) limiting procedure as done in Eq. (11.6). The time-dependent GF is then obtained from the time-independent GF (11.6) via the Fourier transform (14.3), i.e.

$$g^{\pm}(\mathbf{r}, \mathbf{r}', t - t' = \tau) \quad = \quad \frac{1}{2\pi} \int_{-\infty}^{\infty} d\omega' e^{-i\omega'\tau} G^{\pm}(\mathbf{r}, \mathbf{r}', \omega'); \tag{14.4}$$

$$g^{-}(\mathbf{r}, \mathbf{r}', \tau) \quad = \quad \left[g^{+}(\mathbf{r}', \mathbf{r}, -\tau) \right]^{*}. \tag{14.5}$$

Since all singularities of $G^{\pm}(\mathbf{r}, \mathbf{r}', \omega)$ (and $g^{\pm}(\mathbf{r}, \mathbf{r}', w)$) occur on the real w axis we can express the GF $g^{\pm}$ in terms of the discontinuity

$$\tilde{g}(\tau) = g^{+}(\tau) - g^{-}(\tau),$$

namely

$$g^{\pm}(\mathbf{r}, \mathbf{r}', \tau) = \pm \Theta(\pm\tau) \tilde{g}(\tau). \tag{14.6}$$

From Eq. (11.13) it follows for the propagator $\tilde{g}(\tau)$

$$\begin{aligned}
\tilde{g}(\tau) &= \frac{1}{2\pi} \int_{-\infty}^{\infty} d\omega' e^{-i\omega'\tau} \left[G^{+}(\omega') - G^{-}(\omega') \right], \\
&= -i \sum_{\nu} \int_{-\infty}^{\infty} d\omega' e^{-i\omega'\tau} |\alpha_{\nu}\rangle \, \delta(\omega' - \lambda_{\nu}) \, \langle\alpha_{\nu}|, \tag{14.7} \\
&= -i \sum_{\nu} e^{-i\lambda_{\nu}\tau} |\alpha_{\nu}\rangle \langle\alpha_{\nu}|, \\
&= -i e^{-iH(t-t')} = -i U(t - t'), \tag{14.8}
\end{aligned}$$

where $U(t - t')$ is the time-development operator. The solution of the inhomogeneous Schrödinger equation (14.1) is the sum of the general solution $\phi_0(\mathbf{r}, t)$ of the homogeneous equation and the particular solution given by the Green's function, i.e.

$$\phi(\mathbf{r}, t) \quad = \quad \phi_0(\mathbf{r}, t) + \int d\mathbf{r}' dt' g^{+}(\mathbf{r}, \mathbf{r}', t - t') f(\mathbf{r}', t'), \tag{14.9}$$

$$\phi_0(\mathbf{r}, t) \quad = \quad i \int d\mathbf{r}' \tilde{g}(\mathbf{r}, \mathbf{r}', t - t') \phi_0(\mathbf{r}', t'). \tag{14.10}$$

In Eq. (14.9) we rely on the causality principle to exclude the solution involving g^{-} (the response at the time t occurs due to the action of the source at $t' < t$). For this reason the GF g^{+} (g^{-}) is often referred to as the single-particle, retarded (advanced) Green's function g_1^r (g_1^a), because it describes the retarded (advanced) response of the system to an external perturbation. Thus, from Eq. (14.7) one deduces for the non-local spectral density (per unit volume) (cf. the time-independent case (11.12))

$$g_1^r(\tau) - g_1^a(\tau) \quad = \quad -i \sum_{\nu} \int_{-\infty}^{\infty} d\omega' e^{-i\omega'\tau} |\alpha_{\nu}\rangle \, \delta(\omega' - \lambda_{\nu}) \, \langle\alpha_{\nu}|, \tag{14.11}$$

$$g_1^r(\omega) - g_1^a(\omega) \quad = \quad -i 2\pi \sum_{\nu} |\alpha_{\nu}\rangle \, \delta(\omega - \lambda_{\nu}) \, \langle\alpha_{\nu}|. \tag{14.12}$$

Eq. (14.9) makes evident the role of the GF: the wave function ϕ at the position $\mathbf{r}$ and time t emerges from the wave function at a previous time $t' < t$ and all positions $\mathbf{r}'$ with a weight factor determined by the amplitude $\tilde{g}$.

The matrix elements of the single particle Green's function (14.8) reads

$$
\begin{aligned}
\langle \alpha | \tilde{g}(\tau) | \beta \rangle &= -i \langle \alpha | e^{-iH(t-t')} | \beta \rangle \\
&= -i \langle 0 | c_\alpha e^{-iH(t-t')} c_\beta^\dagger | 0 \rangle,
\end{aligned}
\tag{14.13}
$$

where c_j ($c_j^\dagger$) are the fermionic annihilation (creation) operators and $|0\rangle$ is the vacuum state. The GF (14.13) contains information on kinetic quantities such as particle density (cf. (11.13), (11.12)) and is the analog to the so-called "greater" (correlation) function $g^>$ (Eqs. (A.35)), which will be introduced in the context of many-body system (cf. appendix A.5).

Equivalently to Eq. (14.13), one can express the retarded and advanced GF (14.6) as the expectation value of an operator describing a single particle creation, its propagation and subsequent annihilation, i.e.

$$
\langle \alpha | g^\pm(t-t') | \beta \rangle = \mp i \Theta(\pm(t-t')) \langle 0 | c_\alpha e^{-iH(t-t')} c_\beta^\dagger | 0 \rangle.
\tag{14.14}
$$

To derive an expression of these matrix elements in the frequency space one employs the integral representation of the Θ function $\Theta(t-t') = \frac{i}{2\pi} \lim_{\eta \to 0+} \int d\omega' \frac{\exp[-i\omega'(t-t')]}{\omega' + i\eta}$ and obtains

$$
\langle \alpha | g^\pm(\omega) | \beta \rangle =: g^\pm(\alpha, \beta, \omega) = \langle 0 | c_\alpha \frac{1}{\omega - H \pm i\eta} c_\beta^\dagger | 0 \rangle.
\tag{14.15}
$$

This relation offers a way for a systematic perturbational treatment. In case H can be decomposed as $H = H_0 + V$ we can use the relation $(A-B)^{-1} = A^{-1} + A^{-1}B(A-B)^{-1}$ which is valid for any two operators A and B and deduce from (14.15) the Dyson equation

$$
\begin{aligned}
g^\pm(\alpha, \beta, \omega) &= \langle 0 | c_\alpha \frac{1}{\omega - H_0 \pm i\eta} c_\beta^\dagger | 0 \rangle + \\
&\quad + \sum_{\gamma\delta} \langle 0 | c_\alpha \frac{1}{\omega - H_0 \pm i\eta} c_\gamma^\dagger | 0 \rangle \langle \gamma | V | \delta \rangle \langle 0 | c_\delta \frac{1}{\omega - H \pm i\eta} c_\beta^\dagger | 0 \rangle.
\end{aligned}
\tag{14.16}
$$

14.2   Green's function approach to excitations in extended electronic systems

For strongly correlated systems or for multiple high excitations in extended systems (such as one-electron or one-photon double electron emission, i. e. $(e, 2e)$ [356, 357, 358, 359, 360] or $(\gamma, 2e)$ [112]) methods are needed that go beyond a mean field or a first-order perturbational treatment of the interaction between the excited particles. This is because the (highly) excited particles have access to a large manifold of degenerate states and will thus adjust their motion according to their mutual interactions. This is contrast to a ground-state behaviour or to the case of small perturbations where the effective mean field created by all the particles plays a dominant role. For the latter case there are a number of theories available, like the hole-line expansion or the coupled-cluster methods [106, 52, 53], however these theories can not cope with high excitations. For the treatment of correlated excited states the Green's function approach is well suited, however the method as introduced in previous sections, becomes intractable with increasing N since in this case one works within the first quantization scheme, i. e. states have to be (anti)symmetrized. In addition, as discussed in the previous chapter, the reduction of the N-body Green's function to quantities that are amenable to calculations becomes increasingly difficult for large N. An attractive and a power method for the treatment of extended systems within Green's function theory has been put forward by Migdal and Galitskii as well as by Martin and Schwinger [107, 108]. Applying methods of field theory they developed [107, 108] a technique that connects, by means of Feynman diagrams, the single particle (sp) propagator to higher-order propagators. In addition, a systematic approximation scheme is available [52, 84, 85] that allows the efficient evaluations of the physical observables. The system symmetry in this case enters though (anti)commutation relations of the operators [52, 84, 85]. This route has found extensive applications in various fields of physics. Here, we mention the aspects that are of an immediate relevance to multiple excitations in electronic systems.

14.2.1   Single-particle Green's functions for extended systems

In a many-body system the single-particle Green's function $g(\alpha t, \beta t')$ is defined as an expectation value of the time-ordered product of two operators evaluated with respect to the correlated, exact (normalized) ground-state $|\Psi_0\rangle$ of the N electron system (see appendix A.2 for a brief introduction)

$$g(\alpha t, \beta t') = -i\langle \Psi_0 | T[a_{H\alpha}(t)\, a^\dagger_{H\beta}(t')] | \Psi_0 \rangle, \tag{14.17}$$

where T is the time ordering operator. This definition is in analogy to the single-particle case (14.13). The fermionic creation and annihilation operators $a^\dagger_{H\beta}(t')$ and $a_{H\alpha}(t)$ are given in the Heisenberg picture, e.g.

$$a_{H\alpha}(t) = e^{iHt} a_\alpha e^{-iHt}. \tag{14.18}$$

The operators $a^\dagger_{H\beta}(t')$ and $a_{H\alpha}(t)$ are represented by an appropriate basis, the members of which are characterized by α and β. If the system possesses translational symmetry it is advantageous to employ the momentum eigenstates $\{k\}$ as the basis states, as done below. The time-ordering operator has the action

$$T[a_{H\alpha}(t)\, a^\dagger_{H\beta}(t')] = \begin{cases} a_{H\alpha}(t)\, a^\dagger_{H\beta}(t') & (t > t') \\ -a^\dagger_{H\beta}(t')a_{H\alpha}(t) & (t < t') \end{cases}. \tag{14.19}$$

The effect of the chronological operator T can be described in terms of the step function $\Theta(t-t')$, in which case the Green's function is given by

$$\begin{aligned} ig(k, t-t') &= \Theta(t-t')\langle \Psi_0 | a_{Hk}(t) a^\dagger_{Hk}(t')|\Psi_0\rangle - \Theta(t'-t)\langle \Psi_0 | a^\dagger_{Hk}(t')a_{Hk}(t)|\Psi_0\rangle \\ &= \Theta(t-t')\sum_\gamma e^{-i[E_\gamma^{(N+1)}-E_0^{(N)}](t-t')} \left| \langle \Psi_\gamma^{(N+1)}|a^\dagger_k|\Psi_0\rangle \right|^2 \\ &\quad -\Theta(t'-t)\sum_\delta e^{-i[E_0^{(N)}-E_\delta^{(N-1)}](t-t')} \left| \langle \Psi_\delta^{(N-1)}|a_k|\Psi_0\rangle \right|^2. \tag{14.20} \end{aligned}$$

$\Psi_\gamma^{(N+1)}$ and $\Psi_\delta^{(N-1)}$ stand for a complete set of eigenstates of the $(N+1)$- and the $(N-1)$-particle system, respectively. The energy of the correlated ground state of the N particle system is denoted by $E_0^{(N)}$, whereas $E_\gamma^{(N+1)}$ and $E_\delta^{(N-1)}$ refer to the energies for the excited correlated states of, respectively, the $(N+1)$ and the $(N-1)$ particle systems. The exponentials with the energy arguments in Eq. (14.20) stem from the Hamiltonians in the exponential functions in the definition (14.18) of the Heisenberg operators. Recalling the integral representation of the step function $\Theta(t) = -\lim_{\eta\to 0}\frac{1}{2\pi i}\int_{-\infty}^\infty d\omega \frac{e^{-i\omega t}}{\omega + i\eta}$, the Green's function in the energy space can be obtained via Fourier transforming the time difference $t - t'$ to the energy variable ω (homogeneity of the time-space is assumed). This yields the spectral or Lehmann representation of the sp Green's function [109],

$$
g(k, k', \omega) = \lim_{\eta \to 0} \left[\langle \Psi_0 | a_k^\dagger \frac{1}{\omega - [H - E_0^{(N)}] + i\eta} a_{k'}^\dagger | \Psi_0 \rangle \right.
$$

$$
\left. + \langle \Psi_0 | a_{k'}^\dagger \frac{1}{\omega - [E_0^{(N)} - H] - i\eta} a_k^\dagger | \Psi_0 \rangle \right], \tag{14.21}
$$

$$
g(k, k', \omega) = \lim_{\eta \to 0} \left[\sum_\gamma \frac{z_{k\gamma}^* z_{k'\gamma}}{\omega - \omega_\gamma^+ + i\eta} + \sum_\delta \frac{\bar{z}_{k\delta} \bar{z}_{k'\delta}^*}{\omega - \omega_\delta^- - i\eta} \right], \tag{14.22}
$$

$$
z_{k\gamma} = \langle \Psi_\gamma^{(N+1)} | a_k^\dagger | \Psi_0 \rangle, \quad \omega_\gamma^+ = E_\gamma^{(N+1)} - E_0^{(N)},
$$

$$
\bar{z}_{k\delta} = \langle \Psi_\delta^{(N-1)} | a_k | \Psi_0 \rangle, \quad \omega_\delta^- = E_0^{(N)} - E_\delta^{(N-1)}, \tag{14.23}
$$

$$
g(k, \omega) = \lim_{\eta \to 0} \left[\sum_\gamma \frac{\left| \langle \Psi_\gamma^{(N+1)} | a_k^\dagger | \Psi_0 \rangle \right|^2}{\omega - [E_\gamma^{(N+1)} - E_0^{(N)}] + i\eta} \right.
$$

$$
\left. + \sum_\delta \frac{\left| \langle \Psi_\delta^{(N-1)} | a_k | \Psi_0 \rangle \right|^2}{\omega - [E_0^{(N)} - E_\delta^{(N-1)}] - i\eta} \right]. \tag{14.24}
$$

Equation (14.21) is in complete analogy with the single-particle equation (14.15), except for the second part of (14.21) which describes the propagation of the hole (naturally absent in the single particle case). Relations (14.24, 14.21) highlight the significance of the sp Green's function for measurable physical quantities. The poles of g correspond to the change ($\omega_\gamma^+ = E_\gamma^{(N+1)} - E_0^{(N)}$) in energy (with respect to $E_0^{(N)}$) if a particle have been added to the system and occupies the excited state γ or if one particle is removed ($\omega_\delta^- = E_0^{(N)} - E_\delta^{(N-1)}$) from the reference ground state with N interacting particles, leaving the $N - 1$ particle system in the excited state δ. The residua of these poles are given by the *spectroscopic factors* $z_{k\delta}$ and $\bar{z}_{k\delta}$, meaning that the measurable probabilities of adding and removing one particle with wave vector $\mathbf{k}$ to produce the specific state γ (δ) of the residual system. Clearly, the latter probability is of a direct relevance to the (e,2e) experiments [356, 357, 358, 359, 360]. The infinitesimal quantity η in Eq. (14.24) shifts the poles below the Fermi energy [the states of the $(N - 1)$ system] to slightly above the real axis and those above the Fermi energy [the states of the $(N + 1)$ system] to slightly below the real axis.

Useful quantities for the study of the influence of correlation are the so-called the hole and the particle spectral functions which are obtained from the diagonal elements (14.24) of the spectral representation of the single-particle Green's function as

$$
\begin{aligned}
S_{\mathrm{h}}(k,\omega) &= \frac{1}{\pi}\Im\, g(k,\omega) \\
&= \sum_{\gamma}\left|\langle\Psi_{\gamma}^{(N-1)}|a_k|\Psi_0\rangle\right|^2 \delta(\omega-(E_0^{(N)}-E_{\gamma}^{(N-1)})),\quad \text{for } \omega\le\epsilon_F^-;
\end{aligned}
$$

(14.25)

$$
\begin{aligned}
S_{\mathrm{p}}(k,\omega) &= \frac{1}{\pi}\Im\, g(k,\omega) \\
&= \sum_{\gamma}\left|\langle\Psi_{\gamma}^{(N+1)}|a_k^{\dagger}|\Psi_0\rangle\right|^2 \delta(\omega-(E_{\gamma}^{(N+1)}-E_0^{(N)})),\quad \text{for } \omega>\epsilon_F^+,
\end{aligned}
$$

(14.26)

where

$$
\epsilon_F^+ = E_0^{(N+1)} - E_0^{(N)};\quad \epsilon_F^- = E_0^{(N)} - E_0^{(N-1)}.
$$

The diagonal part of the sp Green's function can thus be represented as an integral over all single hole and single particle excitations, i. e.

$$
g(k,\omega) = \lim_{\eta\to 0}\left(\int_{-\infty}^{\epsilon_F} d\omega'\,\frac{S_{\mathrm{h}}(k,\omega')}{\omega-\omega'-i\eta} + \int_{\epsilon_F}^{\infty} d\omega'\,\frac{S_{\mathrm{p}}(k,\omega')}{\omega-\omega'+i\eta}\right).
$$

(14.27)

In general (and in particular for finite systems) ϵ_F^+ and ϵ_F^- are different. For extended "normal" systems[1] ϵ_F^+ and ϵ_F^- are equal within an error of N^{-1}.

A quantity of a fundamental importance is the single-particle density matrix $\rho_{\beta\alpha}$ which derives from the sp propagator by means of the Lehmann representation (14.21) as

$$
\begin{aligned}
\rho_{\beta\alpha} &= -\frac{i}{2\pi}\int d\omega\, e^{i\omega\eta} g(\alpha,\beta,\omega) = \sum_{n}\langle\Psi_0|a_{\beta}^{\dagger}|\Psi_n^{(N-1)}\rangle\langle\Psi_n^{(N-1)}|a_{\alpha}|\Psi_0\rangle \\
&= \langle\Psi_0|a_{\beta}^{\dagger}a_{\alpha}|\Psi_0\rangle
\end{aligned}
$$

(14.28)

Furthermore, the expectation value for *any* single-particle operator $\mathcal{O}$ is obtained as $\langle\hat{\mathcal{O}}\rangle = \sum_{\alpha\beta}\langle\alpha|\mathcal{O}|\beta\rangle\rho_{\alpha\beta}$, where $\langle\alpha|\mathcal{O}|\beta\rangle$ is the matrix representation of $\mathcal{O}$ in the basis $|\alpha\rangle$. The

---

[1] Normal systems are those for which there exists a discontinuity at the Fermi momentum k_F in the momentum distribution. For non-interacting normal Fermi liquids this discontinuity is 1 and vanishes for systems with pairing correlations, such as superconductors and super fluids.

occupation (depletion) number $n(\alpha)$ $(d(\alpha))$ of the single particle state α can be evaluated from the hole (particle) spectral functions as

$$n(\alpha) = \langle \Psi_0 | a_\alpha^\dagger a_\alpha | \Psi_0 \rangle = \sum_n \left| \langle \Psi_n^{(N-1)} | a_\alpha | \Psi_0 \rangle \right|^2 = \int_{-\infty}^{\epsilon_F^-} d\omega S_h(\alpha, \omega),$$

$$d(\alpha) = \int_{\epsilon_F^+}^{+\infty} d\omega S_p(\alpha, \omega),$$

$$n(\alpha) + d(\alpha) = 1. \tag{14.29}$$

The last relation follows from the anticommutation relation for $a_\alpha^\dagger$ and a_α. As shown by Galitskii and Migdal [363] the ground state energy (for a system with two-body interactions) can as well be obtained from the hole spectral function and the expectation value of the kinetic energy part T_{kin} (i.e. $E_0^{(N)} = \frac{1}{2} \sum_{\alpha,\beta} \int_{-\infty}^{\epsilon_F^-} d\omega \, S_h(\alpha, \beta, \omega) \left[\langle \alpha | T_{kin} | \beta \rangle + \omega \delta_{\alpha,\beta} \right]$).

The above relations highlight the importance of the hole and the particle spectral functions as well as the significance of the single-particle removal or addition spectroscopies, such as single photoemission [110, 111] and (e,2e) processes [114, 116]. In fact it has been documented that the high-energy transmission mode (e,2e) spectroscopy is capable of mapping out the hole spectral functions for a variety of condensed matter systems [364].

14.2.2   The self-energy concept

It is useful to split the total Hamiltonian H in a non-interacting (single particle) part H_0 (with known spectrum ϵ_α, i.e. $H_0 = \sum_\alpha \epsilon_\alpha a_\alpha^\dagger a_\alpha$) and a part V containing two-particle correlations. The Lehmann-representation for the uncorrelated single-particle Green's function $g_0(\alpha, \beta, \omega)$ associated with H_0 is readily deduced as

$$g_0(\alpha, \beta, \omega) = \delta_{\alpha\beta} \left\{ \frac{\Theta(\epsilon_\alpha - \epsilon_F)}{\omega - \epsilon_\alpha + i\eta} + \frac{\Theta(\epsilon_F - \epsilon_\alpha)}{\omega - \epsilon_\alpha - i\eta} \right\}. \tag{14.30}$$

Employing the equation of motion (in the Heisenberg picture) for the operators $a_\alpha^\dagger$ and a_α, e.g. $i\partial_t a_{H\alpha} = [a_{H\alpha}, H]$ and Eq. (14.30) one deduces the relation [84]

$$g(\alpha, \beta, t) = g_0(\alpha, \beta, t) + g_0(\alpha, \gamma, t - t') \left[\Sigma(\gamma, \delta, t' - t'') \right] g(\delta, \beta, t''), \tag{14.31}$$

$$g(\alpha, \beta, \omega) = g_0(\alpha, \beta, \omega) + g_0(\alpha, \gamma, \omega) \Sigma(\gamma, \delta, \omega) g(\delta, \beta, \omega), \tag{14.32}$$

Here and unless otherwise specified, the following convention is used hereafter. For product terms one sums over all repeated indices and integrate over the whole range of repeated continuous (time) variables. The operator Σ is called the mass operator or the irreducible self-energy. From (14.32) it is clear that all the effects of the (non-interacting) residual part H_0 are encapsulated in g_0, whereas all correlation effects induced by the interaction V are accounted for by Σ. It should be noted in this context that (14.32) is still a single particle equation, i.e. Σ does not contain full information on many-body processes in the system, such as double and triple excitations (described by the two and three particle Green's function). Σ describes (exactly) however all (integral) many-body effects that are relevant for the *single particle* Green's function. The (simplest) first order approximation to Σ is the so-called Hartree-Fock (static) self energy

$$\Sigma^{HF}(k, k') = -\frac{i}{2\pi} \lim_{\eta \to 0+} \int d\omega' \, e^{i\omega' \eta} V_{k\alpha, k'\beta} \, g(\alpha, \beta, \omega'), \tag{14.33}$$

where $V_{k\alpha, k'\beta}$ are the matrix elements of the two-particle interaction including the direct and the exchange term.

14.2.3   Hedin equations and the GW approximation

The role of Σ is elucidated by inserting the Lehmann representations for g_0 and g into the Dyson equation (14.32) and inspecting the residua of the poles ω_γ^+ and ω_δ^-. By doing so one derives eigenvalue equations for the spectroscopic factors $z_{k\gamma}$ and $\bar{z}_{k\delta}$ (14.23), e.g. for the hole pole ω_δ^- one obtains

$$\sum_{k'} [\langle k|H_0|k' \rangle + \Sigma(k, k', \omega)] \, \bar{z}_{k'\delta} \bigg|_{\omega=\omega_\delta^-} = \omega \bar{z}_{k\delta}. \tag{14.34}$$

Expanding the Dyson equation in terms of ω in the neighborhood of the pole ω_δ^- and taking only first order terms of this expansion one obtains the magnitude of the spectroscopic factors as [365, 366]

$$\sum_k |\bar{z}_{k\delta}|^2 = 1 + \sum_{k, k'} (\bar{z}_{k\delta})^* \left(\partial_\omega \Sigma(k, k', \omega)|_{\omega=\omega_\delta^-} \right) \bar{z}_{k'\delta}. \tag{14.35}$$

Eq. (14.34) makes clear that the self-energy plays the role of an effective, energy-dependent (non-local) potential. In fact the (integral) Dyson equation converted into a differential form reads

$$[i\partial_t - H_0] \, g(\alpha t, \beta t') - \sum_\gamma \int dt'' \Sigma(\alpha t, \gamma t'') g(\gamma t'', \beta t') = \delta(\alpha - \beta)\delta(t - t').$$

$$\tag{14.36}$$

As shown by Hedin [367] the self-energy is related to the Green's function g by a set of coupled integral equations, referred to as the Hedin equations (the symbol $y = x^+$ means $y = \lim_{\eta \to 0^+} (x + |\eta|)$)

$$g(1,2) = g_0(1,2) + \int d(3,4) g_0(1,3) \Sigma(3,4) g(4,2), \tag{14.37}$$

$$\Sigma(1,2) = i \int d(3,4) W(1,3^+) g(1,4) \Gamma(3,2,4), \tag{14.38}$$

$$W(1,2) = V(1,2) + \int d(3,4) P(3,4) W(4,2), \tag{14.39}$$

$$P(1,2) = -i \int d(3,4) g(1,3) \Gamma(3,4,2) g(4,1^+), \tag{14.40}$$

$$\Gamma(1,2,3) = \delta(1-2)\delta(2-3) + \int d(4,5,6,7) \frac{\delta \Sigma(1,2)}{\delta g(4,5)} g(4,6) g(7,5) \Gamma(6,7,3). \tag{14.41}$$

In these equations the variables $(\alpha_1 t_1)$ are grouped into a single index 1. These equations reveal in a transparent way how the various physical quantities are interrelated: The polarization function P, which describes the response of the system to an external perturbation as a (de)excitation of a particle-hole pair, re-normalizes the bare interaction V to result in the screened interaction W (14.39). The screened interaction W and the vertex function Γ, which describes the interactions between the holes and the particles, are the essential ingredients for the determination of Σ. On the other hand the change in the effective potential following the excitation is decisive in determining Γ, as given by Eq. (14.41).

Obviously the practical solution of the Hedin equations is complicated and hence one has to resort to approximation schemes. A widely used approximation is the so-called GW approximation [367] in which one neglects the second term in (14.41), i.e. one sets

$$\Gamma(1,2,3) \approx \delta(1-2)\delta(2-3), \tag{14.42}$$

i.e. one discards the vertex corrections. The polarization function is then given by the

$$P(1,2) = -ig(1,2)g(2,1^+) \tag{14.43}$$

which amounts to the random-phase approximation to the polarization propagator [84]. The self-energy attains the form

$$\Sigma^{GW}(1,2) = ig(1,3)W(3,1), \tag{14.44}$$

and hence the name GW for this approximate procedure. Comparing Eqs. (14.33, 14.44) one sees that the GW scheme can be viewed as an extension of the HF scheme in that a screened interaction W is used instead of the bare interaction $V_{k\alpha,k'\beta}$. This is due to the influence of the fluctuations of the medium (described by P) on the two-body interaction V. In contrast to Σ^{HF} the self-energy Σ^{GW} acts as a dynamical effective potential. Such dynamical screening effects can as well be incorporated within the HF scheme but on the expense of calculating the particle-hole excitation, e.g. by means of RPA [368, 21]. In principle, the Hedin equations, even within the GW approximation have to be performed self-consistently. It turns out however, that the first iteration provides useful results for a number of physical quantities (cf. e.g. Refs. [369, 370] and references therein), whereas the second iteration does not yield a systematic improvement. The (e,2e) experiments measuring the hole spectral functions of aluminum have been reproduced by the GW fairly well near the quasi-particle peak but substantial deviations have been observed for the satellite structures (plasmons) [371]. To circumvent this shortcoming of the GW the cumulant expansion[2] [372] turns out to be a computationally tractable method [371].

14.3 Two-body Green's functions

As mentioned above the self-energy carries (integral) information on many-body excitations that are relevant for the sp Green's function. These excitations are naturally described by higher order Green's function, which hints on the interrelation between the various many-particle GF. In fact, by considering the equation of motion for the sp propagator one can derive the relation [85]

$$i\partial_t g(\alpha, \beta, t - t') = \delta(t - t') + \epsilon_\alpha g(\alpha, \beta, t - t') +$$
$$+ \frac{1}{2} \sum_{\eta\eta'\eta''} V_{\alpha\eta,\eta'\eta''} (-i) \langle \Psi_0^{(N)} | \mathcal{T} \left[a_{H\eta}^\dagger(t) a_{H\eta''}(t) a_{H\eta'}(t) a_{H\beta}^\dagger(t') \right] | \Psi_0^{(N)} \rangle.$$

(14.45)

---

[2]In the cumulant expansion the hole spectral function is written as $S_h(k,\omega) = \frac{n_k}{2\pi} \int_{-\infty}^{\infty} d\tau \, e^{i\omega\tau} \cdot e^{-i\epsilon_k\tau + C(k,\tau)}$, where n_k is the occupation number of the state characterized by the wave vector k and the energy ϵ_k. $C(k,\tau)$ is the cumulant operator containing only boson-type diagrams describing the emission and absorption of plasmons. Interactions between the hole and the particle-hole pairs are not accounted for. In practice one equates the cumulant expansion $G(k,\tau) = e^{-i\epsilon_k\tau + C(k,\tau)} = G_0(k,\tau) \left[1 + C(k,t) + C^2(k,t) + \cdots \right]$ with the iteration of the Dyson equation ($G = G_0 + G_0\Sigma G_0 + \cdots$), where $\Sigma \equiv \Sigma^{GW}$ is used (without self-consistency). This yields for the cumulant $G_0 C = G_0 \Sigma G_0$.

Obviously the second term is a two-body Green's function $(g_2(\beta t_1, \beta' t_1', \gamma t_2, \gamma' t_2'))$ with a special time ordering describing the propagation of a particle-hole pair, meaning that the sp and the two-particle GFs are interrelated. Equation (14.45) is in fact the first cycle in a hierarchy that links the N-particle propagator to the $(N+1)$-particle propagator [108, 107].

Here we briefly discuss the two-body Green's function which is of a direct relevance to the two-particle spectroscopy, such as two-electron emission upon the impact of photons or charged particles (for a review see Ref. [214] and references therein) . Generally the two-body (four-point) Green's function is defined as

$$g_2(\beta t_1, \beta' t_1', \gamma t_2, \gamma' t_2')$$
$$= -i\langle \Psi_0^{(N)} | \mathcal{T} \left[a_{H\beta'}(t_1') a_{H\beta}(t_1) a_{H\gamma}^\dagger(t_2) a_{H\gamma'}^\dagger(t_2') \right] | \Psi_0^{(N)} \rangle. \quad (14.46)$$

As seen from this definition, depending on the time ordering the Green's function g_2 describes particle-hole, particle-particle and hole-hole excitations. Generally, the (four-time) two-body Green's function can be split into a non-interacting part consisting of two noninteracting single-particle (dressed) propagators [cf. Eq. (14.32)] and a second part involving the vertex function Γ [52]

$$g_2(\alpha t_1, \alpha' t_1', \beta t_2, \beta' t_2')$$
$$= i \Big[g(\alpha\beta, t_1 - t_2)\, g(\alpha'\beta', t_1' - t_2') - g(\alpha\beta', t_1 - t_2')\, g(\alpha'\beta, t_1' - t_2)$$
$$- \int dt_a\, dt_b\, dt_c\, dt_d \left[\sum_{abcd} g(\alpha a, t_1 - t_a)\, g(\alpha' b, t_1' - t_b) \right.$$
$$\left. + \langle ab|\Gamma(t_a, t_b; t_c, t_d)|cd\rangle\, g(c\beta, t_c - t_2)\, g(d\beta', t_d - t_2') \right]. \quad (14.47)$$

According to equation (14.47) the kernel function Γ can be considered as an effective coupling between dressed particles. In fact, within the ladder approximation, Γ can be expressed in terms of the two-body interaction V, as done below.

14.3.1   The polarization propagator

Assuming homogeneity of time space the Green's function (14.46) can be reduced to the two-time polarization propagator as [52]

$$\Pi(\beta, \beta', \gamma, \gamma', t, t' = 0)$$

$$= g_2(\beta t^+, \beta' 0^+, \gamma t, \gamma' 0) - \langle \Psi_0^{(N)} | a_{\beta'}^\dagger a_\beta | \Psi_0^{(N)} \rangle \langle \Psi_0^{(N)} | a_\gamma^\dagger a_{\gamma'} | \Psi_0^{(N)} \rangle. \quad (14.48)$$

Fourier transforming this equation into frequency space, as done in the case of the sp GF, one obtains the Lehmann-representation of Π as

$$\Pi(\alpha\beta, \gamma\delta; \Omega) = \sum_{n \neq 0} \frac{\langle \Psi_0^{(N)} | a_\beta^\dagger a_\alpha | \Psi_n^{(N)} \rangle \langle \Psi_n^{(N)} | a_\gamma^\dagger a_\delta | \Psi_0^{(N)} \rangle}{\Omega - [E_n^{(N)} - E_0^{(N)}] + i\eta}$$

$$- \sum_{m \neq 0} \frac{\langle \Psi_0^{(N)} | a_\gamma^\dagger a_\delta | \Psi_m^{(N)} \rangle \langle \Psi_m^{(N)} | a_\beta^\dagger a_\alpha | \Psi_0^{(N)} \rangle}{\Omega - [E_0^{(N)} - E_m^{(N)}] - i\eta}. \quad (14.49)$$

From this relation it is evident that the polarization function describes all particle-hole excitations within the N body system, i.e. it accounts for the fluctuations with respect to the ground state. While Π does not describe particle removal or addition to the N particle system it still may have a strong influence on such processes. This is because Π plays an essential role in the determination of the self-energy which in turn dictates the behaviour of the sp Green's function and the spectral functions. For the electron removal from C_{60}, this has been demonstrated explicitly [368] using the random phase approximation[3] for Π.

14.3.2   Particle-particle and hole-hole propagators

As discussed above the Π describes particle-hole excitations without changing the number of particles in the system. For the treatment of processes that involve the propagation of two particles or holes different type of Green's functions are needed. These are obtained

---

[3]The RPA for Π is obtained by considering the free ph propagator $\Pi_0(\omega)$, which couples two excitations of a particle and a hole type that do not interact with each other. In the language of Feynman diagrams this corresponds to two dressed sp lines propagating in time in opposite directions, i.e.

$$\Pi_0(\alpha\beta, \gamma\delta, \omega) = \frac{i}{2\pi} \int d\omega' ig(\alpha, \gamma, \omega + \omega') ig(\delta, \beta, \omega').$$

Interaction between the hole and the particle lines can then be included by means of an integral equation of the algebraic form $\Pi = \Pi_0 + \Pi_0 V \Pi$. g entering the definition of Π_0 stands for the (dressed) single particle propagator. If the uncorrelated sp Green's function is used $g \approx g_0$ one obtains the random-phase approximation as well as the Tamm-Dancoff approximation [52].

from $g_2(\beta t_1, \beta' t'_1, \gamma t_2, \gamma' t'_2)$ (14.46) using certain time ordering, e.g. the (two-time) particle-particle propagator $g^{II}(\alpha, \beta, \gamma, \delta, t, t' = 0)$ is obtained in the limit

$$g^{II}(\alpha, \beta, \gamma, \delta, t, t' = 0) = g_2(\alpha t, \beta t^+, \gamma 0^+, \delta 0). \tag{14.50}$$

The non-interacting part of g_2 (14.47) results in a similar contribution g_f^{II} to g^{II} consisting of products of dressed sp propagators. Using the spectral representation of the sp Green's function is it readily deduced that

$$g_f^{II}(\alpha\beta, \gamma\delta; \Omega) = \frac{i}{2\pi} \int d\omega \left[g(\alpha, \gamma; \omega) g(\beta, \delta; \Omega - \omega) - g(\alpha, \delta; \omega) g(\beta, \gamma; \Omega - \omega) \right]$$

$$= \sum_{mm'} \frac{\langle\Psi_0^{(N)}|a_\alpha|\Psi_m^{(N+1)}\rangle\langle\Psi_m^{(N+1)}|a_\gamma^\dagger|\Psi_0^{(N)}\rangle\langle\Psi_0^{(N)}|a_\beta|\Psi_{m'}^{(N+1)}\rangle\langle\Psi_{m'}^{(N+1)}|a_\delta^\dagger|\Psi_0^{(N)}\rangle}{\Omega - \{[E_m^{(N+1)} - E_0^{(N)}] + [E_{m'}^{(N+1)} - E_0^{(N)}]\} + i\eta}$$

$$- \sum_{nn'} \frac{\langle\Psi_0^{(N)}|a_\gamma^\dagger|\Psi_n^{(N-1)}\rangle\langle\Psi_n^{(N-1)}|a_\alpha|\Psi_0^{(N)}\rangle\langle\Psi_0^{(N)}|a_\delta^\dagger|\Psi_{n'}^{(N-1)}\rangle\langle\Psi_{n'}^{(N-1)}|a_\beta|\Psi_0^{(N)}\rangle}{\Omega - \{[E_0^{(N)} - E_n^{(N-1)}] + [E_0^{(N)} - E_{n'}^{(N-1)}]\} + i\eta}$$

$$- (\gamma \longleftrightarrow \delta). \tag{14.51}$$

Proceeding along the same lines, followed for the derivation of the Lehmann-representation of the sp GF, one obtains the Lehmann representation of the (full) particle-particle Green's function in terms of energies and states of the systems with N and $N \pm 2$ particles [the $(N-2)$ particle state of the system is achieved upon a $(\gamma, 2e)$ reaction], i.e.

$$g^{II}(\alpha\beta, \gamma\delta; \Omega) = \sum_n \frac{\langle\Psi_0^{(N)}|a_\beta a_\alpha|\Psi_n^{(N+2)}\rangle\langle\Psi_n^{(N+2)}|a_\gamma^\dagger a_\delta^\dagger|\Psi_0^{(N)}\rangle}{\Omega - [E_n^{(N+2)} - E_0^{(N)}] + i\eta}$$

$$- \sum_m \frac{\langle\Psi_0^{(N)}|a_\gamma^\dagger a_\delta^\dagger|\Psi_m^{(N-2)}\rangle\langle\Psi_m^{(N-2)}|a_\beta a_\alpha|\Psi_0^{(N)}\rangle}{\Omega - [E_0^{(N)} - E_m^{(N-2)}] - i\eta}. \tag{14.52}$$

Eqs. (14.47, 14.50) indicate the possibility of expressing g^{II} in terms of g_f^{II} and the vertex function. This is achieved *via* the ladder approximation to the particle-particle propagator which states that

$$g_L^{II}(\alpha\beta, \gamma\delta; \Omega) = g_f^{II}(\alpha\beta, \gamma\delta; \Omega) + \frac{1}{4}\sum_{\epsilon\eta\theta\zeta} g_f^{II}(\alpha\beta, \epsilon\eta; \Omega)\langle\epsilon\eta|V|\theta\zeta\rangle g_L^{II}(\theta\zeta, \gamma\delta; \Omega). \tag{14.53}$$

V stands for the naked two-body interaction. The integral equation (14.53) can be iterated yielding a set of ladder diagrams. The corresponding ladder sum for the effective interaction

Γ, as it appears in [cf. Eq. (14.47)], can be deduced from this result as

$$\langle \alpha_1 \beta_2 | \Gamma_{\mathrm{L}}(\Omega) | \alpha_1' \beta_2' \rangle \;\; = \;\; \langle \alpha_1 \beta_2 | V | \alpha_1' \beta_2' \rangle$$
$$+ \frac{1}{4} \sum_{\epsilon \eta \theta \zeta} \langle \alpha_1 \beta_2 | V | \epsilon \eta \rangle g_{\mathrm{f}}^{\mathrm{II}}(\epsilon \eta, \theta \zeta; \Omega) \langle \theta \zeta | \Gamma_{\mathrm{L}}(\Omega) | \alpha_1' \beta_2' \rangle.$$

$$(14.54)$$

From this relation it is evident that Γ plays the role of the T matrix in the single particle case.

The ladder approximation (14.53) for the two-particle Green's function can be employed to define the self-energy Σ [85, 107] which can then be used to obtain the single-particle Green's function via Eq. (14.32). On the other hand, this Green's function enters in the definition of the two-particle Green's function, as clear, e. g. from Eqs. (14.51, 14.53). Thus, in principle, the Dyson Eqs. (14.32) and (14.53) for the one-body and two-body Green's functions have to be solved in a self-consistent manner.

For the single particle Green's function we discussed the relation between the (hole) spectral representation to the high-energy (e,2e) experiments. Similarly one can relate g^{II} to the $(\gamma,2e)$ measurements by means of Eq. (14.52): g^{II} shows poles at energies (relative to the ground state) corresponding to adding (the part containing $[E_n^{(N+2)} - E_0^{(N)}]$) two particles to the excited two-particle state n. g^{II} also describes the process of removing (the part involving $[E_0^{(N)} - E_n^{(N-2)}]$) two particles from the unperturbed ground state leaving the system in the excited state n. The residua of these poles are related to the measurable spectroscopic factors for the addition or removal of the two particles, e. g., as done in $(\gamma,2e)$ experiment. For the independent particle part of g^{II} one can establish [113] an exact relation between these spectroscopic factors and the single particle spectral functions.

Utilizing the above GF formalism calculations have been performed [115, 373] for the two-particle photocurrent generated from surfaces in a $(\gamma, 2e)$ experiments [112]. The next section provides the basic formulas for the calculations of the photocurrent and contrasts with the Fermi golden rule (5.10), which is widely applied for photoexcitation processes in few-body systems (see Ref. [239] for a review and further references).

In the context of comparing theory with experiment it should be noted that the GF formalism outlined above is valid at absolute zero temperature. The finite temperature case poses no obstacle due to the formal equivalence of the statistical operator and the analytically continued evolution operator [277] (see appendix A.4 for some details on the finite temperature (Matsubara) Green's function).

In contrast, the treatment of non-equilibrium processes, such as the influence of a strong time-dependent external perturbation, requires the use of the non-equilibrium (Keldysh) Green's function approach [278, 374, 375] (a brief summary of the basic elements of the non-equilibrium (Keldysh) GF is provided in appendix A.5). As shown below the results for the (single and many-particle) photocurrent can be expressed [113, 376] in terms of GFs that we can handle formally and computationally.

14.3.3   Single and many-particle photoelectron currents

In a seminal work Caroli *et al.* [376] showed how the Keldysh non-equilibrium Green's function formalism can be applied to the process of photoelectron emission from extended systems. Subsequently, based on Caroli's ideas efficient schemes have been developed for the calculations of the single particle photocurrent (see Refs. [377, 378, 100, 111, 379] for more details and further references). Here we sketch the main steps of the theory for a monochromatic photon field described by the vector potential introduced in chapter 5 by Eq. (5.1) (on page 49). At first let us re-consider the case of chapter 5 for a system consisting of a single electron subject to the confining potential $V(\mathbf{r})$, i.e. Eq. (5.4) reduces to $\mathcal{H} = H + W(\mathbf{r}, t)$, where (cf. Eq. (5.6))

$$H|i\rangle = \left[\frac{\mathbf{p}^2}{2} + V(\mathbf{r}) \right] |i\rangle = \epsilon_i |i\rangle,$$

and the time-dependent perturbation potential has the form

$$W(\mathbf{r}, t) = \frac{1}{2c} [\mathbf{p} \cdot \mathbf{A}(\mathbf{r}, t) + \mathbf{A}(\mathbf{r}, t) \cdot \mathbf{p}] + \frac{\mathbf{A}(\mathbf{r}, t)^2}{2c^2}. \tag{14.55}$$

The properties of H and W have been discussed at length in chapter 5.

Upon the photoabsorption process a photoelectron is generated which is collected by a detector positioned at a distance $\mathbf{r}_0$ from the specimen, and it resolves the wave vector $\mathbf{k}$ of the photoelectron. The quantity of interest for the calculation of the photoelectron spectra is the current density $\mathbf{j}$ generated by the photo-excited state $|\Psi\rangle$ at the position $\mathbf{r}_0$. For a particle described by the state $|\Psi\rangle$ the current density is generally evaluated using the formula

$$\langle \Psi | \mathbf{j}(\mathbf{r}_0) | \Psi \rangle = \frac{i}{2} \left. (\Psi^\star \boldsymbol{\nabla} \Psi - \Psi \boldsymbol{\nabla} \Psi^\star) \right|_{\mathbf{r}=\mathbf{r}_0}. \tag{14.56}$$

Now one proceeds as discussed in the appendix A.3 (in the context of the adiabatic hypothesis) by switching on adiabatically the time-dependent perturbation $W(\mathbf{r}, t)$. The unperturbed state

$|i\rangle$ at the $t = -\infty$ develops in time to the state $|\Psi_i(t)\rangle$ under the action of the perturbation W. To a first order in the perturbation W the time-development of the state $|\Psi_i(t)\rangle$ is given by

$$|\Psi_{i\mathrm{I}}(t)\rangle = |i\rangle - \mathrm{i} \int_{-\infty}^{t} dt' W_{\mathrm{I}}(t') |i\rangle. \tag{14.57}$$

The current density at $\mathbf{r}_0$ associated with the state that had developed from $|i\rangle$, is

$$\langle \mathbf{j}_i(\mathbf{r}_0)\rangle = \langle \Psi_{i\mathrm{I}}(t)|\mathbf{j}_{\mathrm{I}}(\mathbf{r}_0, t)|\Psi_{i\mathrm{I}}(t)\rangle. \tag{14.58}$$

Neglecting the diamagnetic term $(\mathbf{A}^2/2)$ and going over into the Schrödinger picture one obtains the current in terms of GFs as

$$\langle \mathbf{j}_i(\mathbf{r}_0)\rangle = \langle i|O g_1^{\mathrm{a}}(\omega_i + \omega)\mathbf{j}(\mathbf{r}_0)g_1^{\mathrm{r}}(\omega_i + \omega)O|i\rangle. \tag{14.59}$$

The operator O is defined as $O = (\mathbf{p} \cdot \mathbf{A}_0 + \mathbf{A}_0 \cdot \mathbf{p})/2c$, where $\mathbf{A}_0$ has already been introduced in Eq. (5.1). The retarded and the advanced single-particle GFs are defined by Eqs. (11.6, 14.4). Recalling that the non-local spectral density operator $\sum_n |n\rangle \delta(\omega - \omega_n)\langle n|$ is derived from the retarded GF according to (11.11) and inserting Eq. (14.56) into Eq. (14.59) one obtains the following expression for the photocurrent

$$r_0^2 \langle j(\mathbf{r}_0)\rangle = -\frac{k}{32\pi^2 c^2} \langle \Phi(\omega_k + \omega)|\Delta [\Im g_1^{\mathrm{r}}(\omega_k)]\Delta^\dagger|\Phi(\omega_k + \omega)\rangle. \tag{14.60}$$

Here the dipole operator is denoted by Δ, whereas the magnitude of the momentum k of the photoelectron is $k = \sqrt{2(\omega_k + \omega)}$, i.e. the initial particle distribution at the energy ω_k is elevated upon the photoabsorption by the photon energy ω. The state vector $|\Phi\rangle$ is the final state of the photoelectron obtained by back-propagating in time the asymptotic (detector) state $|\Phi_0\rangle$, i.e.

$$|\Phi\rangle = g_1^{\mathrm{r}}(\omega_k + \omega)|\Phi_0\rangle.$$

Using this relation we can re-write Eq. (14.60) to obtain the well-known expression for the photocurrent

$$j \sim - \langle \Phi_0|g_1^{\mathrm{a}}(\omega_k + \omega)\Delta [\Im g_1^{\mathrm{r}}(\omega_k)]\Delta^\dagger g_1^{\mathrm{r}}(\omega_k + \omega)|\Phi_0\rangle. \tag{14.61}$$

According to this equation the single photoelectron current can be evaluated along the following lines: At first, beginning at the time when the photoelectron reaches the detector (we may refer to this time instance as $t = -\infty$), the known detector state $|\Phi_0(\omega_k + \omega)\rangle$ is back propagated using the retarded Green's function g_1^r to the instance of the photoabsorption (one may refer to this time as $t = +\infty$). During the back-propagation the external perturbation (dipole operator) is switched on adiabatically. The adjunct dipole photon operator $\Delta^\dagger$ causes the de-excitation to an initial distribution of states that is described by the non-local spectral density operator $\Im g_1^r(\omega_k)$. The de-excited state is then re-propagated by means of g_1^a to $t = -\infty$, where the detector state is reached. This final state forms the conventional outgoing photo-electron state. This (time) path propagation is to be contrasted with the time-loop contour used for the definition of the Keldysh GF, as discussed in appendix A.5. In fact, in case of a many-body target the (non-equilibrium) photoemission process can be described by means of the non-equilibrium Green's function method [376]. In this case the same formula (14.61) for the photoelectron current is derived with the exception that the retarded and the advanced Green's function g^r and g^a are employed. These GF's have been introduced by respectively Eq. (A.38) and Eq. (A.39). On the other hand, for the single particle case one concludes [113] that Eq. (14.61) reduces to the Fermi-golden rule (5.10) established in chapter 5.

For the one-photon induced two-particle emission the photocurrent j_{II} can be evaluated using the relation [113]

$$ j_{II} \propto - \left\langle \Phi_{II0} \left| g^{IIa}(\omega_{k_1,k_2} + \omega) \Delta_{II} \left[\Im g^{rII}(\omega_{k_1,k_2}) \right] \Delta_{II}^\dagger g^{IIr}(\omega_{k_1,k_2} + \omega) \right| \Phi_{II0} \right\rangle . $$

$$ (14.62) $$

Here $\Delta_{II} = \Delta_1 + \Delta_2$ is the two-particle dipole operator, where Δ_j is the dipole operator acting on the particle j. The (quasi) two-particle energy ω_{k_1,k_2}, before the photon has been observed, is generally a complex quantity (due to the finite life time of the quasi particles). This fact distinguishes the present photocurrent approach from the methods relying on the Fermi golden rule (where ω_{k_1,k_2} is real). The state $|\Phi_{II0}\rangle$ is the detector two-particle state that can be constructed from two single particle detector states. The GF g^{II} is the particle-particle Green's function (14.50). Its first order approximation (14.51) and its ladder expansion terms (14.53) can be constructed from the (dressed) single particle Green's function. The latter sp GF is needed for the calculation of the single particle photocurrent (14.61). Thus, the single and the two-particle photocurrents should be (if tractable) calculated simultaneously. This has been done in Ref. [373] using an approximate model for (14.50). The details of the calculations and an analysis of the theoretical and experimental results will be presented in the second volume of this work.

A Appendices

A.1 Tensor Operators

A spherical tensor T_{JM} of rank J and components M is defined according to its transformation properties under rotations generated by the angular momentum operator $\mathbf{J}$. A rotation can be specified by a rotation angle ψ with respect to an axis $\hat{n}$. Generally, the state vectors are rotated upon the action of the unitary operator $U_{\hat{n}}(\psi) = e^{-i\psi\hat{n}\cdot\mathbf{J}}$. The rotated angular momentum state $U\,|jm\rangle$ is an eigenstate of $\mathbf{J}^2$ with the same eigenvalue, because $\mathbf{J}^2$ commutes with any function of its components J_l. The transform O' of the operator O is obtained via the unitary transformation

$$
\begin{aligned}
O' &= U^\dagger O U = e^{-i\psi\hat{n}\cdot\mathbf{J}}Oe^{i\psi\hat{n}\cdot\mathbf{J}} \\
&= \lim_{N\to\infty}\left\{\left[1+i\frac{\psi}{N}\hat{n}\cdot\mathbf{J}\right]^N O\left[1-i\frac{\psi}{N}\hat{n}\cdot\mathbf{J}\right]^N\right\} \\
&= \sum_{N=0}^{\infty}\frac{1}{N!}\,[i\psi\,\hat{n}\cdot\mathbf{J},O]_N\,.
\end{aligned}
\tag{A.1}
$$

The commutator $[A,B]_N$ is defined as

$$
[A,B]_0 = B, \quad [A,B]_1 = AB - BA, \quad \cdots, \quad [A,B]_\nu = \left[A,[A,B]_{\nu-1}\right].
$$

Spherical tensors T_{JM} are defined according to the special transformation properties

$$
[J_\mu,T_{JM}] = \sqrt{J(J+1)}\,\langle JM1\mu\mid JM+\mu\rangle\,T_{J,M+\mu},
\tag{A.2}
$$

where J_μ are the spherical components of the operator $\mathbf{J}$, i.e.

$$
J_\mu = \begin{cases} -\frac{1}{\sqrt{2}}(J_x + iJ_y) & \text{for} \quad \mu = 1, \\ J_z & \text{for} \quad \mu = 0, \\ \frac{1}{\sqrt{2}}(J_x - iJ_y) & \text{for} \quad \mu = -1. \end{cases}
\tag{A.3}
$$

Here J_x, J_y and J_z are the cartesian components of $\mathbf{J}$. The relation (A.2) can also be formu-

lated as

$$[J_\mu, T_{JM}] = \sum_{M'} T_{JM'} \langle JM' | J_\mu | JM \rangle, \tag{A.4}$$

which means that the relation

$$[i\psi\, \hat{\mathbf{n}} \cdot \mathbf{J}, T_{JM}] = \sum_{M'} T_{JM'} \langle JM' | i\psi\, \hat{\mathbf{n}} \cdot \mathbf{J} | JM \rangle \tag{A.5}$$

applies and we can furthermore write for the N fold commutator

$$[i\psi\, \hat{\mathbf{n}} \cdot \mathbf{J}, T_{JM}]_N = \sum_{M'} T_{JM'} \langle JM' | (i\psi\, \hat{\mathbf{n}} \cdot \mathbf{J})^N | JM \rangle. \tag{A.6}$$

Combining this relation with Eq. (A.1) we conclude that spherical tensors transform under U

as follows

$$U T_{JM} U^\dagger = e^{-i\psi\, \hat{\mathbf{n}} \cdot \mathbf{J}} T_{JM} e^{+i\psi\, \hat{\mathbf{n}} \cdot \mathbf{J}} = \sum_{M'} T_{JM'} \langle JM' | U | JM \rangle = \sum_{M'} T_{JM'} D^J_{M'M}. \tag{A.7}$$

The matrix elements $D^J_{M'M} := \langle JM' | U | JM \rangle$ are the called the Wigner functions [227].

Employing the Euler angles $(\alpha_e, \beta_e, \gamma_e)$ to characterize the most general rotation, one finds

$D^J_{M'M}(\alpha_e, \beta_e, \gamma_e) = \langle JM' | U_z(\alpha_e) U_y(\beta_e) U_z(\gamma_e) | JM \rangle = e^{-i(M'\alpha_e + M\gamma_e)} \langle JM' | e^{-i\beta_e J_y} | JM \rangle =$

$e^{-i(M'\alpha_e + M\gamma_e)} d^J_{M'M}(\beta_e)$.

A.1.1   Wigner-Eckart theorem

The Wigner-Eckart theorem is indispensable for the evaluation and the analysis of the matrix

elements $\langle j'm' | T_{KQ} | jm \rangle$ of the spherical tensor operators T_{KQ} in an angular momentum

basis. Some examples we encountered in chapter (5). The Wigner-Eckart theorem states that

$$\langle j'm' | T_{KQ} | jm \rangle = \langle jmKQ \mid j'm' \rangle \langle j' \| T_K \| j \rangle \tag{A.8}$$

with the reduced matrix elements being given by the relation

$$\langle j' \| T_K \| j \rangle = \sum_{mQ} \langle jmKQ \mid j'm' \rangle \langle j'm' | T_{KQ} | jm \rangle. \tag{A.9}$$

The proof of this theorem can be found in standard books on angular momentum theory,

e.g. [236, 144].

A.1.2 Tensor products of spherical tensors

For two spherical tensors S_{k_1} and T_{k_2} we define the tensor product P_k as

$$P_{kq} = [S_{k_1} \otimes T_{k_2}]_{kq} = \sum_{q_1 q_2} \langle k_1 q_1 k_2 q_2 \mid kq \rangle S_{k_1 q_1} T_{k_2 q_2}. \tag{A.10}$$

The quantity P_k is as well a spherical tensor. This can be seen as follows.

At first we note that

$$
\begin{aligned}
U S_{k_1 q_1} U^\dagger &= \sum_{q_1'} S_{k_1 q_1'} D^{k_1}_{q_1' q_1}, \\
U T_{k_2 q_2} U^\dagger &= \sum_{q_2'} T_{k_2 q_2'} D^{k_2}_{q_2' q_2}.
\end{aligned}
\tag{A.11}
$$

Furthermore, since $S_{k_1 q_1}$ and $T_{k_2 q_2}$ are both spherical tensors the transformation applies

$$
\begin{aligned}
U S_{k_1 q_1} T_{k_2 q_2} U^\dagger &= U S_{k_1 q_1} U\, U^\dagger T_{k_2 q_2} U^\dagger \\
&= \sum_{q_1' q_2'} S_{k_1 q_1'} T_{k_2 q_2'} D^{k_1}_{q_1' q_1} D^{k_2}_{q_2' q_2} \\
&= \sum_{q_1' q_2'} S_{k_1 q_1'} T_{k_2 q_2'} \sum_{k} \langle k_1 q_1' k_2 q_2' \mid kq_1' + q_2' \rangle \langle k_1 q_1 k_2 q_2 \mid kq_1 + q_2 \rangle D^{k}_{q_1' + q_2', q_1 + q_2}
\end{aligned}
\tag{A.12}
$$

Multiplying this expression with $\langle k_1 q_1 k_2 q_2 \mid kq \rangle$ and summing over q_1, q_2 we find

$$
\sum_{q_1 q_2} \langle k_1 q_1 k_2 q_2 \mid kq \rangle U S_{k_1 q_1} T_{k_2 q_2} U^\dagger
$$

$$
= \sum_{q'} \left\{ \sum_{q_1' q_2'} \langle k_1 q_1' k_2 q_2' \mid kq' \rangle S_{k_1 q_1'} T_{k_2 q_2'} \right\} D^{k}_{q' q}. \tag{A.13}
$$

With this finding we deduce the tensorial transformation behaviour of the product P_{kq} (A.10) as

$$
U [S_{k_1} \otimes T_{k_2}]_{kq} U^\dagger = \sum_{q'} [S_{k_1} \otimes T_{k_2}]_{kq'} D^{k}_{q' q}. \tag{A.14}
$$

A.2 Time ordering and perturbation expansion

The Green's function of a many body system is defined as the expectation value of a time-ordered product with respect to the exact correlated ground state $|\Psi_0\rangle$ which is generally unknown. In fact, the computation of $|\Psi_0\rangle$ is one of the purposes of the Green's function theory. Thus, it is imperative for any progress in practical application to express the GF in terms of known quantities, such as the ground state of a non-interacting system. Also, it is of great advantage to develop a systematic scheme for obtaining approximate expressions. This appendix sketches the main steps to achieve this goal, more extensive details can be found in the standard books on many-body theory, e.g. [84, 52, 85].

Let us consider a system described by the Hamiltonian

$$H = H_0 + V,$$

where H_0 is a reference (non-interacting) Hamiltonian and V is the interacting part of H. The time evolution of the system is described within the Schrödinger picture via the time-dependence of the state vector

$$|\Psi(t)\rangle = e^{-iHt}|\Psi(0)\rangle = U(t, 0)|\Psi(0)\rangle,$$

whereas any observable $\mathcal{O}$ is time-independent. In the Heisenberg picture the wave functions do not change in time ($|\Psi_H(t)\rangle = |\Psi_H(0)\rangle$), whereas the time dependence of the operator $\mathcal{O}$ is given by

$$\mathcal{O}_H = e^{iHt}\mathcal{O}e^{-iHt}.$$

In the interaction picture the operators evolve in time under the influence of H_0, i.e.

$$\mathcal{O}_I(t) = e^{iH_0 t}\mathcal{O}e^{-iH_0 t}, \quad i\partial_t \mathcal{O}_I(t) = [\mathcal{O}_I(t), H_0].$$

The time development of the state vectors in the interaction picture is given by

$$|\Psi_I(t)\rangle = e^{-iH_0 t}|\Psi(t)\rangle = e^{-iH_0 t}e^{-iHt}|\Psi(0)\rangle = U_I(t, t_0)|\Psi_I(t_0)\rangle.$$

Using this equation and noting that H and H_0 do not commute it is straightforward to show that the time-evolution operator[1] in the interaction picture satisfies

$$U_I(t, t_0) = e^{iH_0 t}e^{-iH(t-t_0)}e^{-iH_0 t_0}.$$

---

[1] The following steps can as well be formulated in terms of the scattering (S) operator which is defined as

$$S(t, t') = U(t)U^\dagger(t'). \tag{A.15}$$

Further important (group) properties of the evolution operator are

$$
\begin{align}
U_{\mathrm{I}}^{\dagger}(t,t_0)U_{\mathrm{I}}(t,t_0) &= U_{\mathrm{I}}(t,t_0)U_{\mathrm{I}}^{\dagger}(t,t_0) = \mathbb{1}, \ U_{\mathrm{I}}^{-1}(t,t_0) = U_{\mathrm{I}}^{\dagger}(t,t_0), \tag{A.16}\\
U_{\mathrm{I}}(t_1,t_2)U_{\mathrm{I}}(t_2,t_3) &= U_{\mathrm{I}}(t_1,t_3), \quad U_{\mathrm{I}}(t_1,t_2)U_{\mathrm{I}}(t_2,t_1) = \mathbb{1}, \tag{A.17}\\
i\partial_t U_{\mathrm{I}}(t,t') &= V_{\mathrm{I}}(t)U_{\mathrm{I}}(t,t'). \tag{A.18}
\end{align}
$$

Since $U_{\mathrm{I}}(t,t) = \mathbb{1}$ one obtains thus

$$
U_{\mathrm{I}}(t,t_0) = \mathbb{1} - i \int_{t_0}^{t} dt' V_{\mathrm{I}}(t')U_{\mathrm{I}}(t',t_0).
$$

Iterating this integral equation one obtains the perturbative expansion in powers of V_{I}:

$$
\begin{align}
U_{\mathrm{I}}(t,t_0) = & \ \mathbb{1} + (-i) \int_{t_0}^{t} dt_1 \, V_{\mathrm{I}}(t_1) + \\
& + (-i)^2 \int_{t_0}^{t} dt_1 \, V_{\mathrm{I}}(t_1) \int_{t_0}^{t_1} dt_2 \, V_{\mathrm{I}}(t_2) + \dots. \tag{A.19}
\end{align}
$$

This relation can be rearranged [84] using the time ordering operator $\mathcal{T}$ as follows

$$
U_{\mathrm{I}}(t,t_0) = \sum_{n=0}^{\infty} \frac{(-i)^n}{n!} \int_{t_0}^{t} dt_1 \dots \int_{t_0}^{t} dt_n \, \mathcal{T}\left[V_{\mathrm{I}}(t_1)\dots V_{\mathrm{I}}(t_n)\right]. \tag{A.20}
$$

The time ordering operator $\mathcal{T}$ moves the operators with the latest time farthest to the left.

A.3 Adiabatic switching of interactions

Adiabatically switching the interaction V is the decisive step in making use of the uniqueness of the ground state to obtain quantities related to the exact correlated ground state from those associated with the non-interacting Hamiltonian using a perturbative expansion in V.

Operating within the Schrödinger picture the perturbation V is "switched off" at times $t = -\infty$ and $t = +\infty$, in which case H reduces to H_0. As the time $t = 0$ is approached, the interaction V is turned on adiabatically. This is can be formulated mathematically by writing the time-dependent Hamiltonian H in the form

$$
H_\epsilon(t) = H_0 + e^{-\epsilon|t|} H_1, \tag{A.21}
$$

where ϵ is an infinitesimally small positive real number. Thus, the eigenstates of the hamiltonian H_ϵ coincide with the eigenstates of the unperturbed hamiltonian H_0 at $|t| = \infty$ and evolve to the fully correlated eigenstates of the total H as t approaches zero. Since the explicit time-dependence introduced through the definition (A.21) commutes with all relevant terms occurring in the expansion (A.20) one finds for the time evolution operator $U_{I\epsilon}(t, t')$ associated with the hamiltonian H_ϵ the corresponding equation

$$U_{I\epsilon}(t, t_0) = \sum_{n=0}^{\infty} \frac{(-i)^n}{n!} \int_{t_0}^{t} dt_1 \dots \int_{t_0}^{t} dt_n e^{-\epsilon(|t_1|+|t_1|+\dots+|t_n|)} \, \mathcal{T}\left[V_I(t_1) \dots V_I(t_n)\right].$$

(A.22)

Denoting the time-independent (non-degenerate) ground state of H_0 by Φ_0 one obtains by means of (A.22) the exact fully correlated ground state of H at time $t = 0$ according to

$$|\Psi_0(t = 0)\rangle = \lim_{\epsilon \to 0} |\Psi_{0\epsilon}\rangle = \lim_{\epsilon \to 0} U_{I\epsilon}(0, -\infty)|\Phi_0\rangle\,,$$

(A.23)

where $|\Psi_{0\epsilon}(t = 0)\rangle$ is an eigenstate of $H_\epsilon(t)$. The Gell-Mann and Low theorem [361] states that if the limit

$$\frac{|\Psi_0\rangle}{\langle\Phi_0|\Psi_0\rangle} = \lim_{\epsilon \to 0} \frac{|\Psi_{0\epsilon}\rangle}{\langle\Phi_0|\Psi_{0\epsilon}\rangle} = \lim_{\epsilon \to 0} \frac{U_{I\epsilon}(0, -\infty)|\Phi_0\rangle}{\langle\Phi_0|U_{I\epsilon}(\infty, 0)U_{I\epsilon}(0, -\infty)|\Phi_0\rangle}$$

(A.24)

exists for all orders in the perturbation theory then the state $|\Psi_0\rangle/(\langle\Phi_0|\Psi_0\rangle)$ is an eigenstate of H. This result is in so far important as in general the numerator and denominator of Eq. (A.24) do not exist separately. The practical importance of the Gell-Mann Low theorem is that the expectation value of any observable with respect to the correlated ground state can now be expressed in terms of the uncorrelated (asymptotic) state $|\Phi_0\rangle$. In particular one employs the expansion (A.22) for the time evolution operator to prove that the matrix elements for any time-ordered product of two Heisenberg operators a_H and b_H are evaluated using $|\Phi_0\rangle$

$$\frac{\langle\Psi_0|\mathcal{T}\left(a_H(t)b_H(t')\right)|\Psi_0\rangle}{\langle\Psi_0|\Psi_0\rangle} = \lim_{\epsilon \to 0}\left[\frac{1}{\langle\Psi_0|\Psi_0\rangle}\sum_{n=0}^{\infty}\frac{(-i)^n}{n!}\right.$$

$$\times \int_{-\infty}^{\infty} dt_1 \dots \int_{-\infty}^{\infty} dt_n \, e^{-\epsilon(|t_1|+\dots|t_n|)}$$

$$\left.\times\langle\Phi_0|\mathcal{T}\left[V_I(t_1)\dots V_I(t_n)a_I(t)b_I(t')\right]|\Phi_0\rangle\right].$$

(A.25)

This reduces the problem of evaluating $|\Psi_0\rangle$ and then the matrix elements of the operators to the calculations of matrix elements of time-ordered products with respect to the ground state $|\Phi_0\rangle$ of the known reference hamiltonian H_0. For the evaluation of the matrix elements of time-ordered products of field operators one utilizes the Wick's theorem [362, 84] and visualizes the various contributions using Feynman diagrams. Wick [362] showed that the time-ordered product of any set of operators $a_{I1}, \cdots, a_{In}$ is equal to the sum of their normal products with all possible contractions. The normal product (symbolized by $: a_{I1}, \cdots, a_{In} :$) means all destruction operators are ordered to the right of the construction operators[2]. The contraction (also called pairing and symbolized by $\overbrace{a_{Im} a_{In}}$) is the difference between the time and the normal-ordered product, i.e.[3]

$$\overbrace{a_{Im} a_{In}} = \mathcal{T}\left[a_{Im} a_{In}\right] - : a_{Im} a_{In} : .$$

A particularly important example for (A.25) is the single-particle Green's function

$$g(\alpha t, \beta t') = -i \langle \Psi_0 | \mathcal{T}\left[a_{H\alpha}(t) a_{H\beta}^\dagger(t')\right] | \Psi_0 \rangle , \tag{A.26}$$

which is the expectation value for the time-ordered product of two operators calculated for the exact ground-state and can thus be expressed according to Eq. (A.25).

A.4   Finite-temperature equilibrium (Matsubara) Green's function

Originally, the Green's function theory has been developed for equilibrium systems at zero temperatures taking advantage of the uniqueness of the ground state. When attempting to apply this technique to excited states one is faced with the problem of how to average over the highly degenerate excited states. This question is addressed in the next section for non-equilibrium systems.

For systems in thermal equilibrium and at a finite temperature T Matsubara [277] pointed out that the formal equivalence of the statistical operator and the analytically continued evolution operator can be exploited to develop a Green's function theory valid at finite temperatures

---

[2]Destruction operators are those operators that yield zero when acting on the unperturbed ground state; their conjugates are the construction operators. Therefore, as a matter of definition, the ground state average of the normal product of the operators $a_{I1}, \cdots, a_{In}$ vanishes, i.e.

$$\langle \Phi_0 | : a_{I1}, \cdots, a_{In} : | \Phi_0 \rangle = 0.$$

[3]If a_{Im} and a_{In} are both annihilation or both creation operators the contraction of a_{Im} and a_{In} vanishes. In addition, due to the simple time-dependence of the Heisenberg operators, the contraction of an operator with its conjugate is a number.

and has the desirable properties of the zero-temperature GF. Here we sketch briefly the main ideas of this approach.

The normalized, equilibrium statistical operator is given by $\rho = e^{-H/T}$ (T is the temperature), whereas the time evolution operator has the form $U(t', t'' = 0) = e^{-iHt'}$ (t' is the time). ρ satisfies the Bloch equation

$$\partial_\tau \rho(\tau) = -H\rho(\tau), \tag{A.27}$$

where the formal variable τ has been introduced and its range is restricted to

$$0 < \tau < 1/T. \tag{A.28}$$

Using the (Wick) complex rotation $t \leftrightarrow -i\tau$ the Bloch equation (A.27) can be mapped onto a Schrödinger-type equation, i.e.

$$i\partial_t \rho(it) = H\rho(it), \quad 0 > t > -i\tau,$$

which explicitly shows the formal equivalence between the structure of the evolution operator U and the analytically continued ρ. Considering the time as a complex variable and rotating by 90° ($t \to it = \tau$) the Heisenberg creation and annihilation operators become

$$a^\dagger_{H\alpha}(\tau) = e^{H\tau} a^\dagger_\alpha e^{-H\tau} =: a^\dagger_{M\alpha}(\tau), \quad a_{H\alpha}(\tau) = e^{H\tau} a_\alpha e^{-H\tau} =: a_{M\alpha}(\tau).$$

The finite temperature sp (Matsubara) Green's function [277] is then defined as [4]

$$\begin{aligned} g_M(\alpha\tau, \beta\tau') &= -\left\langle \mathcal{T}_\tau \left[a_{M\alpha}(\tau) a^\dagger_{M\beta}(\tau') \right] \right\rangle \\ &= -\mathrm{tr} \left\{ e^{-(H-\Omega)/T} \left(\mathcal{T}_\tau \left[a_{M\alpha}(\tau) a^\dagger_{M\beta}(\tau') \right] \right) \right\}, \end{aligned} \tag{A.29}$$

where $\mathcal{T}_\tau$ is the (inverse) temperature ordering operator. It orders the field operators depending on τ in the same way as $\mathcal{T}$ acts in the time space. In Eq. (A.29) Ω refers to the grand potential. Higher order (two-particle, three-particle,...) finite temperature GFs can be defined in a similar manner as done for the case $T = 0$.

As for the zero temperature case, the Matsubara GF (A.29) depends only on the difference $\bar\tau = \tau - \tau'$ [84]. On the other hand, due to (A.28), $\bar\tau$ can change only from $-1/T$ to $1/T$ which means that the Matsubara GF is a periodic function in $\bar\tau$ with a $2/T$ period. Consequently,

---

[4] We recall that the statistical average $< \mathcal{O} >$ of any observable $\mathcal{O}$ is given by $< \mathcal{O} >= \mathrm{tr}(e^{-(H-\Omega)/T}\mathcal{O})$. For the zero temperature sp GF this means $< g(\alpha t, \beta t') >= -i\,\mathrm{tr}\left\{ e^{-(H-\Omega)/T} \left(\mathcal{T} \left[a_{I\alpha}(t) a^\dagger_{I\beta}(t') \right] \right) \right\}$.

in frequency-space, the Matsubara GF can be expressed in terms of discrete frequencies, the Matsubara frequencies

$$\omega_n = n\pi/(1/T),$$ (A.30)

as the Fourier sum

$$g_M(\alpha, \beta, \bar{\tau}) = T\sum_{-\infty}^{\infty} g_M(\alpha, \beta, \omega_n)e^{-i\omega_n\bar{\tau}}.$$

For fermions (bosons) the integer number n in (A.30) takes on odd (even) values only, i. e. $n = 1 + 2\nu$; $(n = 2\nu)$, $n, \nu \in \mathbb{Z}$.

As for the zero-temperature GF, for the Matsubara GF an efficient diagrammatic technique based on perturbation theory has been developed [84].

A.5   Non-equilibrium Green's function: The Keldysh formalism

The Gell-Mann and Low theorem (A.24), which is of a central importance for the evaluation of the GF, requires that at long times ($t = +\infty$) the system will return to its (known) initial state (at $t = -\infty$). This assumption, however does not hold true for a variety of important physical non-equilibrium processes. For example, an ion impinging on the surface may well be neutralized and the resulting atom can be detected at large times. Thus, the initial ($t = -\infty$) and the final states ($t = +\infty$) are very different. To circumvent this difficulties one avoids the use of the final state $|\Psi(+\infty)\rangle$ reached at $t = +\infty$ by propagating back [374] to the initial time $t = -\infty$ where the state is well defined. This time-loop method introduces several Green's functions that are briefly mentioned in this appendix, further details can be found in Ref. [84] and references therein.

The single-particle casual non-equilibrium GF is defined as (for simplicity, statistical averaging is suppressed)

$$g^{--}(\alpha t, \beta t') = -i\langle\Psi|\mathcal{T}[a_{H\alpha}(t)\, a_{H\beta}^{\dagger}(t')]|\Psi\rangle,$$ (A.31)

which is in complete formal analogy to the equilibrium case (A.26), except for the fact that the expectation value is taken with respect to an arbitrary fully correlated (unknown) state $|\Psi\rangle$, which may differ substantially from the ground state, e.g. due to the presence of external sources. The main consequence of this difference to the equilibrium case is that the Gell-Mann

and Low theorem (A.24) does not apply and a diagrammatic perturbation expansion in terms
of an uncorrelated (known) state $|\Phi_0\rangle$ is not possible. Nevertheless, one can propagate the
initial state $|\Phi_0\rangle$ specified at $t = -\infty$ (in absence of the interactions) to obtain

$$g^{--}(\alpha t, \beta t') = -i\langle\Psi(t = +\infty)|\mathcal{T}S(\infty, -\infty)[a_{H\alpha}(t)\, a^{\dagger}_{H\beta}(t')]|\Phi_0(t = -\infty)\rangle,$$

$$\text{(A.32)}$$

where $\langle\Psi(t = +\infty)| \neq \langle\Phi_0(t = -\infty)|$ and Eqs. (A.32, A.24) are thus different. Therefore,
considering $\Psi(t = +\infty)|$, one reverses the time propagation from $+\infty$ back to $t = -\infty$,
where the state of the system is well defined, i.e. one utilizes the relation $|\Psi(t = +\infty)\rangle =
S_I(+\infty, -\infty)|\Phi_0\rangle$ (S is the S operator given by Eq. (A.15)). This introduces a time ordering
along a (single) contour running from $t = -\infty$ to enclose $t = +\infty$ (this part is called the
$-$ branch) and back to $t = -\infty$ (this part is referred to as the $+$ branch of the contour).
The mapping of the state $|\Psi(t = +\infty)\rangle$ onto the known state $|\Phi_0\rangle$ renders possible the use
of the Wick theorem. The time-ordering operator $\mathcal{T}_t$ orders the field operators on the entire
time-loop with earliest times occurring first. More specifically, let us consider the operators
$a_{H\alpha}(t)$ and $a^{\dagger}_{H\beta}(t')$. If t and t' belong both to the $+$ branch, where the time direction and
the contour direction coincide, the action of $\mathcal{T}_t$ is identical to the conventional time-ordering
$\mathcal{T}$. On the other hand if t and t' are on the $-$ branch, the operator $\mathcal{T}_t$ acts as a conventional
anti-time-ordering operator $\tilde{\mathcal{T}}$. In the case t and t' are on different branches then the operators
are automatically ordered.

Therefore, we obtain four Green's functions g^{--}, g^{-+}, g^{+-} and g^{++} defined depending
on the pairing of the operators on the $+$ and $-$ branches, i.e.

$$g^{--}(\alpha t, \beta t') \;=\; \langle\mathcal{T}\left[a_{H\alpha}(t)\, a^{\dagger}_{H\beta}(t')\right]\rangle =: g_c(\alpha t, \beta t'), \quad \text{called casual GF,}$$

$$\text{(A.33)}$$

$$g^{-+}(\alpha t, \beta t') \;=\; -i\langle a_{H\alpha}(t)\, a^{\dagger}_{H\beta}(t')\rangle =: g^{>}(\alpha t, \beta t'), \quad \text{greater function,}$$

$$\text{(A.34)}$$

$$g^{+-}(\alpha t, \beta t') \;=\; i\langle a^{\dagger}_{H\beta}(t')\, a_{H\alpha}(t)\rangle =: g^{<}(\alpha t, \beta t'), \quad \text{lesser function,}$$

$$\text{(A.35)}$$

$$g^{++}(\alpha t, \beta t') \;=\; \langle\tilde{\mathcal{T}}\left[a_{H\alpha}(t)\, a^{\dagger}_{H\beta}(t')\right]\rangle, \qquad\qquad\qquad\qquad\quad \text{(A.36)}$$

$$\left[g^{++}(\beta t', \alpha t)\right]^{*} \;=\; -g^{--}(\alpha t, \beta t'). \qquad\qquad\qquad\qquad\quad\; \text{(A.37)}$$

The brackets indicate average with respect to the ground state of the interacting system including, if necessary, statistical average. The Keldysh GF is then defined as

$$g_K(\alpha t, \beta t') = g^{-+}(\alpha t, \beta t') + g^{+-}(\alpha t, \beta t') = g^{--}(\alpha t, \beta t') + g^{++}(\alpha t, \beta t').$$

Furthermore, the retarded g^r and the advanced g^a GF are obtained as

$$g^r \;=\; g^{--} - g^{-+} = -(g^{++} - g^{+-}), \qquad\qquad\qquad\text{(A.38)}$$

$$g^a \;=\; g^{--} - g^{+-} = -(g^{++} - g^{-+}), \qquad\qquad\qquad\text{(A.39)}$$

$$g^r - g^a \;=\; g^> - g^<. \qquad\qquad\qquad\qquad\qquad\qquad\text{(A.40)}$$

These quantities are of direct relevance to physical observables and processes, e. g. the particle density is given by $-ig^<$, whereas the spectral density $A(1, 2)$ is described by (note the
analogy to Eq. (14.12) for the single particle case)

$$A(1, 2) = i\left[g^>(1, 2) - g^<(1, 2)\right] = i\left[g^r(1, 2) - g^a(1, 2)\right]. \qquad\qquad\text{(A.41)}$$

Bibliography

[1] E. Merzbacher, *Quantum Mechanics*, (John Wiley and Sons, 1998); A. Messiah, *Quantum Mechanics*, (North Holland Publishing Co., 1961); L.I. Schiff, *Quantum Mechanics*, (McGraw-Hill Inc., 1968).

[2] W. Lenz, Z. Physik **24**, 197 (1924).

[3] E.G. Kalnius, W. Miller jr., P. Winternitz, J. Appl. Math. **30**, 630 (1976).

[4] W. Pauli, Z. Physik **36**, 336 (1926).

[5] V. Fock, Z. Physik **98**, 145 (1935).

[6] V. Bargmann, Z. Physik **99**, 576 (1936).

[7] J. Schwinger, J. Math. Phys. **5**, 1606 (1964).

[8] S.P. Alliluev, JETP **33**, 200 (1957); Soviet Phys. JETP **6**, 156 (1958).

[9] M. Engelfield, *Group Theory and the Coulomb Problem*, (Wiley, N. Y., 1972).

[10] O. Klein, footnote on page 22 of L. Hulten, Z. Physik **86**, 21 (1933).

[11] A. Erdelyi, W. Magnus, Oberhettinger and F.G. Triconi, *Transcendental Functions*, (McGraw-Hill,1953).

[12] W. Magnusand, F. Oberhettinger, *Formulas and Theorems for the Functions of Mathematical Physics*, (New York, Chelsea, 1949).

[13] F. Calogero, *Variable Phase Approach to Potential Scattering*, (Academic Press, NY, 1967).

[14] V. Babikov, *Phase-amplitude method in quantum mechanics*, (Nauka, Moscow, 1967).

[15] F. Calogero, N. Cimento **28**, 66 (1963); Calogero F., Nuovo Cimento **33**, 352 (1964).

[16] V. Babikov, *Preprint* **2005**, 28–32 (Moscow, Dubna: JINR 1964).

[17] K.L. Baluja, A. Jain, Phys. Rev. A **46**, 1279 (1992); R. Raizada, K.L. Baluja, *Ind. J. Phys.* **71B** 33 (1997).

[18] J.L. Bohn, Phys. Rev. A **49**, 3761 (1994).

[19] A.S. Alexandrov, C.J. Dent, J. Phys.: Condensed Matter C **13**, L417 (2001).

[20] M.E. Portnoi, I. Galbraith, Phys. Rev. B **60**, 5570 (1999); Solid State Commun. 103 325 (1997).

[21] O. Kidun, J. Berakdar, *Surf. Science* **507** 662 (2002).

[22] H. Faxén and J. Holtsmark, Zeitschrift für Physik **45** 307 (1927).

[23] O. Kidun, N. Fominykh, and J. Berakdar, Journal of Physics A: Mathematical and General 35 9413-9424 (2002).

[24] J. Schwinger Phys. Rev. **72**, 742 (1947).

[25] D.R. Hartree, Proc. Cambridge Philos. Soc. **24**, 89 (1928).

[26] Fock, V., Z. Phys. **61**, 126 (1930).

[27] I. Shavitt, *The method of configuration interaction, in Methods of Electronic Structure Theory*, edited by H.F. Schaefer, pages 189-275, (Plenum Press, New York, 1977).

[28] P. Fulde, *Electron Correlation in Molecules and Solids*, Springer Series in Solid-State Sciences, Vol. 100, (Springer Verlag, Berlin, Heidelberg, New York, 1991).

[29] Recent Advances in Computational Chemistry - Vol. 3 *Recent Advances In Coupled-Cluster Methods* edited by Rodney J. Bartlet (World Scientific, 2002).

[30] J. Cizek, Adv. Chem. Phys. 1435 (1969).

[31] J. Gauss, D. Cremer *Analytical Energy Gradients in Møller-Plesset Perturbation and Quadratic Configuration Interaction Methods: Theory and Application*, Advances in quantum chemistry Vol. **23**, 205–299, (1992).

[32] S. Gasiorowicz, *Quantum Physics*, chap. 16, p. 255ff. (John Wiley & Sons, New York, 1974).

[33] M.E.J. Newman and G.T. Barkema, *Monte Carlo Methods in Statistical Physics*, (Oxford University Press, 1999).

[34] M. Born and R. Oppenheimer. Zur Quantentheorie der Molekeln. Ann. Phys. (Leipzig) 84 (20), 457 (1927).

[35] J.C. Tully, Nonadiabatic processes in molecular collisions, in Dynamics of Molecular Collisions, edited by W.H. Miller, pages 217-267. Plenum, New York, 1976; A. Szabo and N.S. Ostlund, Modern Quantum Chemistry: Introduction to Advanced Electronic

Structure Theory. McGraw-Hill, New York, 1989; J.I. Steinfeld, Molecules and Radiation: An Introduction to Modern Molecular Spectroscopy. MIT Press, Cambridge, MA., second edition, 1985.

[36] R. Car and M. Parrinello. Unified approach for molecular dynamics and density-functional theory. Phys. Rev. Lett. **55**, (22), 2471 (1985).

[37] T.A. Arias, M.C. Payne, and J.D. Joannopoulos. Phys. Rev. B **45**, (4), 1538 (1992).

[38] A. Mauger and M. Lannoo, Phys. Rev. B **15**, 2324 (1977).

[39] N. Metropolis and S. Ulam, J. Am. Stat. Assoc. 44, 335 (1949).

[40] P.J. Reynolds, D.M. Ceperley, B.J. Alder and W.A. Lester, J. Chem. Phys. **77**, 5593 (1982).

[41] P. Hohenberg, and W. Kohn, Phys. Rev. **136**, B 864 (1964).

[42] T.L. Gilbert, Phys. Rev. B **12**, 2111 (1975).

[43] W. Kohn, and L. Sham, Phys. Rev. **140**, A1133 (1965).

[44] John P. Perdew, Mel Levy, Phys. Rev. B. **31**, 6264 (1985).

[45] D.M. Ceperley and B.J. Alder, Phys. Rev. Lett. **45**, 566 (1980).

[46] R.O. Jones and O. Gunnarsson, Rev. Mod. Phys. **61**, 689 (1989).

[47] R.Q. Hood, M.Y. Chou, A.J. Williamson, G. Rajagopal, and R.J. Needs, Phys. Rev. Lett. **78**, 3350 (1997), Phys. Rev. B **57**, 8972 (1998).

[48] R.N. Schmid, E. Engel, R.M. Dreizler, P. Blaha and K. Schwarz, Adv. Quant. Chem. **33**, 209 (1999).

[49] T.-C. Li and P.-Q. Tong, Phys. Rev. A **34**, 529 (1986).

[50] D. Pines, Phys. Rev., **92**, 626 (1953);
E. Abrahams, Phys. Rev., **95**, 839 (1954);
D. Pines, P. Nozieres, *The Theory of Quantum Liquids*, Addison-Wesley, Reading, MA 1966, and references therein.

[51] A. Gonis, *Theoretical materials science: tracing the electronic origins of materials behavior*, (Warrendale, Materials Research Society 2000).

[52] A.L. Fetter, Walecka J.D., *Quantum theory of many-particle systems*, (New York, McGraw-Hill, 1971).

[53] I. Lindgren, J. Morrison, *Atomic Many-Body Theory* (Springer Verlag, Berlin 1982).

[54] R.M. Dreizler and E.K.U. Gross, *Density Functional Theory* (Springer, Berlin, 1995).

[55] H. Eschrig, *The Fundamentals of Density Functional Theory* (Teubner, Stuttgart, 1996).

[56] R.G. Parr and W. Yang, *Density-Functional Theory of Atoms and Molecules* (Oxford University Press, Oxford, 1989).

[57] J.F. Dobson, G. Vignale, and M.P. Das (eds.), *Electron Density Functional Theory, Recent Progress and New Directions*, (Plenum Press, 1998).

[58] M. Levy, Proc. Nat. Acad. Sci. 76, 6062 (1979); M. Levy and J. P. Perdew, in *Density Functional Methods in Physics* (eds. R.M. Dreizler and J. da Providencia), p. 11ff. (Plenum Publishing Corporation, New York, 1985).

[59] J.P. Perdew and Y. Wang, Phys. Rev. B **33**, 8800 (1986) (PW86); (Erratum: Phys. Rev. B **40**, 3399 (1989); J.P. Perdew, K. Burke and Y. Wang, Phys. Rev. B **54**, 10533 (1996) (PW91).

[60] R.T. Sharp and K.G. Horton, Phys. Rev. 90, 317 (1953); J.D. Talman and W.F. Shadwick, Phys. Rev. A **14**, 36 (1976); B.A. Shadwick, J.D. Talman and M.R. Norman, Comp.Phys.Commun. 54, 95 (1989).

[61] A. Görling and M. Levy, Phys. Rev. A **50**, 196 (1994);

[62] E. Engel and R.M. Dreizler, J. Comp. Chem. **20**, 31 (1999).

[63] A.K. Rajagopal and J. Callaway, Phys. Rev. B **7**, 1912 (1973); A.K. Rajagopal, J. Phys. C 11, L943 (1978); A.H. MacDonald and S.H. Vosko, J. Phys. C12, 2977 (1979).

[64] O. Gunnarsson and B.I. Lundqvist, Phys. Rev. B **13**, 4274 (1976).

[65] M.P. Grumbach, D. Hohl, R.M. Martin, and R. Car. J. Phys. Cond. Mat., 6:1999, (1994).

[66] L.N. Oliveira, E.K.U. Gross, and W. Kohn, Phys. Rev. Lett. **60**, 2430 (1988).

[67] E. Runge and E.K.U. Gross, Phys. Rev. Lett. **52**, 997 (1984).

[68] E.K.U. Gross and W. Kohn, Phys. Rev. Lett. **55**, 2850 (1985).

[69] A. Henne, A. Toepfer, H.J. Lüdde and R.M. Dreizler, J. Phys. B: 19, L361 (1986).

[70] M. Petersilka, U.J. Gossmann and E.K.U. Gross, Phys. Rev. Lett. 76, 1212 (1996).

[71] L. Rosenberg, Phys. Rev. D **8**, 1833 (1973).

[72] M. Brauner, Briggs, J.S., Klar, H., J. Phys. B, **22**, 2265 (1989).

[73] J.S. Briggs, Phys. Rev. A, **41**, 539 (1990).

[74] J. Berakdar, (thesis) University of Freiburg i.Br. (unpublished) 1990; Klar, H., Z. Phys. D, 16, 231 (1990).

[75] E.O. Alt, A.M. Mukhamedzhanov, Phys. Rev. A, **47**, 2004 (1993); E.O. Alt, M. Lieber, *ibid*, **54**, 3078 (1996).

[76] J. Berakdar, J.S. Briggs, Phys. Rev. Lett, **72**, 3799 (1994); J. Phys. B **27**, 4271 (1994).

[77] J. Berakdar, Phys. Rev. A **53**, 2314 (1996).

[78] J. Berakdar, Phys. Rev. A **54**, 1480 (1996).

[79] Sh.D. Kunikeev, V.S. Senashenko, 1996, Zh. E′ksp. Teo. Fiz., 109, 1561; [1996, Sov. Phys. JETP, 82, 839]; 1999, Nucl. Instrum. Methods B, 154, 252.

[80] D.S. Crothers, J. Phys. B **24**, L39 (1991).

[81] G. Gasaneo, F.D. Colavecchia, C.R. Garibotti, J.E. Miraglia, P. Macri, Phys. Rev. A, **55**, 2809 (1997); G. Gasaneo, F.D. Colavecchia, C.R. Garibotti, Nucl. Instrum. Methods B, **154**, 32 (1999).

[82] J. Berakdar, Phys. Lett. A, **220**, 237 (2000); *ibid*, **277**, 35 (1997).

[83] J. Berakdar, Phys. Rev. Lett., **78**, 2712 (1997).

[84] G.D. Mahan, *Many-Particle Physics*, second ed., (Plenum Press, London 1993).

[85] A.A. Abrikosov, L.P. Korkov, I.E. Dzyaloshinski 1975, *Methods of Quantum Field Theory in Statistical Physics*, (Dover).

[86] M.C. Gutzwiller, Rev. Mod. Phys. **70**, 589 (1998).

[87] M.J. Holman, N.W. Murray, Astron. J. **112**, 1278 (1996).

[88] H.A. Bethe, E.E. Salpeter, *Quantum Mechanics of One- and Two-electron Atoms*, Springer Verlag, Berlin, New York, 1957.

[89] J. Berakdar, Aust. J. Phys., **49**, 1095 (1996).

[90] G. Wannier, Phys. Rev. **90**, 817 (1953).

[91] J.H. Macek, S.Yu. Ovchinnikov, Phys. Rev. Lett, **74**, 4631 (1995).

[92] P.G. Burke and W.D. Robb, Adv. At. Mol. Phys. **11**, 143 (1975).

[93] I. Bray and A.T. Stelbovics, Phys. Rev. A **46**, 6995 (1992); I. Bray and D.V. Fursa, ibid. **54**, 2991 (1996); I. Bray, Phys. Rev. Lett. **78**, 4721 (1997); A. Kheifets and I. Bray, J. Phys. B **31**, L447 (1998).

[94] D. Proulx and R. Shakeshaft, Phys. Rev. A **48**, R875 (1993); M. Pont and R. Shakeshaft, J. Phys. B **28**, L571 (1995).

[95] J. Colgan, M.S. Pindzola, and F. Robicheaux, J. Phys. B L457 (2001).

[96] T.N. Rescigno, M. Baertschy, W.A. Isaacs, and C. McCurdy, Science **286**, 2474 (1999).

[97] L. Malegat, P. Selles, and A. Kazansky, Phys. Rev. Lett. **85**, 4450 (2000).

[98] J. Berakdar, P.F. O'Mahony, Mota Furtado, F., Z. Phys. D, **39**, 41 (1997).

[99] M. Abramowitz, and I. Stegun, *Pocketbook of Mathematical Functions*, (Verlag Harri Deutsch, Frankfurt 1984).

[100] *Solid-State Photoemission and Related Methods: Principles and Practices*, Michel A. van Hove (Editor), Wolfgang M. Schattke (Editor) (Wiley, Berlin, 2003).

[101] J. Berakdar, Phys. Rev. Lett., 85, 4036 (2000).

[102] C.N. Yang, T.D. Lee, Phys. Rev. **97**, 404 (1952); *ibid*, **87**, 410 (1952).

[103] S. Grossmann, W. Rosenhauer, Z. Phys., **207**, 138 (1967); *ibid*, **218**, 437 (1969).

[104] M.N. Barber, *Phase Transitions and Critical Phenomena*, eds. Domb, C., Lebowity, J.L., pp. 145-266, 1983.

[105] J. Berakdar, O. Kidun, A. Ernst 2002, in *Correlations, Polarization, and Ionization in Atomic Systems*, eds. D. H. Madison, M. Schulz (AIP, Melville, New York), pp. 64-69.

[106] K.A. Brueckner, Phys. Rev., **97**, 1353 (1955); H.A. Bethe, Ann. Rev. Nucl. Sci., **21**, 93 1971; J.P. Jeukenne, A. Legeunne, C. Mahaux, Physics Reports, **25**, 83 (1976).

[107] A.B. Migdal, *Theory of finite Fermi Systems*, New York 1967.

[108] P.C. Martin, J. Schwinger, Phys. Rev., **115**, 1342 (1959).

[109] H. Lehmann, Nuovo Cimento A **11**, 342 (1954).

[110] S.D. Kevan (Ed.), *Angle-Resolved Photoemission: Theory and Current Application*, Studies in Surface Science and Catalysis Elsevier 1992.

[111] S. Hüfner, *Photoelectron Spectroscopy*, No. 82 in Springer Series in Solid-State Science, Springer Verlag 1995.

[112] R. Herrmann S. Samarin, H. Schwabe, J. Kirschner, Phys. Rev. Lett., **81**, 2148 (1998); J. Phys. (Paris) IV, **9**, 127 (1999).

[113] N. Fominykh, J. Henk, J. Berakdar, P. Bruno, H. Gollisch, R. Feder, Solid State Commun. **113**, 665 (2000).

[114] J. Berakdar, H. Gollisch, R. Feder Solid State Commun. **112**, 587 (1999).

[115] N. Fominykh *et al.*, in *Many-particle spectroscopy of Atoms, Molecules, Clusters and Surfaces*, Eds. J. Berakdar, J. Kirschner, Kluwer Acad/Plenum Pub 2001.

[116] N. Fominykh, J. Berakdar, J. Henk, and P. Bruno Phys. Rev. Lett. **89**, 086402, (2002).

[117] J.P. Perdew and A. Zunger, Phys. Rev. B **23**, 5048 (1981).

[118] J.B. Krieger, Y. Li, and G.J. Iafrate, Phys. Rev. A **45**, 101 (1992).

[119] U. Fano, Phys. Rev. **126**, 1866 (1961).

[120] U. Fano, J.W. Cooper, Phys. Rev., **127**, 1364 (1965).

[121] U. Fano, J.H. Macek, Rev. Mod. Phys. **45**, 553 (1973).

[122] M.R.H. Rudge and M.J. Seaton Proc. Roy. Soc. A, **283**, 262 (1965).

[123] R.K. Peterkop *Theory of Ionization of Atoms by the Electron Impact* (Colorado associated university press: Boulder).

[124] S. Jetzke and F.H.M. Faisal J. Phys. B. **25**, 1543 (1992).

[125] R. Peterkop and A. Liepinsch, J. Phys. B. **14**, 4125, (1981).

[126] S.P. Merkuriev, Theor. and Math. Phys. (USA) **32**, 680, (1977).

[127] U. Fano and A.R.P. Rau *Atomic Collisions and Spectra* (Academic Press: New York) p.342.

[128] U. Fano Rep. Prog. Phys. **46**, 97 (1983).

[129] C.D. Lin Phys. Rep. **257**, No.1 1-83 (1995).

[130] G. Garibotti and J.E. Miraglia, Phys. Rev. A *21* 572 (1980)

[131] R.K. Peterkop J. Phys. B. **4**, 513 (1971).

[132] A.R.P. Rau Phys. Rev. A. **4**, 207 (1971).

[133] H. Klar J. Phys. B. **14**, 3255 (1981).

[134] A.R.P. Rau Phys. Rep. **110**, 369 (1984).

[135] J.M. Feagin J. Phys. B. **17**, 2433 (1984).

[136] A. Huetz, P. Selles, D. Waymel and J. Mazeau J. Phys. B. **24**, 1917 (1991).

[137] A.K. Kazansky and V.N. Ostrovsky Phys. Rev. A. **48**, R871 (1993).

[138] J-M. Rost, J. Phys. B. **28**, 3003 (1995); 1998, Phys. Rep. 297, 271.

[139] Y.L. Luke *Mathematical Functions and their Approximations* (Academic Press Inc. 1975).

[140] T. Kato Commun. Pure Appl. Math. **10**, 151 (1957).

[141] C.R. Myers, C.J. Umrigar, J.P. Sethna and III J.D. Morgan 1991 Phys. Rev. A **44**, 5537 (1991).

[142] S. Ward and J.H. Macek, Phys. Rev. A **49**, 1049 (1994).

[143] D.M. Brink, G.R. Satchler, *Angular momentum*, 2nd Edn. (Oxford Clarendon Press, 1968).

[144] D.A. Varshalovich, A.N. Moskalev, and V.K. Khersonskii, *Quantum Theory of Angular Momentum* (World Scientific, Singapore,1988).

[145] K. Blum, *Density Matrix Theory and Applications*, (Plenum, New York, 1981).

[146] H. Klar and H. Kleinpoppen, J. Phys. B. 15933-50 (1982).

[147] H. Klar, M. Fehr, Z. Phys. D, **23**, 295 (1992).

[148] J. Berakdar, H. Klar, A. Huetz, and P. Selles, J. Phys. B **26**, 1463 (1993).

[149] N.M. Kabachnik, and V. Schmidt, J. Phys. B **28**, 233 (1995).

[150] N. Chandra and S. Sen, Phys. Rev. A **52**, 2820-2828 (1995).

[151] N.L. Manakov, S.I. Marmo, and A.V. Meremianin, J. Phys. B **29**, 2711 (1996).

[152] A.W. Malcherek and J.S. Briggs, J. Phys. B **30**, 4419 (1997).

[153] L. Malegat, P. Selles, and A. Huetz, J. Phys. B **30**, 251 (1997).

[154] L. Malegat, P. Selles, P. Lablanquie, J. Mazeau and A. Huetz, J. Phys. B **30**, 263 (1997).

[155] P.J. Feibelman, Surf. Sci. **46**, 558 (1974).

[156] H.J. Levinson, E.W. Plummer, and P.J. Feibelman Phys. Rev. Lett. **43**, 942 (1979).

[157] J.B. Pendry, *Photoemission and the Electronic Properties of Surfaces* ed. B. Feuerbacher, B. Fitton and R. Willis (New York, Wiley) p. 87 (1978).

[158] L.D. Landau and E.M. Lifschitz, *Lehrbuch der theoretischen Physik*, Vol. III, Quantenmechanik, Akademie Verlag, Berlin (1979).

[159] C.N. Yang, Phys. Rev. **74**, 764 (1948).

[160] D.P. Dewangan, J. Phys. B **30**, L467 (1997).

[161] S. Flügge, *Lehrbuch der theoretischen Physik* (VI), Springer-Verlag, Berlin 1964.

[162] U. Fano, *Atomic Physics* vol. 1, ed. B. Bederson, V.W. Cohen and F.M.J. Pichanick (New York, Plenum) pp. 209-325 (1969), D. Herrick, Adv. Chem. Phys. **52**, 1 (1983); U. Fano, Rep. Prog. Phys. **29**, 32 (1984); C.D. Lin, Adv. At. Mol. Phys. **22**, 77 (1986); R.S. Berry and J.L. Krause, Adv. Chem. Phys. **70**, 35 (1988);Rau, A.R.P., Phys. Rep. **110**, 369 (1984); U. Fano and A.R.P. Rau, *Atomic Collisions and Spectra* (Academic Press, New York 1986).; S.J. Buckman, and C.W. Clark, Rev. Mod. Phys. **66**, 539 (1994); M. Aymar, C.H. Greene, and E. Luc-Koenig, Rev. Mod. Phys. **68**, 1015 (1996). J.M. Rost, Phys. Rep. **297**, 271 (1998); G. Tanner, K. Richter, J.M. Rost, Rev. Mod. Phys. **72**, 497, (2000).

[163] R.P. Madden and K. Codling, Phys. Rev. Lett. **10**, 516 (1963).

[164] J.H. Macek, J. Phys. B **1**, 831 (1968).

[165] H. Klar and M. Klar, J. Phys. B **13**, 1057 (1980).

[166] T. Atsumi, T. Isihara, M. Koyama, and M. Matsuzawa, Phys. Rev. A **42**, 6391 (1990).

[167] O.I. Tolstikhin, S. Watanabe, and M. Matsuzawa, Phys. Rev. Lett. **74**, 3573 (1995).

[168] C. Wulfman, K. Sukeyuki, J. Chem. Phys., **23**, 367, (1973).

[169] O. Sinanoglu, and D.E. Herrick, J. Chem. Phys. **62**, 886 (1975).

[170] D.R. Herrick, Phys. Rev. A **12**, 413 1975a. D.R. Herrick, J. Math. Phys. **16**, 281 1975b. D.R. Herrick, and M.E. Kellman, Phys. Rev. A **21**, 418 (1980). D.R. Herrick, M.E. Kellman, and R.D. Poliak, Phys. Rev. A **22**, 1517 (1980). D.R. Herrick, Adv. Chem. Phys. **52**, 1 (1983).

[171] M.E. Kellman, D.R. Herrick, Phys. Rev. A **22**, 1536 (1980).

[172] J.M. Feagin, and J.S. Briggs, Phys. Rev. Lett. **57**, 984 (1986). J.M. Feagin, Nucl. Inst. Meth. Phys. Res. B **24/25** 261 (1987). J.M. Feagin, and J.S. Briggs, Phys. Rev. A **37**, 4599 (1988). J.M. Rost, and J.S. Briggs, J. Phys. B **23**, L339 (1990). J.M. Rost, and J.S. Briggs, J. Phys. B **24**, 4293 (1991). J.M. Rost, J.S. Briggs, and J.M. Feagin, Phys. Rev. Lett. **66**, 1642 (1991). J.M. Rost, R. Gersbacher, K. Richter, J.S. Briggs, and D. Wintgen, J. Phys. B **24**, 2455 (1991).

[173] G.S. Ezra, K. Richter, G. Tanner, and D. Wintgen, Phys. B **24**, L413 (1991).

[174] K. Richter and D. Wintgen, J. Phys. B **23**, L197 1990a. K. Richter, and D. Wintgen, Phys. Rev. Lett. **65**, 1965 1990b. K. Richter, Semiklassik von Zwei-Elektronen-

Atomen, Ph.D. thesis, Universität Freiburg, 1991. Rost, J.M., and G. Tanner, in Classical, Semiclassical and Quantum Dynamics in Atoms, edited by H. Friedrich and B. Eckhardt (Lecture Notes in Physics 485, Springer, Berlin 1997), p. 274. G. Tanner, K. Richter, J.M. Rost, Rev. Mod. Phys. **72**, 497 (2000).

[175] H.R. Sadeghpour and C.H. Greene, Phys. Rev. Lett. **65**, 313 (1990).

[176] A. Vollweiter, J.M. Rost, and J.S. Briggs, J. Phys. B **24**, L155 (1991).

[177] B.C. Gou, Z. Chen, and C.D. Lin, Phys. Rev. A **43**, 3260 (1991).

[178] C.D. Lin, Phys. Rev. Lett. **51**, 1348 (1983); Phys. Rev. A **27**, 22 (1983); Adv. At. Mol. Phys. **22**, 77 (1986).

[179] S. Watanabe, C.D. Lin, Phys. Rev. A **34**, 823 (1986);

[180] Slater, J.C., *Quantum Theory of Matter* (Krieger, Huntington 1977).

[181] Esséen H., Int. J. Quant. Chem. **12**, 721 (1977).

[182] J.M. Rost and J.S. Briggs, J. Phys. B: At. Mol. Opt. Phys. 23 L339 (1990); J. Phys. B: At. Mol. Opt. Phys. **24**, 4293 (1991).

[183] A.K. Bhatia and A. Temkin, Phys. Rev. A **11**, 2018 (1975); **29**, 1895 (1984).

[184] A. Macias and A. Riera, Phys. Lett. A **119**, 28 (1986).

[185] D.H. Oza, Phys. Rev. A **33**, 824 (1986).

[186] J. Tang, S. Watanabe, and M. Matsuzawa, Phys. Rev. A **46**, 2437 (1992).

[187] C. Froese Fischer and M. Idrees, J. Phys. B **23**, 679 (1990).

[188] J. Nuttall, Bull. Am. Phys. Soc. **17**, 598 (1972).

[189] G.D. Doolen, M. Hidalgo, J. Nuttall, and R.W. Stagat, in Atomic Physics, edited by S.J. Smith and G.K. Walters Plenum, New York, 1973, p. 257.

[190] G.D. Doolen, J. Nuttall, and R.W. Stagat, Phys. Rev. A **10**, 1612 (1974).

[191] Y.K. Ho, Phys. Rep. **99**, 1 (1983).

[192] W. Reinhardt, Annu. Rev. Phys. Chem. **33**, 223 (1982).

[193] Y.K. Ho, J. Phys. B **12**, 387 (1979); Phys. Rev. A **23**, 2137 (1981); **34**, 4402 (1986).

[194] K.T. Chung and B.F. Davie, Phys. Rev. A **26**, 3278 (1982).

[195] P. Froelich and S.A. Alexander, Phys. Rev. A **42**, 2550 (1990).

[196] E. Lindroth, Phys. Rev. A **49**, 4473 (1994).

[197] A. Buergers, D. Wintgen, and J.-M. Rost, J. Phys. B **28**, 3163 (1995).

[198] J. Aguilar and J.M. Combes, Commun. Math. Phys. **22**, 269 (1971).

[199] E. Balslev and J.M. Combes, Commun. Math. Phys. **22**, 280 (1971).

[200] B. Simon, Commun. Math. Phys. **27**, 1 1972; Ann. Math. 97, 247 (1973).

[201] A. Herzenberg and F. Mandl, Proc. R. Soc. London Ser. A **274**, 256 (1963); J.N. Bardsley and B.R. Junker, Phys. B **5**, 178 (1972).

[202] T.N. Rescigno, V. McKoy, Phys. Rev. A **12**, 522 (1975).

[203] P.M. Morse and H. Feshbach, *Methods of Theoretical Physics* (New York: McGraw Hill) p. 1730, 1953.

[204] V.A. Fock, Kongelige Norske Videnskaber Selskap Forhandliger 31, 145 (1958).

[205] F. Smith J. Math. Phys. **3**, 735 (1962).

[206] C.D. Lin Phys. Rev. A **10**, 1968 (1974).

[207] H. Klar, Phys. Rev. Lett. **57**, 66 (1986).

[208] G. Tanner, K. Richter, J.M. Rost, Rev. Mod. Phys. **72**, 497, (2000).

[209] K. D Granzow 1963 J. Math. Phys. **4**, 897-900.

[210] H. Klar, W. Schlecht, J. Phys. B **9**, 1699 (1976).

[211] K.A. Poelstra, J.M. Feagin, H. Klar J. Phys. B, **27**, 781-94 (1994).

[212] E.P. Wigner, Phys. Rev. A **73**, 1002 (1948).

[213] D.L. Knirk J. Chem. Phys. **60**, 66 (1974).

[214] J. Berakdar, A. Lahmam-Bennani, C. Dal Cappello, Phys. Rep. **374**, 91 (2003).

[215] W. Walter, *Gewöhnliche Differentialgleichungen*, (Berlin: Springer-Verlag 1976).

[216] J.H. Bartlett, Phys. Rev. **51**, 661-9 (1937); J.H. Bartlett, J.J. Gibbons and C.G. Dunn, Phys. Rev. **47**, 679-80 (1935).

[217] K. Frankowski and C.L. Pekeris, Phys. Rev. **146**, 46-9 (1966).

[218] J.H. Macek, Phys. Rev. **160**, 170-4 (1967).

[219] D.L. Knirk, J. Chem. Phys. **60**, 66-80 (1974).

[220] A.M. Ermolaev 1958 Vest. Len. Univ. 14 48-64; 1961 Vest. Len. Univ. 16 19-33; 1968 Sov. Phys.-Dokl. 12 1144-6; A. M Ermolaev and G. B Sochilin 1964 Sov. Phys.-Dokl. 9 292-5; 1968 Int. J. Quantum Chem. 2 333-9.

[221] Y.N. Demkov and A.M. Ermolaev 1959 Sov. fPhys.-JETP 9 633-5.

[222] J.D. Morgan 1977 J. Phys. A : Math. Gen. 10 L91-3; 1986 Theor. Chim. Acta 69 181-223.

[223] P.C. Abbott and E.N. Maslen, J. Phys. A: At. Mol. Phys. 17 L489-92 (1984).

[224] J.M. Feagin, J. Macek, and A.F. Starace, Phys. Rev. A **32**, 3219-30 (1985).

[225] J. Leray, *Methods of Functional Analysis and Theory of Elliptic Operators* (Naples: Universita di Napoli 1982); *Bifurcation Theory, Mechanics and Physics* ed C.P. Bruter et al (Dordrecht: Reidel 1983) pp. 99-108; Lecture Notes in Physics vol 195 (Beerlin: Springer) pp. 235-47, 1984.

[226] R.T. Pack and W. Byers-Brown, J. Chem. Phys. **45**, 556 (1966).

[227] E.P. Wigner, *Group Theory and its Application to the Quantum Mechanics of Atomic Spectra* (Academic, New York, 1959).

[228] A.V. Meremianin, J.S. Briggs, Phys. Rev. Lett, **89** 200405-1 (2002).

[229] G. Breit, Phys. Rev. **35**, 569-78 (1930).

[230] E.A. Hylleraas, Z. Phys. **54**, 347-66 (1929).

[231] Hylleraas, Skrift. Utg. Norske Vidensk. Akad. Oslo no 6 (1932).

[232] F. Smith, Phys. Rev. **120**, 1058 (1960).

[233] U. Fano, D. Green, J.L. Bohn and T.A. Heim, J. Phys. B: At. Mol. Opt. Phys. **32**, R1-R37 (1999).

[234] Yu. V. Popov and L.U. Ancarani, Phys. Rev. A **62** 042702-1 (2000).

[235] H. Klar, J. Phys. B **34**, 2725-2730 (2001).

[236] L.C. Biedenharn, and J.D. Louck, *Angular Momentum in Quantum Physics: Theory and Application.* Reading, MA: Addison-Wesley, 1981. pp. 314.

[237] M. Rotenberg, R. Bivens, N. Metropolis, and J.K. Wooten, *The 3-j and 6-j symbols.* Cambridge, MA: MIT Press, (1959).

[238] I.S. Gradstein and M.I. Ryshik, 1981 *"Summen-, Produkt- und Integral-Tafeln"*, Verlag Harri Deutsch, Thun.

[239] J.S. Briggs and V. Schmidt, J. Phys. B **33** (2000) R1.

[240] J. Berakdar and H. Klar, Physics Reports **340**, 473 - 520 (2001).

[241] P. Harrison, *Quantum Wells, Wires and Dots*, John Wiley & sons, UK, 2000.

[242] E. Wigner, Phys. Rev. **46**, 1002 (1934).

[243] D. Ceperley, Rev. Mod. Phys. **67**, 279 (1995).

[244] I. Taouil, A. Lahmam-Bennani, A. Duguet, L. Avaldi, Phys. Rev. Lett. **81**, 4600 (1998).

[245] Lahmam-Bennani, A. Duguet, A.M. Grisogono and M. Lecas, J. Phys. B **25**, 2873 (1992).

[246] Lahmam-Bennani, I. Taouil, A. Duguet, M. Lecas, L. Avaldi and J. Berakdar, Phys. Rev. A **59**, 3548 (1999).

[247] B. El Marji, C. Schröter, A. Duguet, A. Lahmam-Bennani, M. Lecas and L. Spielberger, J. Phys. B **30**, 3677 (1997).

[248] C. Schröter, B. El Marji, A. Lahmam-Bennani, A. Duguet, M. Lecas, L. Spielberger J. Phys. B **31**, 131 (1998).

[249] A. Lahmam-Bennani, A. Duguet, M.N. Gaboriaud, I. Taouil, M. Lecas, A. Kheifets, J. Berakdar, C. Dal Cappello, J. Phys. B **34**, 3073 (2001).

[250] S. Cvejanović, and F.H. Read, (1974). J. Phys. B **7**, 1841; S. Cvejanović, R.C. Shiell, and T.J. Reddish, (1995). J. Phys. B **28**, L707; J.M. Feagin, (1984). J. Phys. B **17**, 2433. Klar, H. (1981). J. Phys. B **14**, 3255; K. Kossmann, V. Schmidt, and T. Andersen, (1988). Phys. Rev. Lett **60**. 1266. P. Lablanquie, K. Ito, P. Morin, I. Nenner, and J.H. D. Eland, (1990). Z. Phys. D **16**, 77.

[251] B. El-Marji, J.P. Doering, J.H. Moore, M.A. Coplan, Phys. Rev. Lett. **83**, 1574 (1999).

[252] R.W. van Boeyen, J.P. Doering, J.H. Moore, M.A. Coplan and J.W. Cooper J. Phys. B **35**, (2002) L97.

[253] J.H. Moore, M.A. Coplan, J.W. Cooper, J.P. Doering, and B. El Marji, AIP Conf. Proc. 475 96 (1999).

[254] M.J. Ford, B. El-Marji, J.P. Doering, J.H. Moore, M.A. Coplan, and J.W. Cooper, Phys. Rev. A **57**, 325 (1998).

[255] M.J. Ford, J.H. Moore, M.A. Coplan, J.W. Cooper, and J. P. Doering, Phys. Rev. Lett. **77**, 2650 (1996).

[256] M.J. Ford, J.P. Doering, J.H. Moore, and M.A. Coplan, Rev. Sci. Instrum. 66 3137 (1995).

[257] Robert S. Freund, Robert C. Wetzel, Randy J. Shul, and Todd R. Hayes, Phys. Rev. A **41**, 3575 (1990).

[258] A. Dorn, R. Moshammer, C. Schröter, T.J.M. Zouros, W. Schmitt, H. Kollmus, R. Mann, J. Ullrich, Phys. Rev. Lett. **82**, 2496 (1999).

[259] A. Dorn, A. Kheifets, C.D. Schröter, B. Najjari, C. Hö hr, R. Moshammer, J. Ullrich, Phys. Rev. A **65**, 032709 (2002).

[260] A. Dorn, A. Kheifets, C.D. Schröter, B. Najjari, C. Hö hr, R. Moshammer, Phys. Rev. Lett. **86**, 3755 (2001).

[261] R. Wehlitz, M.-T. Huang, B.D. DePaola, J.C. Levin, I.A. Sellin, T. Nagata, J.W. Cooper, and Y. Azuma, Phys. Rev. Lett. **81** (1998) 1813.

[262] L.H. Andersen, P. Hvelplund, H. Knudsen, S.P. Møller, J. O.P. Pedersen, S. Tang-Petersen, E. Uggerhøj, K. Elsener, and E. Morenzoni, Phys. Rev. A **40**, 7366 (1989).

[263] M. Unverzagt, R. Moshammer, W. Schmitt, R.E. Olson, P. Jardin, V. Mergel, J. Ullrich, and H. Schmidt-Böcking, Phys. Rev. Lett. 76 1043 (1996).

[264] B. Bapat, S. Keller, R. Moshammer, R. Mann and J. Ullrich J. Phys. B **33**, 1437 (2000).

[265] M. Schulz, R. Moshammer, W. Schmitt, H. Kollmus, R. Mann, S. Hagmann, R.E. Olson and J. Ullrich, J. Phys. B **32**, (1999) L557.

[266] R. Moshammer, J. Ullrich, H. Kollmus, W. Schmitt, M. Unverzagt, O. Jagutzki, V. Mergel, H. Schmidt-Böcking, R. Mann, CJ. Woods, RE. Olson, Phys. Rev. Lett. **77** 1242 (1996).

[267] J. Moxom, D.M. Schrader, G. Laricchia, Jun Xu, and L.D. Hulett, Phys. Rev. A **60**, 2940 (1999).

[268] F. Melchert, E. Salzborn, D.B. Uskov, L.P. Presnyakov, Catalina Gonzalez, and J.H. McGuire, Phys. Rev. A **59**, 4379 (1999).

[269] R. Egger, W. Häusler, C.H. Mak, and H. Grabert, Phys. Rev. Lett. **82**, 3320 (1999).

[270] A.V. Filinov, M. Bonitz, and Yu.E. Lozovik, Phys. Rev. Lett. vol. 86, p. 3851 (2001).

[271] B.A. Lippmann, J. Schwinger Phys. Rev. **79**, 469 (1950).

[272] C. Møller, K. Dan. Vidensk. Selsk. Mat. Fys.Medd. 23, 1 (1945).

[273] B. Podolsky and L. Pauling, Phys. Rev., 34:109, (1929).

[274] J. Avery, Hyperspherical Harmonics, Applications in Quantum Theory. Kluwer Academic, Dordrecht, The Netherlands, 1989.

[275] J. Avery, Hyperspherical Harmonics and Generalized Sturmians. Kluwer Academic Publishers, Dordrecht, The Netherlands, 2000.

[276] T. Shibuya and C.E. Wulfman, Am. J. Phys. **33**, 570 (1965).

[277] T. Matsubara, Prog. Theor. Phys. (Kyoto) 14, 315 (1955).

[278] L.P. Keldysh, Sov. Phys. JETP 20, 1018 (1965).

[279] E.N. Economou, *Green's Functions in Quantum Physics*, Springer Series in Solid-State Siences 7, (Springer Verlag, Berlin, 1983).

[280] M. Rotenberg, Ann. Phys., NY 19 262-78 (1962); Adv. At. Mol. Phys. **6**, 233-68 (1970).

[281] S. Okubo, D. Feldman, Phys. Rev. **117**, 292 (1960).

[282] J. Macek, Phys. Rev. A **37**, 2365 (1988).

[283] E.H. Wichmann, C.H. Woo, J. Math. Phys. **2**, 178 (1961).

[284] M. and C. Itzyckson, Rev. Mod. Phys. **38**, 330-45 (1966).

[285] A.M. Perelomov and V.S. Popov, Sov. Phys. JETP, 23, 118 (1966).

[286] R. Finkelstein, D. Levy, J. Math. Phys., 8, 2147 (1967).

[287] A.K. Rajagopal and C.S. Shastry, J. Math. Phys. **12**, 2387-94 (1971).

[288] L. Hostler 1964 J. Math. Phys. **5**, 591-611; J. Math. Phys. **5**, 1235-40 (1964).

[289] J.C.Y. Chen and A.C. Chen, Adv. At. Mol. Phys. **8** 71-129 (1972), Proc. Phys. Soc. London 4, L102 (1971); J.C.Y. Chen, A.C. Chen, A.K. Rajagopal, Phys. Rev. A **5**, 2686 (1972).

[290] W.F. Ford, J. Math. Phys. **7**, 626-33 (1966).

[291] M. Lieber, Phys. Rev. **174**, 2037-54 1968; Relativistic, *Quantum Electrodynamic and Weak Interaction Effect in Atoms* (AIP Conf. Proc. W. Johnson, P. Mohr and J. Sucher (New York: AIP) pp. 445-59 1989.

[292] N.L. Manakov, V.D. Ovsiannikov and L.P. Rapoport Phys. Rep. **141**, 319-33 (1986).

[293] H. Van Haeringen, *Charged Particle Interactions. Theory and Formulas* (Leyden: Coulomb 1980).

[294] A. Maquet, V. Véniard, T.A. Marian, J. Phys. B **31**, 3743-3764 (1998).

[295] J.C.Y. Chen, A.C. Chen, P.J. Kramer 1971 Phys. Rev. A, **4**, (1982).

[296] G.L. Nutt, J. Math. Phys. **9**, 796 (1968).

[297] G. Wentzel, Z. Phys. **40**, 590 (1927).

[298] J.R. Oppenheimer, Z. Phys. **43**, 413 (1927).

[299] W. Gordon, Z. Phys. **48**, 180 (1928).

[300] N.F. Mott, Proc. Roy. Soc. Ser. A **118**, 542 (1928).

[301] R.H. Dalitz, Proc. Roy. Soc. Ser. A **206**, 509 (1951).

[302] C. Kacser, Nuovo Cimento **13**, 303 (1959).

[303] S. Weinberg, Phys. Rev.B **140**, 516 (1965).

[304] L.D. Faddeev, Soviet Phys. JETP **12**, 1014 (1961); Soviet Phys. JETP 39, 1459, (1960).

[305] L.D. Faddeev *Mathematical Aspects of the Three-Body Problem* (Davey, New York, 1965); *Mathematical Problems of the Quantum Theory of Scattering for a Three-Particle System* Publication of the Steklov Mathematical Institute, Leningrad 1963.

[306] B.A. Lippmann, Phys. Rev. **102**, 264 (1956).

[307] L.L. Foldy and W. Tobocman, Phys. Rev. **105**, 1099 (1957).

[308] F. Smithies, *Integral Equations* (Cambridge University press, New York 1958).

[309] S. Weinberg, Phys. Rev. A **133**, B232 (1964).

[310] R.D. Amado, Phys. Rev. **132** 485 (1963).

[311] C. Lovelace, Phys. Rev. **135** B1225 (1964).

[312] L. Rosenberg, Phys. Rev. **135** B715 (1964).

[313] L.R. Dodd, K.R. Greider, Phys. Rev. **146** 146 (1966).

[314] E.O. Alt, P. Grassberger, W. Sandhas, Nucl. Phys. B2 167 (1967).

[315] C.J. Joachain, *The Quantum Collision Theory* (North-Holland Publishing Company 1983).

[316] O.A. Yakubowsky. Sov. J. Nucl. Phys. 5, 937, (1967).

[317] W. Glöckle, *The Quantum mechanical Few-Body Problem*, Springer Verlag, 1983.

[318] A.N. Mitra, Nucl. Phys. **32**, (1962) 529

[319] J.P. Taylor, 1972 *Scattering Theory,* John Wiley & Sons, Inc.

[320] C.R. Chen, G.L. Payne, J.L. Friar, and B.F. Gibson, Phys. Rev. C 39, 1261 (1989).

[321] W. Glöckle, H. Witala, D. Hüber, H. Hamada, and J. Golak, Phys. Rep. 274, 107 (1996).

[322] E. Nielsen, D.V. Fedorov, A.S. Jensen, E. Garrido Phys. Rep. 347 (2001) 373-459; D.V. Fedorov, A.S. Jensen, Phys. Rev. Lett. **71**, (1993) 4103.

[323] J.V. Noble, Phys. Rev. **161**, 945 (1967).

[324] S.P. Merkuriev, C. Gignoux, and A. Laverne, Ann. Phys. N.Y. 39, 30 (1976).

[325] L.D. Faddeev and S.P. Merkuriev, *Quantum Scattering Theory for Several Particle Systems* (Kluwer, Dordrecht, 1993).

[326] E.O. Alt, in Few-Body Methods: Principles and Applications, edited by T.K. Lim et al. World Scientific, Singapore, 1986, p. 239.

[327] C. Chandler, Nucl. Phys. A463, 181c (1987).

[328] Gy. Bencze, P. Doleshall, C. Chandler, A.G. Gibson, and D. Walliser, Phys. Rev. C **43**, 992 (1991).

[329] Gy. Bencze, Nucl. Phys. A196, 135 (1972).

[330] E.O. Alt, W. Sandhas, and H. Ziegelmann, Phys. Rev. C 17, 1981 (1978).

[331] Z. Papp, Phys. Rev. C 55, 1080 (1997).

[332] H. Kleinert, *Path Integrals* (World Scientific, Teaneck, NJ, 1995).

[333] C. Grosche and F. Steiner, *Handbook of Feynman Path Integrals* (Springer, Berlin, 1998).

[334] M.F. Trotter, Proc. Am. Math. Soc. **10**, 545 (1959).

[335] R.P. Feynman, A.R. Hibbs : *Quantum Mechanics and Path Integrals* New York: McGraw Hill 1965.

[336] D.M. Ceperly, Phys. Rev. Lett. **69**, 331 (1992)
E.L. Pollock, D.M. Ceperly, Phys. Rev. **B 30**, 2555 (1984)
E.L. Pollock, D.M. Ceperly, Phys. Rev. **B 36**, 8343 (1987).

[337] M. Takahashi, M. Imada, J. Phys. Soc. Jpn. **53**, 963 (1984); M. Takahashi, M. Imada, J. Phys. Soc. Jpn. **53**, 3765 (1984).

[338] N. Metropolis, A. Rosenbluth, M.N. Rosenbluth, A.H. Teller, E. Teller, J. Chem. Phys. **21**, 1087 (1953).

[339] D. Ceperley, Phys. Rev. Lett. **69**, 331 (1992).

[340] A. Lyubartsev and P. Vorontsov-Velayaminov, Phys. Rev. A **48**, 4075 (1993).

[341] I. Morgenstern, Z. Phys. B **77**, 267 (1989).

[342] W.H. Newman and A. Kuki, J. Chem. Phys. **96**, 356 (1992).

[343] C. Mak, R. Egger, and H. Weber-Gottschick, Phys. Rev. Lett. **81**, 4533 (1998).

[344] J. Harting, O. Mülken and P. Borrmann, Phys. Rev. B 62, 10207 (2000).

[345] H. Ibach, D.L. Mills, *Electron Energy Loss Spectroscopy and Surface Vibrations*, Academic Press, London, 1982.

[346] M.A. van Hove, W.H. Weinberg, and C.-M. Chan, *Low Energy Electron Diffraction*, Springer Series in Surface Science, Springer, Berlin 1986.

[347] I. Turek, V. Drchal, J. Kudrnovský, M. Šob, and P. Weinberger, *Electronic Structure of Disrodered Alloys, Surfaces, and Interfaces*, Kluwer Academic Pub., Bosten, London, Dordrecht, 1997.

[348] P. Weinberger, *Electron Scattering Theory for Ordered and Disordered Matter*, Clarendon Press, Oxford, 1990.

[349] D.A. King and D.P. Woodruff (Eds.), *Chemical Physics of Solid Surfaces and Heterogeneous Catalysis*, Vol.3. Chemisorption systems. Part A., Elsevier, Amsterdam, 1990.

[350] B.L. Gyorffy, *Fondamenti di fisica dello stato solida*, Scuola Nazionale di Struttura della Materia (1971).

[351] B.L. Gyorffy and M.J. Stott, in: D.J. Fabian and L.M. Watson (eds.), *Bandstructure Spectroscopy of Metals and Alloys*, Academic Press, London, 1973.

[352] I. Mertig, E. Mrosan and P. Ziesche *Multiple scattering theory of point defects in metals: electronic properties* (Teubner Texte zur Physik, Leipzig, 1987).

[353] V. Bortolani, N.H. March and M.P. Tosi (Eds.), *Interaction of Atoms and Molecules with Solid Surfaces*, (Plenum, New York, 1990).

[354] J. Berakdar, *Surf. Rev. and Letters* **7**, 205 (2000).

[355] K. Kouzakov, J. Berakdar, Phys. Rev. B **66**, 235114 (2002); J. Phys.: Condens. Matter **15**, L41-L47 (2003).

[356] S. Iacobucci, L. Marassi, R. Camilloni, S. Nannarone, and G. Stefani, Phys. Rev. B **51**, 10252 (1995).

[357] O.M. Artamonov, S.N. Samarin, and J. Kirschner, Applied Physics a (Materials Science Processing) A65, 535 (1997).

[358] J. Berakdar, S.N. Samarin, R. Herrmann, and J. Kirschner, Phys. Rev. Lett. **81**, 3535 (1998).

[359] R. Feder, H. Gollisch, D. Meinert, T. Scheunemann, O.M. Artamonov, S.N. Samarin, and J. Kirschner, Phys. Rev. B **58**, 16418 (1998).

[360] A.S. Kheifets, S. Iacobucci, A. Ruocco, R. Camiloni, and G. Stefani, Phys. Rev. B **57**, 7360 (1998).

[361] M. Gell-Mann and F.E. Low, Phys. Rev. **84** 350 1951.

[362] G.C. Wick, Phys. Rev. **80** 268 1950.

[363] V.M. Galitskii and A.B. Migdal 1958 Zh Eksp. Teor. Fiz. 34 139 (1958 Sov. Phys.JETP 7 96).

[364] E. Weigold and I.E. McCarthy, *Electron Momentum Spectroscopy* (Kluwer Academic/Plenum Publishers, 1999).

[365] S.D. Yang and T.T.S. Kuo, Nucl. Phys. A **456**, 413 (1986).

[366] W. Hengeveld, W.H. Dickhoff and K. Allaart, Nucl. Phys. A451, 269 (1986).

[367] L. Hedin, Phys. Rev. 139 A **796**, (1965), L. Hedin, Ark. Fys. 30, 231 (1965).

[368] O. Kidun, J. Berakdar 2001, Phys. Rev. Lett. **87**, 263401.

[369] F. Aryasetiawan and O. Gunnarson, Rep. Prog. Phys. **61**, 237 (1998).

[370] G. Onida, L. Reininger, A. Rubio, Rev. Mod. Phys. **74**, 601 (2002).

[371] M. Vos, A.S. Kheifets, E. Weigold, S.A. Canney, B. Holms, F. Aryasetiawan, K. Karlsson, J. Phys. : Condensed matter **11**, 3545 (1999); E. Weigold and M. Vos, in *Many Particle Spectroscopy of Atoms, Molecules, Clusters, and Surfaces*, edited by J. Berakdar and J. Kirschner (Kluwer Academic/Plenum Publishers, New York, 2001).

[372] L. Hedin, Phys. Scr. 21, 477 (1980).

[373] N. Fominykh, J. Berakdar, J. Henk, P. Bruno: Phys. Rev. Lett. 89, 086402 (2002).

[374] J. Schwinger, J. Math. Phys. **2**, (1961), 407.

[375] J. Rammer, H. Smith, Rev. Mod. Phys. **58**, 323 (1986).

[376] C. Caroli C., D. Lederer-Rozenblatt, B. Roulet, D. Saint-James, Phys. Rev. B **8**, 4552 (1973).

[377] M. Cardona, L. Ley (eds.), Photoemission in Solids I, no. 26 in Topics in Applied Physics, Springer, Berlin (1978).

[378] S. V. Kevan (ed.), Angle-resolved Photoemission: Theory and Current Applications, Elsevier, Amsterdam (1992).

[379] J. Henk, *Handbook of Thin Film Materials*, pp. 479, edited by H.S. Nalwa, Vol. 2 (Academic Press, New York, 2002).

Index